Electrical Installation Work

Level 3

Updated in line with the 18th Edition of the Wiring Regulations and written specific... the EAL Diploma in Electrical Installation, this book has a chapter dedicated to each unit of the EAL syllabus, allowing you to master each topic before moving on to the next. This new edition also includes a section on LED lighting. End of chapter revision questions help you to check your understanding and consolidate the key concepts learned in each chapter. A must have for all learners working towards EAL electrical installations qualifications.

Trevor Linsley was formerly Senior Lecturer in Electrical Engineering at Blackpool and the Fylde College of Technology. There he taught subjects at all levels from first year trainee to first year undergraduate courses; and was also head of the multidiscipline NVQ Assessment Centre and responsible for establishing and running the AM2 Electrical Skills Assessment Centre. He has had 27 books published and has also written many bespoke training packages for local SME electrical engineering companies in the North West of England.

Electrical Installation Work

Level 3

EAL Edition

Second Edition

Trevor Linsley

Routledge
Taylor & Francis Group

LONDON AND NEW YORK

Second edition published 2020
by Routledge
2 Park Square, Milton Park, Abingdon, Oxon, OX14 4RN

and by Routledge
52 Vanderbilt Avenue, New York, NY 10017

Routledge is an imprint of the Taylor & Francis Group, an informa business

First edition published by Routledge 2016

British Library Cataloguing-in-Publication Data
A catalogue record for this book is available from the British Library

Library of Congress Cataloging-in-Publication Data
Names: Linsley, Trevor, author.
Title: Electrical installation work. Level 3 / Trevor Linsley.
Description: EAL edition. Second edition. | Abingdon, Oxon ; New York, NY : Routledge, 2019. |
   Includes bibliographical references and index.
Identifiers: LCCN 2019010388| ISBN 9780367195649 (hardback) | ISBN 9780367195632 (pbk.) |
   ISBN 9780429203183 (ebook)
Subjects: LCSH: Electric apparatus and appliances--Installation--Textbooks. | Electric wiring, Interior--Textbooks.
Classification: LCC TK452 .R575 2019 | DDC 621.319/24076--dc23
LC record available at https://lccn.loc.gov/2019010388

ISBN: 978-0-367-19564-9 (hbk)
ISBN: 978-0-367-19563-2 (pbk)
ISBN: 978-0-429-20318-3 (ebk)

Typeset in Helvetica
by Servis Filmsetting Ltd, Stockport, Cheshire

Visit the companion website: **www.routledge.com/cw/linsley**

Printed and bound in Great Britain by
TJ International Ltd, Padstow, Cornwall

Contents

Understand environmental legislation, working practices and the principles of environmental technology systems

EAL Electrical Installation Work – Level 3, 2nd Edition 978 0 367 19564 9
© 2019 T. Linsley. Published by Taylor & Francis. All rights reserved.
https://www.routledge.com/9780367195649

Learning outcomes

When you have completed this chapter you should:

1. Understand the environmental legislation, working practices and principles that are relevant to work activities.
2. Understand how work methods and procedures can reduce material wastage and impact on the environment.
3. Understand how and where environmental technology systems can be applied.

Assessment criteria 1.1

Specify the current, relevant legislation for processing waste

Assessment criteria 1.2

Describe what is meant by the term environment

Environmental laws and regulations

The **environment** describes the world in which we live, work and play; it relates to our neighbourhood and surroundings and the situation in which we find ourselves.

Environmental laws protect the natural environment in which we all live, including: plants, forests and mountains among other elements.

If an offence is identified in the area in which we now think of as 'environmental' it can be of two kinds:

1 An offence in common law, which means damage to property, nuisance or negligence leading to a claim for damages.
2 A statutory offence against one of the laws dealing with the protection of the environment. These offences are nearly always 'crimes' and punished by fines or imprisonment rather than by compensating any individual.

Built environment

Everything we do affects other people or other elements, for instance creating new houses creates jobs but this also impacts the environment given the new amount of materials that have to be provided. We therefore define the materials and resources we use to build things that impact on the environment as the 'built environment'.

Definition

The *environment* describes the world in which we live, work and play. It relates to our neighbourhood, surroundings and the situation in which we find ourselves.

Figure 1.1 Eco buildings such as this one are becoming more and more common today.

However, we also impact on the environment in other ways, which is why we have developed laws and legislation to deal with the environment but devolving responsibility for each part such as: air, water, land noise, radioactive substances. Where organizations' activities impact upon the environmental laws they are increasingly adopting environmental management systems that comply with ISO 14001. Let us now look at some of the regulations and try to see the present picture at the beginning of the new millennium.

Environmental Protection Act 1990

In the context of environmental law, the Environmental Protection Act 1990 was a major piece of legislation. The main sections of the Act are:

Table 1.1 The Environmental Protection Act ensures that companies set up an effective environmental management system

Part 1	Integrated pollution control by HM Inspectorate of Pollution, and air pollution control by Local Authorities
Part 2	Wastes on land
Part 3	Statutory nuisances and clean air
Part 4	Litter
Part 5	Radioactive Substances Act 1960
Part 6	Genetically modified organisms
Part 7	Nature conservation
Part 8	Miscellaneous, including contaminated land

The Royal Commission of 1976 identified that a reduction of pollutant to one medium, air, water or land, then led to an increase of pollutant in another. The need to take an integrated approach to pollution control is therefore stressed.

The processes subject to an integrated pollution control are:

- Air emissions.
- Processes that give rise to significant quantities of special waste, that is, waste defined in law in terms of its toxicity or flammability.
- Processes giving rise to emissions to sewers or 'Red List' substances. These are 23 substances including mercury, cadmium and many pesticides that are subject to discharge consent to the satisfaction of the Environment Agency.

The Inspectorate is empowered to set conditions to ensure that the best practicable environmental option (BPEO) is employed to control pollution. This is the cornerstone of the Environmental Protection Act.

Hazardous Waste Regulations 2005

New Hazardous Waste Regulations were introduced in July 2005 and under these regulations electric discharge lamps and tubes such as fluorescent, sodium, metal halide and mercury vapour are classified as hazardous waste. While each lamp only contains a very small amount of mercury, vast numbers are used and disposed of each year, resulting in a significant environmental threat. The environmentally responsible way to dispose of lamps and tubes is to recycle

Safety first

Disposal of batteries and old fluorescent tubes is regulated through the Hazardous Waste Regulations 2005.

them and this process is now available through electrical wholesalers. Electrical companies produce relatively small amounts of waste and even smaller amounts of special waste. Most companies buy in the expertise of specialist waste companies these days and build these costs into the contract.

Pollution Prevention and Control Regulations 2000

The system of Pollution Prevention and Control replaced that of Integrated Pollution Control established by the Environmental Protection Act 1990, thus bringing environmental law into the new millennium and implementing the European Directive (EC/96/61) on integrated pollution prevention and control. The new system was fully implemented in 2007.

Pollution Prevention and Control is a regime for controlling pollution from certain industrial activities. This regime introduces the concept of Best Available Technique (BAT) for reducing and preventing pollution to an acceptable level. Industrial activities are graded according to their potential to pollute the environment:

- A(1) installations are regulated by the Environment Agency.
- A(2) installations are regulated by Local Authorities.
- Part B installations are also regulated by Local Authorities.

All three systems require the operators of certain industrial installations to obtain a permit to operate. Once an operator has submitted a permit application, the regulator then decides whether to issue a permit. If one is issued it will include conditions aimed at reducing and preventing pollution to acceptable levels. A(1) installations are generally perceived as having the greatest potential to pollute the environment. A(2) installations and Part B installations would have the least potential to pollute.

The industries affected by these regulations are those dealing with petrol vapour recovery, incineration of waste, mercury emissions from crematoria, animal rendering, non-ferrous foundry processes, surface treating of metals and plastic materials by powder coating, galvanizing of metals and the manufacture of certain specified composite wood-based boards.

Control of Pollution Act 1989

This Act concerns making provision with regard to depositing controlled waste, and requires carriers of controlled waste to register with either the Environment Agency in England and Wales or the Scottish Environment Protection Agency (SEPA).

Noise at Work Regulations 1989

The Noise at Work Regulations, unlike the previous vague or limited provisions, apply to all work places and require employers to carry out assessments of the noise levels within their premises and to take appropriate action where necessary. The 1989 Regulations came into force on 1 January 1990 implementing in the United Kingdom the EC Directive 86/188/EEC 'The Protection of Workers from Noise'.

Three action levels are defined by the regulations:

1 The first action level is a daily personal noise exposure of 85 dB, expressed as 85 dB(A).

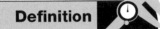

Definition

Remember that there are two basic forms of plastic: thermoplastic and thermosetting. Only thermoplastic can be reheated and reshaped.

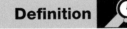

Definition

Burning waste material on site could contravene the Control of Pollution Act 1989.

This Act specifically extends to prosecuting any person including the seizure/disposal of vehicles involved in:

- Knowingly permitting the use of any plant or equipment for the purpose of disposing of controlled waste including sewerage regarding pollution of any relevant water systems such as streams.

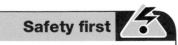

Safety first

The Noise at Work Regulations are intended to reduce hearing damage caused by loud noise.

2 The second action level is a daily personal noise exposure of 90 dB(A).

3 The third defined level is a peak action level of 140 dB(A) or 200 Pa of pressure, which is likely to be linked to the use of cartridge operated tools, shooting guns or similar loud explosive noises. This action level is likely to be most important where workers are subjected to a small number of loud impulses during an otherwise quiet day.

The Noise at Work Regulations are intended to reduce hearing damage caused by loud noise. So, what is a loud noise? If you cannot hear what someone is saying when they are 2 m away from you or if they have to shout to make themselves heard, then the noise level is probably above 85 dB and should be measured by a competent person.

At the first action level an employee must be provided with ear protection (earmuffs or earplugs) on request. At the second action level the employer must reduce, so far as is reasonably practicable, other than by providing ear protection, the exposure to noise of that employee.

Hearing damage is cumulative, it builds up, leading eventually to a loss of hearing ability. Young people, in particular, should get into the routine of avoiding noise exposure before their hearing is permanently damaged. The damage can also take the form of permanent tinnitus (ringing noise in the ears) and an inability to distinguish words of similar sound such as bit and tip. Vibration is also associated with noise.

Figure 1.2 This kind of drilling can cause vibration white finger, and can also damage your hearing if the proper precautions aren't taken.

Direct vibration through vibrating floors or from vibrating tools can lead to damage to the bones of the feet or hands. A condition known as 'vibration white finger' is caused by an impaired blood supply to the fingers, associated with vibrating hand tools.

Employers and employees should not rely too heavily on ear protectors. In practice, they reduce noise exposure far less than is often claimed, because they may be uncomfortable or inconvenient to wear. To be effective, ear protectors need to be worn all the time when in noisy places. If left off for even a short time, the best protectors cannot reduce noise exposure effectively.

Protection against noise is best achieved by controlling it at source. Wearing ear protection must be a last resort. Employers should:

- Design machinery and processes to reduce noise and vibration (mounting machines on shock absorbing materials can dampen out vibration).
- When buying new equipment, where possible, choose quiet machines. Ask the supplier to specify noise levels at the operator's working position.
- Enclose noisy machines in sound-absorbing panels.
- Fit silencers on exhaust systems.
- Install motor drives in a separate room away from the operator.
- Inform workers of the noise hazard and get them to wear ear protection.
- Reduce a worker's exposure to noise by job rotation or provide a noise refuge.

Figure 1.3 Ear defenders protect workers from noise at work.

New regulations introduced in 2006 reduced the first action level to 80 dB(A) and the second level to 85 dB(A) with a peak action level of 98 dB(A) or 140 Pa of pressure. Every employer must make a 'noise' assessment and provide workers with information about the risks to hearing if the noise level approaches the first action level. They must do all that is reasonably practicable to control the noise exposure of their employees and clearly mark ear protection zones. Employees must wear personal ear protection whilst in such a zone.

Packaging (Essential Requirements) Regulations 2003

Definition

The new Packaging Regulations were introduced on 25 August 2003, bringing the UK into harmony with Europe. The regulations deal with the essential requirements of packaging for the storage and transportation of goods.

The new Packaging Regulations were introduced on 25 August 2003 bringing the UK into harmony with Europe. The regulations deal with the essential requirements of packaging for the storage and transportation of goods. There are two essential elements to the regulations:

1 The packaging should be designed and manufactured so that the volume and weight is the minimum amount required to maintain the necessary level of safety for the packaged product.
2 The packaging should be designed and manufactured in such a way that the packaging used is either reusable or recyclable.

The regulations are enforced by the Weights and Measures Authority in Great Britain, the Department of Enterprise, Trade and Investment in Northern Ireland and the Procurator Fiscal or Lord Advocate in Scotland.

Environmental Act 1975

This Act was necessary in order to create specific regulatory bodies to police and protect the environment through proper measures such as pollution control and any element that impacts against water in general. These bodies have already been mentioned and are the EA and SEPA but their responsibilities will also extend to specific codes of practice for recreational duties areas and sites of special scientific interests and repair of contaminated land.

Waste Electrical and Electronic Equipment EU Directive 2007

The Waste Electrical and Electronic Equipment (WEEE) Directive will ensure that Britain complies with its EU obligation to recycle waste from electrical products. The regulations came into effect in July 2007 and from that date any company that makes, distributes or trades in electrical or electronic goods such

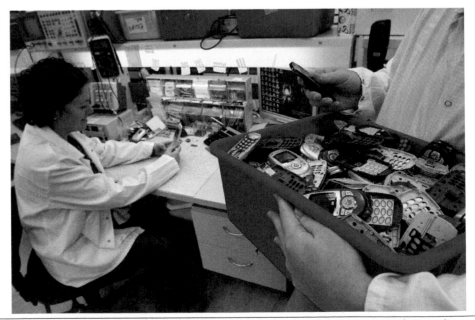

Figure 1.4 Mobile phones contain precious metals and expensive components that can be recycled even if the phones themselves can't be repaired.

as household appliances, sports equipment and even torches and toothbrushes has to make arrangements for recycling these goods at the end of their useful life. Batteries will be covered separately by yet another forthcoming EU directive. Some sectors are better prepared for the new regulations than others. Mobile phone operators, O2, Orange, Virgin and Vodafone, along with retailers such as Currys and Dixons, have already joined together to recycle their mobile phones collectively. In Holland the price of a new car now includes a charge for the recycling costs.

Further information is available on the DTI and DEFRA website under WEEE.

Safety first

Hazardous waste needs to be disposed of through licensed recycle operators.

Assessment criteria 1.3

Describe the ways in which the environment may be affected by work activities

A safe attitude towards health and safety, working conscientiously and neatly, keeping passageways clear and regularly tidying up the workplace is the sign of a good and competent craftsman.

However, the same kind of premise must be levied at those who work in industry in order to reduce the effect on the environment, in particular ensuring that some substances are afforded special consideration and therefore need to be disposed of through licensed operators. Such consideration and applied process is also necessary when burning materials or dealing with contaminated water since, for instance, the use of industrial drains will ensure that any water that contains chemicals is filtered accordingly and does not enter ground water drains. Consideration is also needed to plan any on-going works so as not to affect watercourses as well as implementing pollution action plans and response protocols.

Failure to manage or adopt such systems can lead to, for example, land being affected by contamination, the introduction of damaging and harmful particulates into the Earth's atmosphere, and contamination and pollution of water resources. What happens thereafter is that people can be subjected to disease directly through the environment or indirectly through animals and food crops.

Assessment criteria 1.4

Identify and interpret the requirement for electrical installations as outlined in relevant sections of the Building Regulations and the Code for Sustainable Homes

Building Regulations – Part P 2013 and the self-certification scheme

The Building Regulations (2010 as amended) are one of the most important pieces of legislation controling the installation of environmental technology systems. Part P of the Building Regulations was published on 22 July

2004, bringing domestic electrical installations in England and Wales under Building Regulations control. An amended document was published in an attempt at greater clarity and this came into effect on 6 April 2006. A further amendment occurred in 2013, but it is important to note that this only applies to England.

Part A structure

Part B fire safety

Part C site preparation and resistance to moisture

Part D toxic substances

Part E resistance to the passage of sound

Part F ventilation

Part G hygiene

Part H drainage and waste disposal

Part J combustion appliances and fuel storage systems

Part K protection from falling, collision and impact

Part L conservation of fuel and power

Part M access and facilities for disabled people

Part N glazing – safety in relation to impact, opening and cleaning

Part P electrical safety.

Part Q security – dwellings (England only)

Anyone carrying out domestic electrical installation work must comply with Part P of the Building Regulations. If the electrical installation meets the requirements of the IET Regulations BS 7671, then it will also meet the requirements of Part P of the Building Regulations, so no change there. However, a major change occurred through the concept of 'notification' to carry out electrical work.

Notifiable electrical work

Any work to be undertaken by a firm or individual who is *not* registered under an 'approved competent person scheme' must be notified to the Local Authority Building Control Body before work commences. That is, work that involves:

- the provision of at least one new circuit;
- work carried out in kitchens;
- work carried out in bathrooms;
- work carried out in special locations such as swimming pools and hot-air saunas.

Upon completion of the work, the Local Authority Building Control Body will test and inspect the electrical work for compliance with Part P of the Building Regulations.

Figure 1.5 The Local Authority Building Control Body should be notified before carrying out electrical work within bathrooms.

Non-notifiable electrical work

This is work carried out by a person or firm registered under an authorized Competent Persons Self-Certification Scheme or electrical installation work that does not include the provision of a new circuit. This includes work such as:

- replacing accessories such as socket outlets, control switches and ceiling roses;
- replacing a like for like cable for a single circuit that has become damaged by, for example, impact, fire or rodent;
- refixing or replacing the enclosure of an existing installation component provided the circuit's protective measures are unaffected;
- providing mechanical protection to existing fixed installations;
- adding lighting points (light fittings and switches) to an existing circuit, provided that the work is not in a kitchen, bathroom or special location;
- installing or upgrading the main or supplementary equipotential;
- bonding provided that the work is not in a kitchen, bathroom or special location.

All replacement work is non-notifiable even when carried out in kitchens, bathrooms and special locations, but certain work carried out in kitchens, bathrooms and special locations may be notifiable, even when carried out by an authorized skilled or instructed person. The IET have published a guide called *The Electricians' Guide to the Building Regulations* that brings clarity to this subject. In specific cases the Local Authority building control officer or an approved inspector will be able to confirm whether Building Regulations apply.

Failure to comply with the Building Regulations is a criminal offence and Local Authorities have the power to require the removal or alteration of work that does not comply with these requirements. Electrical work carried out by DIY homeowners will still be permitted after the introduction of Part P. Those carrying out notifiable DIY work must first submit a building notice to the Local Authority before the work begins. The work must then be carried out to the standards set by the IET Wiring Regulations BS 7671 and a building control fee paid for such work to be inspected and tested by the Local Authority.

Top tip

Part P requirements in England have recently changed and no longer recognize kitchens as special locations. This is in contrast to Wales which still recognizes kitchens as well as bathrooms as special locations.

Competent Persons Scheme

The Competent Persons Self-Certification Scheme is aimed at those who carry out electrical installation work as the primary activity of their business. The government has approved schemes to be operated by BRE Certification Ltd, British Standards Institution, ELECSA Ltd, NICEIC Certification Services Ltd (see certsure.com), and Napit Certification Services Ltd. All the different bodies will operate the scheme to the same criteria and will be monitored by the Department for Communities and Local Government, formerly called the Office of the Deputy Prime Minister.

Installers of environmental technology systems must also be registered under the scheme. Those individuals or firms wishing to join the Competent Persons Scheme will need to demonstrate their competence, if necessary, by first undergoing training. The work of members will then be inspected at least once each year. There will be an initial registration and assessment fee and then an annual membership and inspection fee.

Code for sustainable homes

The use of energy to provide heat for central heating and hot water in our homes is responsible for 60% of a typical family's energy bill. Heating accounts for over half of Britain's entire use of energy and carbon emissions. If Britain is to reduce its carbon footprint and achieve energy security, we must revolutionize the way we keep warm in the home.

At present 69% of our home heating comes from burning gas, 11% from oil, 3% from solid fuels such as coal and 14% from electricity, which is mainly generated from these same three fossil fuels. Only 4% is currently provided by renewable sources. If Britain is to meet its clean energy targets, renewable sources will have to increase, and the revolution will have to start in the home because the country's dwellings currently provide more than half of the total demand, almost entirely for hot water and central heating.

There are about 20 million homes in the UK and a review of present buildings has found that about six million homes have inadequately lagged lofts, eight million have uninsulated cavity walls and a further seven million homes with solid walls would benefit from better insulation. If the country is to achieve its reduced carbon emissions targets, these existing homes must be heavily insulated to reduce energy demand and then supplied with renewable heat.

We cannot sustain the present level of carbon emissions without disastrous ecological consequences in the future. Low carbon homes are sustainable homes. HRH The Prince of Wales has entered the debate saying

becoming more sustainable is possibly the greatest challenge humanity has faced and I am convinced that it is therefore, the most remarkable chance to secure a prosperous future for everyone. We must strive harder than ever before to convince people that by living sustainably we will improve our quality of life and our health; that by living in harmony with nature we will protect the intricate, delicate balance of the natural systems that ultimately sustain us.

(Daily Telegraph, 31 July 2010)

The Code for Sustainable Homes, see Fig. 1.6, measures the sustainability of a home against categories of sustainable design, rating the whole home as a complete package, including building materials and services within the building.

The Code uses a one to six star rating to communicate the overall sustainability performance of a new home and sets minimum standards for energy and water use at each level.

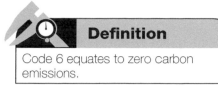

Since May 2008 all new homes are required to have a Code Rating and a Code Certificate. By 2016 all new homes must be built to zero-carbon standards, which will be achieved through step by step tightening of the Building Regulations. If we look at sustainability from a manufacturing point of view, sustainable manufacture is based on the principle of meeting the needs of the current generation without compromising the ability of future generations to meet their needs.

The topics that we will consider in this chapter, together with improved levels of insulation, would all help our homes to be more sustainable.

Date	2010	2013	2016
Energy efficiency improvement of the dwelling compared to 2006 (part L Building Regulations)	2.5%	44%	Zero carbon
Equivalent Standard within the Code	Code level 3 ⌂ THE CODE FOR SUSTAINABLE HOMES ★★★☆☆☆	Code level 4 ⌂ THE CODE FOR SUSTAINABLE HOMES ★★★★★☆	Code level 6 ⌂ THE CODE FOR SUSTAINABLE HOMES ★★★★★★

Figure 1.6 Code for Sustainable Homes.

Assessment criteria 1.5

State materials and products that are classed as hazardous to the environment or recyclable

Assessment criteria 1.6

Describe the organizational procedures for processing materials that are classed as hazardous to the environment or recyclable

New Hazardous Waste Regulations were introduced in July 2005 and all waste material is now coded but any hazardous waste material is marked with an asterisk in order to highlight the danger. Furthermore, the Environment Agency emphasize what are termed as hazardous outright and are coloured red accordingly.

Safety first

- Clean up before you leave the job.
- Put waste in the correct skip.
- Recycle used lamps and tubes.
- Get rid of all waste responsibly.

Under these regulations lamps and tubes are classified as hazardous. While each lamp contains only a small amount of mercury and phosphorous, vast numbers of lamps and tubes are disposed of in the UK every year, resulting in a significant environmental threat.

The environmentally responsible way to dispose of fluorescent lamps and tubes is to recycle them. The process usually goes like this:

- Your employer arranges for the local electrical wholesaler to deliver a plastic used lamp waste container of an appropriate size for the job.
- Expired lamps and tubes are placed whole into the container, which often has a grating inside to prevent the tubes from breaking when being transported.
- When the container is full of used lamps and tubes, you telephone the electrical wholesaler and ask them to pick up the filled container and deliver it to one of the specialist recycling centres.
- Your electrical company will receive a 'Duty of Care Note' and full recycling documents, which should be filed safely as proof that the hazardous waste was recycled safely.
- The charge is approximately 50p for each 1800 mm tube and this cost is passed on to the customer through the final account.

Batteries would also come within this category due to the possibility of leakage from the electrolyte inside.

Controlled Waste Regulations 2012

Under these regulations we have a 'duty of care to handle, recover and dispose of all waste responsibly'. This means that all waste must be handled, recovered and disposed of by individuals or businesses that are authorized to do so under a system of signed Waste Transfer Notes. The Environmental Protection (Duty of Care) Regulations 1991 state that as a business you have a duty to ensure that any waste you produce is handled safely and in accordance with the law. This is the duty of care and applies to anyone who produces, keeps, carries, treats or disposes of waste from business or industry.

You are responsible for the waste that you produce, even after you have passed it on to another party such as a skip hire company, a scrap metal merchant, recycling company or local council. The duty of care has no time limit and extends until the waste has either been finally and properly disposed of or fully recovered. So what does this mean for your company?

- Make sure that waste is only transferred to an authorized company.
- Make sure that waste being transferred is accompanied by the appropriate paperwork showing what was taken, where it was to be taken and by whom.
- Segregate the different types of waste that your work creates.
- Label waste skips and waste containers so that it is clear to everyone what type of waste goes into that skip.
- Minimize the waste that you produce and do not leave waste behind for someone else to clear away.

Remember, there is no time limit on your duty of care for waste. Occupiers of domestic properties are exempt from the duty of care for the household waste that they produce. However, they do have a duty of care for the waste produced by, for example, a tradesperson working at a domestic property.

Special waste is covered by the Special Waste Regulations 1996 and is waste that is potentially hazardous or dangerous and which may, therefore, require special precautions during handling, storage, treatment or disposal. Examples of special waste are asbestos, lead-acid batteries, used engine oil, solvent based paint, solvents, chemical waste and pesticides. The disposal of special waste must be carried out by a competent person, with special equipment and a licence.

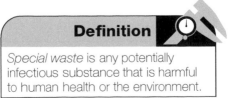

Definition

Special waste is any potentially infectious substance that is harmful to human health or the environment.

Figure 1.7 In order to work with such hazardous materials asbestos removal workers must have a licence.

Assessment criteria 2.1

State installation methods that can help to reduce material wastage

Current government incentives endorse the view that waste should be reduced if possible, but if this is not achievable: reusing it. Consequently, a waste policy looks to protect human health and the impact on the environment and as such a hierarchy of waste management has been produced, which can be seen below.

Waste prevention

The best practice approach to reducing waste is a step-by-step approach and includes the actual design of a building, since judicious use and selection of materials can reduce the waste stream. The process of buying materials is also important and whilst insufficient material stock is never ideal, stock levels should be managed in a way that does not permit excess waste allowance.

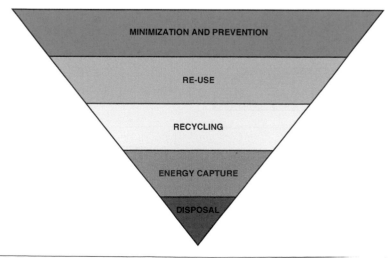

Figure 1.8 Remember that landfill is a last resort.

Re-use or recycle

Re-use

The part coils of cable and any other re-usable left-over lengths of conduit, trunking or tray should be taken back to your employer's stores area. Here it will be stored for future use and the returned quantities deducted from the costs allocated to that job.

Recycle

Recycling involves everyone in all environments including all households in recycling white goods such as microwaves and fridges to plastic bottles and cans.

Energy capture

Traditionally, energy has been generated through waste by burning it, however this has led to phenomena such as acid gases, acid rain and ash. A modern alternative is to create massive 'district heating' systems in use in cities such as Newcastle.

Disposal

But what do you do with the rubbish that the working environment produces? Well, all the packaging material for electrical fittings and accessories usually goes into either your employer's skip or the skip on-site designated for that purpose. All the off-cuts of conduit, trunking and tray also go into the skip. In fact, most of the general site debris will probably go into the skip and the waste disposal company will take the skip contents to a designated local council landfill area for safe disposal.

What goes into the skip for normal disposal into a landfill site is usually a matter of common sense but it is also the least favoured option.

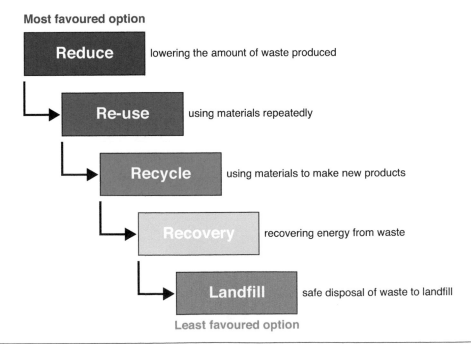

Most favoured option

Reduce — lowering the amount of waste produced

Re-use — using materials repeatedly

Recycle — using materials to make new products

Recovery — recovering energy from waste

Landfill — safe disposal of waste to landfill

Least favoured option

Figure 1.9 Materials should only be sent to landfill as a last resort.

Assessment criteria 2.2

Explain why it is important to report any hazards to the environment that arise from work procedures

If you suspect that an incident or accident has or might affect the environment then it must be reported to your immediate supervisor. The consequences of not reporting such issues could have serious implications, especially when considering exposure to asbestos since it is cumulative – every exposure builds upon the previous ones.

Assessment criteria 2.3

Specify environmentally friendly materials, products and procedures that can be used in the installation and maintenance of electrotechnical systems and equipment

We all have a collective responsibility when it comes to the environment and not just to our families or employer but also to protect future generations through a shift in mindset. There are other benefits, however, since it can reduce both domestic and industrial bills.

Measures include:

- fitting thermostatic control in order to turn down room or working temperature;
- fitting energy efficient replacement items such as compact fluorescent light bulbs or LED fittings;
- installing water meters or smart gas/electricity meters;
- installing insulation;
- repairing leaking taps and pipes;
- fitting light motion sensors in lighting systems or switching off unused rooms, corridors and working areas.

Further development has stemmed from Part L of the Building Regulations and specifically the introduction of Energy Performance Certificates, which grade the environmental efficiency of a building from A being recognized as peak efficiency to G the least efficient. In the electrical installation industry more and more use is being made of installing modular assemblies that have been assembled and tested at the factory. There is therefore less fabrication required on site, which offers a means of waste reduction.

> **Key fact**
>
> Modular or pre-made equipment offers a means of reducing waste.

Assessment criteria 3.1

Describe the fundamental operating principles of environmental technology systems

Assessment criteria 3.2

State the applications and limitations of environmental technology systems

Assessment criteria 3.3

State the Local Authority Building Control requirements that apply to the installation of environmental technology systems

Environmental technology systems and renewable energy

Energy is vital to the modern industrial economy in the UK and Europe. We also need energy in almost every aspect of our lives, to heat and light our homes and offices, to enable us to travel on business or for pleasure, and to power our business and industrial machines.

In the past the UK has benefited from its fossil fuel resources of coal, oil and gas but respectable scientific sources indicate that the fossil fuel era is drawing to a close. Popular estimates suggest that gas and oil will reach peak production in the year 2060 with British coal reserves lasting only a little longer. Therefore we must look to different ways of generating electricity so that:

- the remaining fossil fuel is conserved;
- our CO_2 emissions are reduced to avoid the consequences of climate change;
- we ensure that our energy supplies are secure, and not dependent upon supplies from other countries.

Following the introduction of the Climate Change Act in 2008 the UK and other Member States agreed an EU-wide target of 20% renewable energy by the year 2020 and 60% by 2050. Meeting these targets will mean basing much of the new energy infrastructure around renewable energy, particularly offshore wind power.

The 'Energy Hierarchy' states that organizations and individuals should address energy issues in the following order so as to achieve the agreed targets:

1 Reduce the need for energy – reducing energy demand is cost saving, reduces greenhouse gas emissions and contributes to the security of supply. Reducing the energy loss from buildings by better insulation and switching off equipment when not in use is one way of achieving this target.
2 Use energy more efficiently – use energy-efficient lamps and 'A'-rated equipment.
3 Use renewable energy – renewable energy refers to the energy that occurs naturally and repeatedly in the environment. This energy may come from wind, waves or water, the sun, or heat from the ground or air.
4 Any continuing use of fossil fuels should use clean and efficient technology. Power stations generating electricity from coal and oil (fossil fuel) release a lot of CO_2 in the generating process. New-build power stations must now be fitted with carbon capture filters to reduce the bad environmental effects.

Key fact

Renewable energy is no less reliable than energy generated from more traditional sources.

Funding for environmental technology systems

Renewable energy is no less reliable than energy generated from more traditional sources. Using renewable energy does not mean that you have to change your lifestyle or your domestic appliances. There has never been a better time to consider generating energy from renewable technology than now because grants and funding are available to help individuals and companies.

The Low Carbon Building Programme implemented by the Department of Energy and Climate Change (DECC) provides grants towards the installation of renewable technologies that are available to householders, public non-profit-making organizations and commercial organizations across the UK.

The government's 'Feed-in Tariff' pays a tax-free sum which is guaranteed for 25 years. It is called 'Clean energy cash back' and has been introduced to promote the uptake of small-scale renewable and low-carbon electricity generation technologies. The customer receives a generation tariff from the electricity supplier, whether or not any electricity generated is exported to the national grid, and an additional export tariff when electricity is transported to the electricity grid through a smart meter.

Key fact

The government's 'Feed-in Tariff' pays a tax-free sum which is guaranteed for 25 years.

From April 2010, clean energy generators were paid 41.3p for each kWh of electricity generated. Surplus energy fed back into the national grid earns an extra 3p per unit. However, the scheme has been very popular so the feedback tariffs were reduced to 21p, and from 1 August 2012 the fee fell to only 16p for each kWh. Current rates are 4.32p for each kWh for systems with a capacity of less than 10kW. If you add to this the electricity bill savings, a normal householder could still make some savings. Savings vary according to energy use and type of system used. The Energy Saving Trust at www.energysavingtrust. org.uk, British Gas at www.britishgas.co.uk and Ofgem at www.ofgem.gov.uk/ fits provide an online calculator to determine the cost, size of system and CO_2 savings for PV systems.

Microgeneration technologies

Microgeneration is defined in the Energy Act 2004 Section 82 as the generation of heat energy up to 45 kW and electricity up to 50 kW.

Today, microgeneration systems generate relatively small amounts of energy at the site of a domestic or commercial building. However, it is estimated that by 2050, 30 to 40% of the UK's electricity demand could be met by installing microgeneration equipment to all types of building.

In the USA, the EU and the UK buildings consume more than 70% of the nation's electricity and contribute almost 40% of the polluting CO_2 greenhouse gases. Any reductions which can be made to these figures will be good for the planet, and hence the great interest today in the microgeneration systems. Microgeneration technologies include small wind turbines, solar photovoltaic (PV) systems, small-scale hydro and micro-CHP (combined heat and power) systems. Microgenerators that produce electricity may be used as stand-alone systems, or may be run in parallel with the low-voltage distribution network; that is, the a.c. mains supply.

Key fact

Microgeneration technologies include small wind turbines, solar photovoltaic (PV) systems, small-scale hydro and micro-CHP (combined heat and power) systems.

The April 2008 amendments to the Town and Country Planning Act 1990 brought in 'permitted development' to allow the installation of microgeneration systems within the boundary of domestic premises without obtaining planning permission. However, size limitations have been set to reduce the impact upon neighbours. For example, solar panels attached to a building must not protrude more than 200mm from the roof slope, and stand-alone panels must be no higher than four metres above ground level and no nearer than five metres from the property boundary. See the Electrical Safety Council site for advice on connecting microgeneration systems at www.esc.org.uk/bestpracticeguides.html.

Smart electricity meters

Smart electricity meters are designed to be used in conjunction with micro-generators. Electricity generated by the consumer's microgenerator can be sold back to the energy supplier using the 'smart' two-way meter.

From 2012 the Department of Energy and Climate Change began introducing smart meters into consumers' homes, and this is expected to run until 2020, with the aim being to help consumers reduce their energy bills.

When introducing the proposal Edward Davey, the Energy and Climate Change Secretary, said, 'the meters which most of us have in our homes were designed for a different age, before climate change. Now we need to get smarter with our energy. This is a big project affecting 26 million homes and several million businesses. The project will lead to extra work for electrical contractors through installing the meters on behalf of the utility companies and implementing more energy-efficient devices once customers can see how much energy they are using.'

Figure 1.10 Smart electricity meter.

Figure 1.11 Solar photovoltaic (PV array).

Already available is the Real Time Display (RTD) wireless monitor which enables consumers to see exactly how many units of electricity they are using through an easy to read portable display unit. By seeing the immediate impact in pence per hour of replacing existing lamps with low-energy ones or switching off unnecessary devices throughout the home or office, consumers are naturally motivated to consider saving energy. RTD monitors use a clip-on sensor on the meter tails and include desktop software for PC and USB links.

Let us now look at some of these microgeneration technologies.

Microgeneration technologies

Electricity-producing technologies

1 Solar photovoltaic (PV) microgenerators
2 Wind energy microgenerators
3 Hydro microgenerators

Heat-producing technologies

4 Solar thermal (hot water) microgenerators
5 Ground source heat pump microgenerators
6 Air source heat pump microgenerators
7 Water source heat pump microgenerators
8 Biomass microgenerators

Co-generation technologies

9 Combined heat and power (CHP) microgenerators

Water conservation technologies

10 Rainwater harvesting
11 Grey water recycling

1. Solar photovoltaic (PV) microgenerators

A solar photovoltaic (PV) system is a collection of PV cells known as a PV string, that forms a PV array and collectively are called a PV generator which turns sunlight directly into electricity. PV systems may be 'stand-alone' power supplies or be designed to operate in parallel with the public low-voltage distribution network; that is, the a.c. mains supply.

Stand-alone PV systems are typically a small PV panel of maybe 300 mm by 300 mm tilted to face the southern sky, where it receives the maximum amount of sunlight. They typically generate 12 to 15 volts and are used to charge battery supplies on boats, weather stations, road signs and any situation where electronic equipment is used in remote areas away from a reliable electrical supply.

The developing nations are beginning to see stand-alone PV systems as the way forward for electrification of rural areas beyond the National Grid rather than continuing with expensive diesel generators and polluting kerosine lamps.

The cost of PV generators is falling. The period 2009 to 2010 saw PV cells fall by 30% and with new 'thin-film' cells being developed, the cost is expected to continue downward. In the rural areas of the developing nations they see PV systems linked to batteries bringing information technology, radio and television to community schools. This will give knowledge and information to the next generation which will help these countries to develop a better economy, a better way of life and to have a voice in the developed world.

Stand-alone systems are not connected to the electricity supply system and are therefore exempt from much of BS 7671, the IET Regulations. However, Regulation 134.1.1, 'good workmanship by skilled or instructed persons and

Definition

A *solar photovoltaic* (PV) system is a collection of PV cells known as a PV string, that forms a PV array and collectively are called a PV generator which turns sunlight directly into electricity.

proper materials shall be used in all electrical installations,' will apply to any work done by an electrician who must also pay careful attention to the manufacturer's installation instructions.

PV systems designed to operate in parallel with the public low-voltage distribution network are the type of microgenerator used on commercial and domestic buildings. The PV cells operate in exactly the same way as the stand-alone system described above, but will cover a much greater area. The PV cells are available in square panels which are clipped together and laid over the existing roof tiles as shown in Fig. 1.12, or the PV cells may be manufactured to look just like the existing roof tiles which are integrated into the existing roof.

Figure 1.12 PV system in a domestic situation.

A solar PV system for a domestic three-bedroom house will require approximately 15 to 20 square metres generating 2 to 3 kilowatts of power. To install a 4KWh system, the average costs involved would range between £6000–£7000. On the positive side, a PV system for a three-bedroom house will save around 1,200 kg of CO_2 per year.

These bigger microgeneration systems are designed to be connected to the power supply system and the installation must therefore comply with Section 712 of BS 7671: 2008. Section 712 contains the requirements for protective measures comprising automatic disconnection of the supply wiring systems, isolation, switching and control, earthing arrangements and labelling. In addition, the installation must meet the requirements of the Electricity Safety Quality and Continuity Regulations 2006. This is a mandatory requirement. However, where the output does not exceed 16A per line, they are exempt from some of the requirements, provided that:

- the equipment is type tested and approved by a recognized body;
- the installation complies with the requirements of BS 7671, the IET Regulations;
- the PV equipment must disconnect from the distributor's network in the event of a network fault;
- the distributor must be advised of the installation before or at the time of commissioning.

Installations of less than 16A per phase but up to 5 kilowatt peak (kWp) will also be required to meet the requirements of the Energy Network Assciation's Engineering Recommendation G83/1 for small-scale embedded generators in parallel with public low-voltage distribution networks. Installations generating more than 16A must meet the requirements of G59/1 which requires approval from the distributor before any work commences. The installer must be MCS-accredited if the client is to receive the feedback tariff from the energy supplier.

2. Wind energy microgenerators

Modern large-scale wind machines are very different from the traditional windmill of the last century which gave no more power than a small car engine. Very large structures are needed to extract worthwhile amounts of energy from the wind. Modern large-scale wind generators are taller than electricity pylons, with a three-blade aeroplane-type propeller to catch the wind and turn the generator. If a wind turbine was laid down on the ground, it would be longer and wider than a football pitch. They are usually sited together in groups in what has become known as 'wind energy farms,' as shown in Fig. 1.13.

Each modern grid-connected wind turbine generates about 600 kW of electricity. A wind energy farm of 20 generators will therefore generate 12 MW, a useful contribution to the national grid, using a naturally occurring, renewable, non-polluting source of energy. The Department of Energy and Climate Change considers wind energy to be the most promising of the renewable energy sources.

In 2016 there are approximately 5,220 onshore turbines and 1,465 offshore turbines. Combined they generate 33,940,073 MWh, which in turn powers 8,247,891 homes and reduces CO_2 yearly emissions by 14,594,232 tonnes.

Figure 1.13 Offshore wind farm energy generators.

At one time, the Countryside Commission, the government's adviser on land use, had calculated that to achieve a target of generating 10% of the total electricity supply by wind power will require 40,000 generators of the present size. In 2018 it was reported that wind power contributes approximately 8% of the total of the electricity generated in the UK.

Wind power is an endless renewable source of energy, is safer than nuclear power and provides none of the polluting emissions associated with fossil fuel. If there was such a thing as a morally pure form of energy, then wind energy would be it. However, wind farms are, by necessity, sited in some of the most beautiful landscapes in the UK. Building wind energy farms in these areas of outstanding natural beauty has outraged conservationists. Prince Charles has reluctantly joined the debate saying that he was in favour of renewable energy sources but believed that 'wind farms are an horrendous blot on the landscape'. He believes that if they are to be built at all they should be constructed well out to sea.

The next generation of wind farms will mostly be built offshore, where there is more space and more wind, but the proposed size of these turbines creates considerable engineering problems. From the sea bed foundations to the top of the turbine blade will be up to a staggering 250 metres, three times the height of the Statue of Liberty. Each offshore turbine, generating between 5 and 7 MW, will weigh between 200 and 300 tonnes. When you put large wind forces onto that structure you will create a very big cantilever effect which creates a major engineering challenge.

The 100 turbine 'Thanet' wind farm just off the Kent coast generates enough power to supply 200,000 homes. The Thanet project cost £780 million to build and was the world's largest, when it opened in 2010. The turbines are up to 380 feet high and cover an area as large as 4,000 football pitches. The Thanet project did not retain its title as the world's largest wind farm for long though, because the 'Greater Gabbard' wind farm, off the north-east coast with 140 turbines, opened in 2013. These projects bring Britain's total wind energy capacity above 5,000 megawatts for the first time and more are being built.

The Department of Energy and Climate Change has calculated that 10,000 wind turbines could provide the energy equivalent of eight million tonnes of coal per year and reduce CO_2 emissions. While this is a worthwhile saving of fossil fuel, opponents point out the obvious disadvantages of wind machines, among them the need to maintain the energy supply during periods of calm, which means that wind machines can only ever supplement a more conventional electricity supply.

Small wind microgenerators can be used to make a useful contribution to a domestic property or a commercial building. They can be stand-alone about the size of a tall street lamp. A 12 m-high turbine costs about £24,000 and, with a good wind, will generate 10,000 kWh per year enough for three small domestic homes. However, if you live in a village, town or city you are unlikely to obtain the the Local Authority Building and Planning permissions to install a wind generator because your neighbours will object.

Small wind generators of the type shown in Fig. 1.14 typically generate between 1.5A and 15A in wind speeds of 10 mph to 40 mph.

Definition

Wind power is an endless renewable source of energy, is safer than nuclear power and provides none of the polluting emissions associated with fossil fuel.

Key fact

The UK Government announced in March 2019 that they would double the present capacity of off shore wind farms in order to meet a new target of generating 30% of energy from renewable sources by the year 2030.

Key fact

The Department of Energy and Climate Change has calculated that 10,000 wind turbines could provide the energy equivalent of eight million tonnes of coal per year and reduce CO_2 emissions.

Figure 1.14 Small wind generator on a domestic property.

Hydroelectric power generation

The UK is a small island surrounded by water. Surely we could harness some of the energy contained in tides, waves, lakes and rivers? Many different schemes have been considered in the past 20 years and a dozen or more experimental schemes are currently being tested.

Water power makes a useful contribution to the energy needs of Scotland but the possibility of building similar hydroelectric schemes in England is unlikely chiefly due to the topographical nature of the country.

The Severn Estuary has a tidal range of 15 m, the largest in Europe, and a reasonable shape for building a dam across the estuary. This would allow the basin to fill with water as the tide rises, and then allow the impounded water to

Key fact

Future developments include the Swansea tidal lagoon project which aims to harness the power within tidal surges in the sea in order to drive specialised turbines to generate electricity. The Swansea lagoon could potentially be the first of six installed around Britain's coastline, which potentially could generate 8% of the UK's electricity.

Figure 1.15 Tidal flow water turbines.

flow out through electricity-generating turbines as the tide falls. However, such a tidal barrier might have disastrous ecological consequences upon the many wildfowl and wading bird species by the submerging of the mudflats which now provide winter shelter for these birds. Therefore, the value of the power which might be produced must be balanced against the possible ecological consequences.

France has successfully operated a 240 MW tidal power station at Rance in Brittany for the past 25 years.

Marine Current Turbines Ltd are carrying out research and development on submerged turbines which will rotate by exploiting the principle of flowing water in general and tidal streams in particular. The general principle is that an 11 m diameter water turbine is lowered into the sea down a steel column drilled in the sea bed. The tidal movement of the water then rotates the turbine and generates electricity.

The prototype machines were submerged in the sea off Lynmouth in Devon. In May 2008 they installed the world's first tidal turbine in the Strangford Narrows in Northern Ireland where it is now grid-connected and generating 1.2 MW.

All the above technologies are geared to providing hydroelectric power connected to the national grid, but other micro-hydro schemes are at the planning and development stage.

3. Hydro microgenerators

The use of small hydropower (SHP) or micro-hydropower has grown over recent decades led by continuous technical developments, brought about partly in the UK by the 2010 coalition government's 'feed-in tariff' where green electricity producers are paid a premium to produce electricity from renewable sources.

The normal perception of hydropower is of huge dams, but there is a much bigger use of hydropower in smaller installations. Asia, and especially China, is set to become a leader in hydroelectric generation. Australia and New Zealand are focusing on small hydro plants. Canada, a country with a long tradition of hydropower, is developing small hydropower as a replacement for expensive diesel generation in remote off-grid communities.

Small hydropower schemes generate electricity by converting the power available in rivers, canals and streams. The object of a hydropower scheme is to convert the potential energy of a mass of water flowing in a stream with a certain fall, called the head, into electrical energy at the lower end of the stream where the powerhouse is located. The power generated is proportional to the flow, called the discharge, and to the head of water available. The fundamental asset of hydropower is that it is a clean and a renewable energy source and the fuel used, water, is not consumed in the electricity-generating process.

In the Derbyshire Peak District along the fast-flowing River Goyt there were once 16 textile mills driven by waterwheels. The last textile mill closed in the year 2000 but the Old Torr Mill has been saved. Where once the waterwheel stood is now a gigantic 12-tonne steel screw, 2.4 metres in diameter. The water now drives the Reverse Archimedian Screw, affectionately called 'Archie', to produce 130,000 kWh per year, enough electricity for 40 homes. The electricity-generating

Definition

Small hydropower schemes generate electricity by converting the power available in rivers, canals and streams.

project is owned by the residents of New Mills in a sharing cooperative in which surplus electricity is sold back to the grid. The installation cost was £300,000 in 2008. See Fig. 1.16. See also torreshydro new mills website and manpower for info and an interesting video of fish swimming through the turbine.

The type of turbine chosen for any hydro scheme will depend upon the discharge rate of the water and the head of water available. A Pelton Wheel is a water turbine in which specially shaped buckets attached to the periphery of the wheel are struck by a jet of water. The kinetic energy of the water turns the wheel which is coupled to the generator.

Axial turbines comprise a large cylinder in which a propeller-type water turbine is fixed at its centre. The water moving through the cylinder causes the turbine blade to rotate and generate electricity.

A Francis Turbine or Kaplan Turbine is also an axial turbine but the pitch of the blades can be varied according to the load and discharge rate of the water.

Small water turbines will reach a mechanical efficiency at the coupling of 90%.

Up and down the country, riverside communities must be looking at the relics of our industrial past and wondering if they might provide a modern solution for clean, green, electrical energy. However, despite the many successes and obvious potential, there are many barriers to using waterways for electricity generation in the European countries. It is very difficult in this country to obtain permission from the Waterways Commission to extract water from rivers, even though, once the water has passed through the turbine, it is put back into the river. Environmental pressure groups are opposed to micro-hydro generation because of its perceived local environmental impact on the river ecosystem and the disturbance to fishing. Therefore, once again, the value of the power produced would have to be balanced against the possible consequences.

4. Solar thermal (hot water) microgenerators

Solar thermal hot water heating systems are recognized as a reliable way to use the energy of the sun to heat water. The technology is straightforward and solar thermal panels for a three-bedroomed house cost at the time of going to press between £3,000 and £6,000 for a 3 to 6 m^2 panel and they will save about 260 kg of CO_2 annually.

The solar panel comprises a series of tubes containing water that is pumped around the panel and a heat exchanger in the domestic water cylinder as shown in Fig. 1.17. Solar energy heats up the domestic hot water. A solar panel of about 4 m^2 will deliver about 1,000 kWh per year which is about half the annual water demand of a domestic dwelling. However, most of the heat energy is generated during the summer and so it is necessary to supplement the solar system with a boiler in the winter months. Figure 1.18 shows a photo of an installed solar hot water panel.

If you travel to Germany, you will see a lot of PV and solar thermal panels on the roofs there. In the UK, planning requirements for solar thermal and PV installations have already been made much easier by 'permitted development'. A website detailing planning requirements for solar and wind may be found at www.planningportal.gov.uk/planning.

Figure 1.16 An example of an Archimedian Screw at the River Dart country park, Devon. Credit: WRE Limited.

Key fact

The type of turbine chosen for any hydro scheme will depend upon the discharge rate of the water and the head of water available.

Definition

Solar thermal hot water heating systems are recognized as a reliable way to use the energy of the sun to heat water.

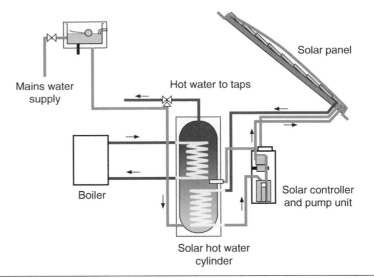

Figure 1.17 Solar-powered hot water system.

Figure 1.18 Solar hot water panel.

Heat pumps

In applications where heat must be upgraded to a higher temperature so that it can be usefully employed, a **heat pump** must be used. Energy from a low-temperature source such as the earth, the air, a lake or river is absorbed by a gas or liquid contained in pipes, which is then mechanically compressed by an electric pump to produce a temperature increase. The high-temperature energy is then transferred to a heat exchanger so that it might do useful work, such as providing heat to a building. For every 1 kWh of electricity used to power the heat pump compressor, approximately 3 to 4 kWh of heating are produced.

How a heat pump works

1   A large quantity of low-grade energy is absorbed from the environment and transferred to the refrigerant inside the heat pump (called the evaporator). This causes the refrigerant temperature to rise, causing it to change from liquid to a gas.

2 The refrigerant is then compressed, using an electrically driven compressor, reducing its volume but causing its temperature to rise significantly.

3 A heat exchanger (condenser) extracts the heat from the refrigerant to heat the domestic hot water or heating system.

4 After giving up its heat energy, the refrigerant turns back into a liquid, and, after passing through an expansion valve, is once more ready to absorb energy from the environment and the cycle is repeated as shown in Fig. 1.19.

A refrigerator works on this principle. Heat is taken out of the food cabinet, compressed and passed onto the heat exchanger or radiator at the back of the fridge. This warm air then radiates by air convection currents into the room. Thus the heat from inside the cabinet is moved into the room, leaving the sealed refrigerator cabinet cold.

5. Ground source heat pump microgenerators

Ground source heat pumps extract heat from the ground by circulating a fluid through polythene pipes buried in the ground in trenches or in vertical boreholes as shown in Fig. 1.20. The fluid in the pipes extracts heat from the ground and a heat exchanger within the pump extracts heat from the fluid. These systems are most effectively used to provide underfloor radiant heating or water heating.

Calculations show that the length of pipe buried at a depth of 1.5 m required to produce 1.2 kW of heat will vary between 150 m in dry soil and 50 m in wet soil. The average heat output can be taken as 28 watts per metre of pipe. A rule of thumb guideline is that the surface area required for the ground heat exchanger should be about 2.5 times the area of the building to be heated.

This type of installation is only suitable for a new-build project and the ground heat exchanger will require considerable excavation and installation. The installer must seek Local Authority Building Control permissions before work commences.

> **Definition**
>
> *Ground source heat pumps* extract heat from the ground by circulating a fluid through polythene pipes buried in the ground in trenches or in vertical boreholes.

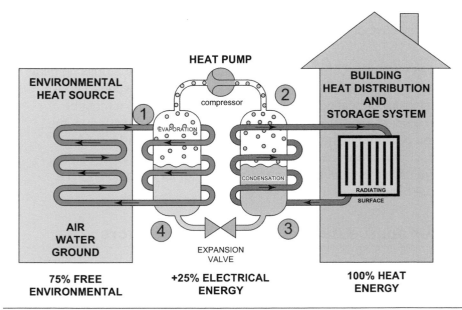

Figure 1.19 Heat pump working principle.

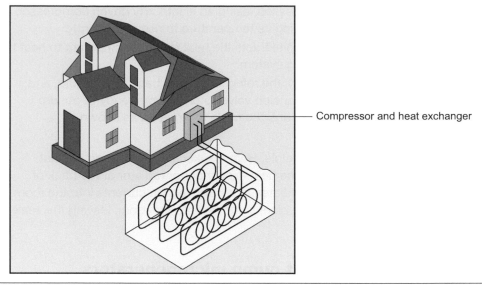

Figure 1.20 Ground source heating system.

Compressor and heat exchanger

6. Air source heat pump microgenerators

The performance and economics of heat pumps are largely determined by the temperature of the heat source and so we seek to use a high-temperature source. The heat sources used by heat pumps may be soil, the air, ground or surface water. Unfortunately all these sources follow the external temperature, being lower in winter when demand is highest. Normal atmosphere is an ideal heat source in that it can supply an almost unlimited amount of heat although unfortunately at varying temperatures, but relatively mild winter temperatures in the UK mean excellent levels of efficiency and performance throughout the year. For every 1 kWh of electricity used to power the heat pump compressor, between 3 and 4 kWh of heating energy is produced. They also have the advantage over ground source heat pumps of lower installation costs because they do not require any groundwork. Figure 1.21 shows a commercial air source heat pump.

If the air heat pump is designed to provide full heating with an outside temperature of 2 to 4 degrees centigrade, then the heat pump will provide approximately 80% of the total heating requirement with high performance and efficiency.

The point at which the output of a given heat pump meets the building heat demand is known as the 'balance point'. In the example described above, the 20% shortfall of heating capacity below the balance point must be provided by some supplementary heat. However, an air-to-air heat pump can also be operated in the reverse cycle which then acts as a cooling device, discharging cold air into the building during the summer months. So here we have a system which could be used for air conditioning in a commercial building.

7. Water source heat pump microgenerators

When we looked at the ground source heat pump at number 5 we said that heat was extracted from the ground by circulating a glycol fluid through polythene pipes buried in the ground. The wetter the soil the more conductive it is to the glycol. Water is an ideal conductor and therefore most suitable as a heat source.

Figure 1.21 Air source heat pump unit.

Connecting the polythene pipe work together to form a raft-type system will hold the collector together. It is then floated out onto the lake using a small dinghy. When in position the collector is filled with glycol which makes it sink to the bottom of the lake. The feeder pipes are connected to the heat pump at the adjacent building and over time the collector becomes covered with silt.

The lake must be close to the building and large enough to provide a stable temperature throughout the seasons so that the pipes do not freeze at the bottom of the lake.

8. Biomass microgenerators

Biomass is derived from plant materials and animal waste. It can be used to generate heat and to produce fuel for transportation. The biomass material may be straw and crop residues, crops grown specially for energy production such as trees or rape seed oil and waste from a range of sources including food production. The nature of the fuel will determine the way that energy can best be recovered from it.

There is a great deal of scientific research being carried out at the moment into 'biomass renewables'; that is, energy from crops. This area of research is at an early stage, but is expected to flourish in the next decade. The first renewable energy plant, which is to be located at Teesport on the River Tees in the north-east of England, has received approval from the Department of Energy and Climate Change for building to commence.

The facility will be one of the largest biomass plants to be built in the world and is scheduled to enter commercial operation in 2020. Young trees will be grown as a crop to produce wood chips. The plant will use 2.5 million tonnes of wood chips each year to produce 300 MW of electrical energy. The plant will operate 24 hours a day, all year round to meet some of the national grid base load. The USA have over 200 biomass plants operating which meet approximately 4% of the total US energy consumption.

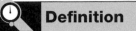 **Definition**

Biomass is derived from plant materials and animal waste. It can be used to generate heat and to produce fuel for transportation.

Definition

'*Biomass renewables*' is energy from crops.

Figure 1.22 Wood pellets being burned in a domestic biomass stove.

Domestic biomass heating can today be found in the increasing popularity of wood stoves. A wide variety of wood-burning appliances are used by households, particularly in rural areas of countries with traditional forest cultures such as Canada, Austria, Finland and Sweden. Since wood fuel was classified as a carbon-neutral fuel under the Kyoto Protocol, most countries that require winter space heating have begun to look again at wood.

Domestic and commercial boilers fuelled by wood logs, wood chips or wood pellets are available. Boiler designs and controls mean that modern boilers can be highly automated. With a water-based heat distribution system already in place, it is relatively simple for any household or commercial operation to remove an existing boiler and replace it with a wood-fuelled one that will do much the same job.

As long as they are locally and sustainably sourced, dry and well-seasoned logs are arguably the greenest of all wood fuels. Apart from drying, they require less processing, and therefore less energy than wood chips or wood pellets.

9. Combined heat and power (CHP) microgenerators

Definition

CHP is the simultaneous generation of usable heat and power in a single process. That is, heat is produced as a by-product of the power-generation process.

CHP is the simultaneous generation of usable heat and power in a single process. That is, heat is produced as a by-product of the power-generation process. A chemical manufacturing company close to where I live has a small power station which meets some of their electricity requirements using the smart meter principle. They use high-pressure steam in some of their processes and so their 100 MW turbine is driven by high-pressure steam. When the steam condenses after giving up its energy to the turbine, there remains a lot of very hot water which is then piped around the offices and some production plant buildings for space heating. Combining heat and power in this way can increase the overall efficiency of the fuel used because it is performing two operations.

CHP can also use the heat from incinerating refuse to heat a nearby school or block of flats. This is called 'district heating', or 'community heating'.

Water conservation

Definition

Conservation is the preservation of something important, especially of the natural environment.

Conservation is the preservation of something important, especially of the natural environment. Available stored water is a scarce resource in England and Wales where there are only 1,400 cubic metres per person per year; very little compared with France, which has 3,100 cubic metres per person per year, Italy which has 2,900 and Spain 2,800. About half of the water used by an average home is used to shower, bathe and wash the laundry, and another third is used to flush the toilet.

At a time when most domestic and commercial properties have water meters installed, it saves money to harvest and re-use water.

The examination board has asked us to look at two methods of water conservation: rainwater harvesting and grey water recycling.

10. Rainwater harvesting

Rainwater harvesting is the collection and storage of rainwater for future use. Rainwater has in the past been used for drinking, water for livestock and water for irrigation. It is now also being used to provide water for car cleaning and garden irrigation in domestic and commercial buildings.

Many gardeners already harvest rainwater for garden use by collecting run-off from a roof or greenhouse and storing it in a water butt or water barrel. However, a 200-litre water butt does not give much drought protection although garden plants much prefer rainwater to fluoridated tap water. To make a useful contribution the rainwater storage tank should be between 2,000 and 7,000 litre capacity. The rainwater-collecting surfaces will be the roof of the building and any hard-paved surfaces such as patios. Downpipes and drainage pipes then route the water to the storage tank situated, perhaps, under the lawn. An electric pump lifts the water from the storage tank to the point of use, possibly a dedicated outdoor tap. The water is then distributed through a hosepipe or sprinkler system to the garden in the normal way.

With a little extra investment, rainwater can be filtered and used inside the house to supply washing machines and WCs. Domestic pipes and interior plumbing can be added to existing homes although it is more straightforward in a new-build home.

With the move towards more sustainable homes UK architects are becoming more likely to specify rainwater harvesting in their design to support alternatives to a mains water supply. In Germany, rainwater harvesting systems are now installed as standard in all new commercial buildings.

> **Definition**
>
> *Rainwater harvesting* is the collection and storage of rainwater for future use.

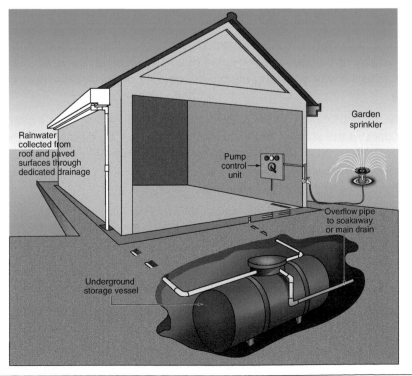

Figure 1.23 Underground rainwater storage vessel.

Figure 1.24 Rainwater harvesting can save a significant amount of water in domestic situations.

11. Grey water recycling

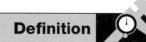

Definition

Grey water is tap water which has already performed one operation and is then made available to be used again instead of flushing it down the drain.

Grey water is tap water which has already performed one operation and is then made available to be used again instead of flushing it down the drain. Grey water recycling offers a way of getting double the use out of the world's most precious resource.

There are many products on the market such as the BRAC system which takes in water used in the shower, bath and laundry, cleans it by filtering and then re-uses it for toilet flushing. It is only a matter of routing the grey waste water drainpipe from the bath, shower and laundry to the filter unit and then plumbing the sanitized grey water to the toilet tank.

These systems are easy to install, particularly in a new-build property. It is only a matter of re-routing the drainpipes. Another option for your grey water is to route it into the rainwater storage tank for further use in the garden.

Planning permission and permitted development

The Town and Country Planning Act 1990 effects control over all building development, its appearance and the layout of the building. The Public Health Acts limit the development of buildings with regard to noise, pollution and public nuisance. The Highways Act 1980 determines the layout and construction of roads and pavements. The Building Act 1984 affects the Building Regulations 2000 which enforce minimum material and design standards.

If you are considering building a new home, office or factory you will definitely require Local Authority planning permission before any work begins. However, under permitted development, small extensions to an existing building may be exempt from a formal planning application. The General Permitted Developments Amendment 2, 2008 said, for example, that porches are exempt from planning permission providing that the floor area was no more than three square metres, the height was no more than three metres high and the porch was more than two metres from the boundary wall of the property.

This same permitted developments amendment also lifted the requirements for planning permission for most domestic microgeneration technologies. However, as with the porch example given above, size limitations have been set to reduce the impact upon neighbours. For example, solar panels attached to a building must not protrude more than 200mm from the roof slope and stand-alone panels must be no higher than four metres above ground level and no nearer than five metres from the property boundary. A website detailing planning requirements may be found at www.planning portal.gov.uk/uploads/hhghouseguide.html.

Installing Microgenerators – the Regulations

Let us look at the regulation requirements for each of the technologies discussed earlier in this chapter.

1. Solar PV. This is probably one of the most popular forms of microgeneration in the UK. Planning permission is not required because the fitting of solar panels comes within permitted development unless the building is listed or in a conservation area.

Building regulations approval may be required because the ability of the roof to carry the weight of the panels and for the panels to remain secure in high winds must be assured. However, this will probably be waived if the installation contractor is MCS-approved because MCS-approved installers must purchase and fit only accredited fixtures, fittings and panels.

2. Small wind generators rated between 1.5 kW and 15 kW and micro wind generators rated up to 1.5 kW. Planning permission is still a genuine issue in the UK for micro and small wind generators despite the introduction in 2008 and 2009 of permitted development rights for renewable technologies. Somehow wind was not included in the permitted development right. This may soon be resolved because further legislation is expected, but it is vital to consult your Local Authority planning office before any work is undertaken.

3. Hydro microgeneration. It is very difficult in this country to obtain permission from the Waterways Commission to extract water from rivers, even though once the water has passed through the turbine it is put back into the river. Environmental pressure groups are opposed to any disturbance of any waterway because of the environmental impact on the river's ecosystem.

Planning permission will be required. The Environment Agency must be consulted and Building Regulations approval must be obtained, in particular with regard to the electrical installation (more details at www.energysavingtrust.org.uk/generate-your-own-electricity).

Figure 1.25 A small-scale wind generator when combined with solar PV panels can produce a considerable amount of energy.

4. Solar thermal. This is another popular form of microgeneration in the UK. Planning permission is not required because the installation is within permitted development unless the building is listed or in a conservation area. Building Regulations may require assurance on the roof's ability to carry the weight and withstand high winds but if an MCS installer carries out the work he must install to the standards required by the permitted development. The electrical installation must of course be installed to the satisfaction of the Building Regulations Part P.

5. Ground source heat pumps. Planning permission is not required unless the building is listed or in a conservation area. The installation of the pipe work which forms the collector will be extensive and necessitate a lot of ground work. The installation will therefore probably require Building Regulations approval.

6. Air source heat pump. Strangely, air source heat pumps currently require planning permission. However, the Government is currently consulting on allowing air source heat pumps as permitted development. So, check with the planning office of your local area before work commences.

7. Water source heat pumps. If the water is a moving watercourse or canal you will require permissions from the Waterways Commission or the Environment Agency. If the water is a big pond or lake on your own land then it probably comes under permitted development but may require Building Regulations approval. Your local planning office must be consulted before any work begins.

8. Biomass microgenerators. If the biomass microgenerator is a giant factory processing crops into fuel or the energy plant located at Teesport which I discussed earlier, then clearly it will require full planning permission and Building Regulations approval. However, if we are looking at a biomass-fuelled appliance such as a log burner in a domestic situation, then planning permission and Building Regulations approval will not be required. Of course the stove must be installed and the chimney checked by an approved installer.

9. Combined heat and power (CHP). If it is proposed to install a CHP microgenerator to provide community heating to an adjacent school or block of flats, then it will require full Planning and Building Regulations permission. However, if it is proposed to install a CHP microgenerator in a domestic situation, then planning will not be required. Building Regulations approval may be required, particularly with regard to the electrical installation and plumbing work in relation to approved documents L1A, L1B, L2A and L2B, which require that reasonable provision has been made to conserve fuel and power and, of course, this new technology must be installed by an MCS-approved installer.

10. Rainwater harvesting. If this water conservation technology is part of a new-build then it will form a part of the Planning and Building Regulations permissions for the whole building. However, if it is to be installed in an existing building, then a very large hole must be excavated to house the storage tank. A 7,000-litre tank is as big as a large skip and will require an extensive hole to bury it underground. Building Regulations may be required for such a large excavation. Check with your local office.

11. Grey water recycling. If this water conservation technology is part of a new build then it will form a part of the Planning and Building Regulations permissions for the whole building. Where it is to be installed in an existing building and incorporating a water filtration unit, then it is little more work than is required to install a washer or dishwasher and permissions will not be required.

The potential to install microgeneration technology – the building location and features

Let us look at each of the technologies discussed earlier.

1. Solar PV. The sun rises in the east and sinks in the west in the northern hemisphere and so solar panels facing south will be the best installation position. Adjacent trees or other buildings which cast a shadow on the panels will reduce the energy generated.

2. Small wind generators 1.5kW to 15kW but not small boat battery chargers. Small wind generators work best in exposed locations where there is a constant high wind speed. For this reason the site of the turbine is most important. Average wind speed for a specific location can be obtained from the internet by entering the post code or ordnance survey map reference. Alternatively the average wind speed can be measured on-site during the planning stage.

A small wind generator should not be fixed to a building because the wind blowing onto the building causes turbulence which reduces the efficiency of generation. Similarly, nearby buildings or trees will reduce efficiency. A turbine fitted to a mast or tower is by far the best method of installation and the higher the turbine the better it will perform.

All turbine towers must be robustly installed to withstand very high winds without becoming a hazard. Every mast should be designed to suit the location in which it is to be installed and for the type of turbine which it will support.

Figure 1.26 Solar powered garden lighting.

3. Hydro microgenerators. The available watercourse must have sufficient flow and head to maintain the flow rate throughout the seasons. However, hydropower is available 24 hours a day, seven days a week, month after month, year after year, unlike solar and wind energy. Hydropower has the most continuous and predictable output of any renewable energy. On-grid it continuously offsets electricity needs while selling any surplus energy back to the grid.

The old reputation of small hydro schemes being unreliable and requiring constant maintenance has been cast aside in the past 20 years, initially with the introduction of automatic electronic load controllers and more recently with the introduction of easy maintenance intake screens, similar to those used in domestic Koi pool installations.

If you have legitimate access to an appropriate water resource and can obtain the necessary permissions, then hydro could well be the best of all renewable energy solutions, despite being the most expensive.

4. Solar thermal. Solar thermal systems use the energy of the sun to heat water which is then stored in the domestic hot water cylinder to be used as required. The solar collector is usually placed on a south-facing roof with a slope of between 20 and 50 degrees for best results. Adjacent buildings or trees which cast a shadow on the panel will reduce the heat energy generated. A four-metre square panel will usually meet the hot water demands of the average family during the summer months.

5. Ground source heat pumps. The ground heat exchanger must be about 2.5 times the area of the building to be heated. The heat exchanger is harnessing the solar energy absorbed by the soil above it and so this area must remain uncovered and in direct sunlight for best results. Shading from other buildings or trees will reduce the energy efficiency of the system. To disguise the area dedicated to the heat exchanger will require some architectural feature such as a large lawn, a five-a-side football pitch, a tennis court or two or a 'paddock to exercise the ponies'. It is not a system to be considered for urban development.

6. Air source heat pump. You can see from Fig. 1.21 that this system can fit into an urban development or a city centre home. A suitable external wall is all that is required and almost all properties have this except high-rise buildings. The essential resource for this system is air within the temperature range 8 °C to 17 °C or above. Air source heat pumps can be used for heating or cooling, making them ideal for domestic or commercial buildings provided that the temperature range is available.

7. Water source heat pump. The heat exchanger will be large. You will need a large pond or watercourse close to the property in which to submerge the heat exchanger coils. The water must be deep enough to avoid freezing at the bottom in winter and of course you must have permission to use the water source.

8. Biomass heating. If the biomass heating system is based upon a wood-burning stove, then the property must have a chimney or flue system or the facility to install a chimney. I see lots of stainless steel chimneys being installed to provide a flue for a wood burner. A wood burner is a desirable feature in many homes these days, even very modern homes. The stove provides a focal point in the living area while the remaining rooms are heated more conventionally with underfloor heating or radiators. Wood-burning stoves may also incorporate a back boiler which was the open coal fire method of heating hot water before central heating became so popular.

The principal fuel is wood which is classified by the Kyoto Protocol as carbon neutral. Dry wood burns best and so you must also build an outside wood store to accommodate, ideally, about two to three tonnes of logs. In this way they will be air-dried before burning.

9. Combined heat and power (CHP). A micro combined heat and power unit can be a direct replacement for a conventional domestic boiler. The combined heat and power unit can use most types of fuel and is therefore described as fuel neutral. This new technology is developing quite quickly.

10. Rainwater harvesting. At a time when many homes are connected to water meters, it makes good economical sense to try to preserve this scarce natural resource. All that is required is a large storage vessel to collect all the rainwater which falls on the roof and flat paved areas around the property.

The ideal place for a large storage vessel is underground and when connected to a submersible pump and outside tap it will provide just the same amenity for garden watering and car washing as a tap connected to the water mains. However, you will not be using fluorinated water which has travelled through the water meter. Water from a storage vessel is also exempt from a summer hosepipe ban.

11. Grey water recycling. At its simplest it is only a matter of connecting the waste pipes from the bath, shower and laundry into the rainwater storage vessel. However, all commercial systems filter the water before reusing it again to, for example, flush the toilet. Commercial car-washing machines also recycle and filter the water for re-use.

Figure 1.27 Small-scale wood burning stove.

Figure 1.28 Outside wood store.

Advantages and disadvantages of microgeneration technology

Our energy bills are set to rise above inflation for the foreseeable future. To reduce the effect of these rises upon household bills we should consider:

- using energy more efficiently;
- insulating our homes more effectively;
- the advantages and disadvantages of using microgeneration technology.

1. Solar PV. Government grants and the feed-in tariffs are available for solar PV even though the price per unit is reducing in value. The installation is simple and straightforward. PV systems up to 5 kW are exempt from planning permission under the permitted development laws. However, the panels must be fixed in a south-facing situation for best results. Home owners should not consider the 'rent a roof scheme' because this may reduce the value of their home. Financial payback is about 30 years if no grants or feedback tariff are available.

2. Small wind generators. People living in isolated hill farms may find benefit from a wind turbine, but not everyone will live or work in a location which has sufficient wind resources. There are issues of local wind turbulence from buildings and trees which will impede the efficient operation of a wind turbine. Lots of people will object to the erection of a tall tower with a wind turbine at the top. Planning permission will be required. Financial payback for a 100 kW installation will be in the region of 15 to 20 years.

3. Hydro microgeneration. Hydropower is considered the renewable energy of choice if the resources are available. The civil works and capital investment required for hydropower look overwhelming when compared with solar or wind power but the benefits of hydro outweigh solar and wind because it will generate power all day every day. However, permissions will be time consuming and installation costs high. Financial payback for a 100 kW turbine installation will be in the region of 15 to 20 years.

4. Solar thermal. Solar thermal can be installed in two days by an MCS installer. The installation is simple and straightforward and requires no planning permission. Panels must be south facing for best results. Financial payback is 8 to 20 years depending upon the site chosen and the utilization of the equipment.

5. Ground source heat pump. This microgenerator is invisible, unlike solar PV, solar thermal, wind generation and hydro. This may therefore be one option for a building in a conservation area. However, a large piece of land must be given over to the heat exchanger. Financial payback is 8 to 15 years.

6. Air source heat pump. This system is not suitable for high-rise buildings. Air temperatures within the range 8 to 17 degrees or above are essential for the efficient operation of the system. Financial payback is 8 to 15 years.

7. Water source heat pump. This is another invisible microgenerator but in this case the heat exchanger must be submerged into a large mass of water. Financial payback time is 8 to 15 years.

8. Biomass heating. A wood burner gives a very cosy focal point to a living space, especially on a cold winter's day. However, unlike gas and electricity heating you must be able to store the fuel on the premises. You must also have a chimney or flue connecting the wood burner to the outside. Financial payback is not applicable here because fuel costs are about the same as coal, oil or gas and the installation of a 7 kW stove and chimney liner if necessary is less than £1,000.

Figure 1.29 Solar powered speed warning sign.

9. Combined heat and power (CHP). A micro CHP domestic unit can be a straight swap for the existing boiler which may save up to 30% on the fuel bill. However, this is very new technology which is improving all the time. Financial payback is three to five years.

10. Rainwater harvesting. At a time when many people are on water meters these water-saving techniques make a lot of sense. The disadvantage is that you must find a place to put the large water storage vessel.

11. Grey water recycling. By simply connecting a few pipes in a different way you can save a precious resource and money. So why wouldn't you? It is very simple to install on a new-build property but less so on an existing property.

Financial payback

The payback period from fuel savings or income is a useful way of assessing if an investment is worthwhile and it also allows you to compare the financial merits of different technology options.

To carry out a simple estimate of the financial payback of a proposed installation you must divide the total installation costs by the annual savings. For example, let us suppose that you intend to replace your existing coal, calor gas or oil-fired heating system with a ground source heat pump because you live in the country and there is a large plot of grassland next to your house. The estimated cost of the new installation is £12,000 and this will save £1,000 in fuel for the old system each year. The financial payback period is therefore 12 years.

The calculations which I have done on the microgeneration technologies at 1 to 9 above take no account of any tax credits, grants or feedback tariffs which may be applicable to your particular microgeneration technology.

Test your knowledge

When you have completed the questions check out the answers at the back of the book.

Note: more than one multiple-choice answer may be correct.

Learning outcome 1

1   Which of the following was set up to protect the environment?
    a.   the Hazardous Waste Regulations 2005
    b.   the Pollution Prevention and Control Act 1999
    c.   the Control of Pollution Act 1989
    d.   the Environmental Act 1995.

2   Which of the following was set up to limit pollution from industry?
    a.   the Hazardous Waste Regulations 2005
    b.   the Environmental Act 1995
    c.   Radio Active Substances Act 1991
    d.   the Pollution Prevention and Control Regulations 2000.

3   Which of the following was set up to control how carriers of waste products operate?
    a.   the Hazardous Waste Regulations 2005
    b.   the Control of Pollution Act 1989
    c.   the Pollution Prevention and Control Act 1999
    d.   the Environmental Act 1995.

4   Our surroundings and the world in which we live is one definition of:
    a.   the Health and Safety at Work Act
    b.   the Building Regulations
    c.   the environment
    d.   the water table.

5   Which of the following is covered under the Environmental Protection Act?
    a.   damage to property
    b.   emissions
    c.   waste management
    d.   disease.

6   The Packaging Regulations tell us that all packaging must be designed and manufactured so that the:
    a.   goods can never be broken
    b.   volume and weight are at a minimum
    c.   packaged goods can be moved with a forklift truck and so avoid manual handling
    d.   used packaging can be recycled and re-used.

7   Any home with zero carbon emissions would be allocated which code?

a.   6

b.   0

c.   1

d.   5.

8   The heights of switches and sockets are covered in which element of the Building Regulations 2000, amended?

a.   Part P

b.   Part M

c.   Part C

d.   Part L.

9   Site preparation and resistance to contaminants and moisture is covered in which element of the Building Regulations 2000, amended?

a.   Part P

b.   Part G

c.   Part C

d.   Part K.

10   The total amount of greenhouse gases produced by an organization, event, product or person is?

a.   green house effect

b.   global warming

c.   carbon footprint

d.   depleted ozone layer.

11   The energy efficiency requirements of a building are covered in which element of the Building Regulations 2000, amended?

a.   Part P

b.   Part L

c.   Part C

d.   Part K.

12   Recycling hazardous waste must be carried out through a:

a.   licensed recycle operator

b.   licensed landfill site

c.   licensed public tip

d.   licensed recycle contractor.

13   Identify the hazardous materials below:

a.   old glass bottles

b.   old fluorescent tubes

c.   used batteries

d.   offcuts of trunking and conduit.

14   Identify the recyclable materials below:

a.   old glass bottles

b.   old fluorescent tubes

c.   used batteries

d.   offcuts of trunking and conduit made from thermoplastic.

15   We know that we have a 'duty of care' to dispose of all waste material that the working environment creates responsibly. So what is the 'responsible' way to dispose of the packaging from boxes of electrical equipment and fittings?
  a.   place it in the general waste skip which will go to landfill
  b.   place it in the skip designated for recycling materials
  c.   have it removed by a specialist waste contractor before your electrical work begins
  d.   recycle through a specialist hazardous waste company.

16   We know that we have a 'duty of care' to dispose of all waste material that the working environment creates responsibly. So what is the 'responsible' way to dispose of asbestos material?
  a.   place it in the general waste skip, which will go to landfill
  b.   place it in the skip designated for recycling materials
  c.   have it removed by a specialist waste contractor before your electrical work begins
  d.   recycle through a specialist hazardous waste company.

17   We know that we have a 'duty of care' to dispose of all waste material which the working environment creates responsibly. So what is the 'responsible' way to dispose of the off cuts of conduit and trunking made from thermosetting material?
  a.   place it in the general waste skip which will go to landfill
  b.   place it in the skip designated for recycling materials
  c.   have it removed by a specialist waste contractor before your electrical work begins
  d.   recycle through a specialist hazardous waste company.

18   We know that we have a 'duty of care' to dispose of all waste material that the working environment creates responsibly. So what is the 'responsible' way to dispose of dozens of old fluorescent tubes and SON lamps?
  a.   place it in the general waste skip which will go to landfill
  b.   place it in the skip designated for recycling materials
  c.   have it removed by a specialist waste contractor before your electrical work begins
  d.   recycle through a specialist hazardous waste company.

Learning outcome 2

19   Which of the following apply to the built environment?
  a.   materials used in building
  b.   all green belt land
  c.   all common land
  d.   product of human labour.

20   Which of the following can reduce the impact on the environment?
  a.   arriving separately on site
  b.   arriving together on site
  c.   carefully planning the installation
  d.   carefully planning your breaks.

21 If a company causes a hazard to the environment they:
    a. should always inform the customer
    b. need to place warning signs
    c. must clean up without notification
    d. have a duty to report the incident.

22 To assist in helping to reduce the environmental impact of an installation, it is advisable to use:
    a. correct isolation procedures
    b. site security systems
    c. inspection and testing equipment
    d. licensed waste companies.

23 When stripping out an old installation, the waste should be:
    a. shipped off site as rapidly as possible unsorted
    b. sorted into different types but not labelled
    c. placed in one waste skip
    d. segregated into the different types and labelled.

24 Anyone disposing of electrical or electronic waste must:
    a. store it on site
    b. record how you got rid of it
    c. have a record to show that it was disposed of in an authorized way
    d. use an authorized agency and rely on them to maintain disposal records.

25 Waste cable being burnt on site is required to be reported as an environmental hazard because:
    a. the copper will be too damaged and cannot be recycled
    b. the PVC sheathing is considered as hazardous waste
    c. it contravenes the Manual Handling Operations Regulations 1992
    d. it will cause air pollution.

Learning outcome 3

26 Environmental technology systems:
    a. are eco friendly
    b. use renewable energy
    c. use fossil fuel
    d. use nuclear energy.

27 In 2008 the Planning Laws were relaxed to allow the installation of small micro-generators on domestic properties within certain guidelines under an act called:
    a. the Health and Safety at Work Act
    b. the Building Regulations
    c. permitted development
    d. the Microgeneration Certification Scheme.

28 Solar photovoltaic systems use which of the following devices?
a. photodiode and transformer
b. diode and transformer
c. photodiode and inverter
d. diode and inverter.

29 Which of the following is not included in grey water recycling?
a. dishwashers
b. showers
c. washing machine
d. toilets.

30 Which of the following uses heat from the sun to work alongside a conventional water heater?
a. photovoltaic
b. air source heat pump
c. solar thermal
d. ground source pump.

31 Solar photovoltaic panels in the northern hemisphere are used to generate electricity by being aligned:
a. east facing
b. north facing
c. south facing
d. west facing.

32 The collection and storage of rainwater for future use is one definition of:
a. grey water recycling
b. brown water harvesting
c. rainwater harvesting
d. brown water recycling.

33 The best type of container to be used for rainwater harvesting in a domestic household would be a:
a. 30 litre bucket
b. 200 litre rainwater butt
c. 5,000 litre underground storage vessel
d. small garden pond.

34 Tap water that has already performed one operation and is then made available to be used again is one definition of:
a. grey water recycling
b. brown water harvesting
c. rainwater harvesting
d. brown water recycling.

35 Which of the following would be most suitable to coastal regions?
a. photovoltaic
b. hydro generation
c. micro hydro
d. micro turbine.

36 Which of the following is a means of generating electricity from a living or recently living organism?
    a. rainwater harvesting
    b. solar water heating
    c. rainwater harvesting
    d. biomass.

37 Identify four things that you use at work that would require to be disposed of as hazardous waste.

38 Identify six pieces of equipment that would require to be disposed of correctly under the WEEE Regulations.

39 How do the Noise at Work Regulations protect workers?

40 Use bullet points to state the basic operating principle of a solar hot water heating system.

41 Use bullet points to state the applications and limitations of a solar hot water heating system.

42 Use bullet points to state the basic operating principle of a solar photovoltaic system.

43 Use bullet points to state the applications and limitations of a solar photovoltaic system.

44 State the advantages and disadvantages of wind energy generation.

45 Very briefly, in three sentences, describe the basic principle of heat pumps.

46 Very briefly, in three sentences, describe the basic principle of CHP systems.

47 In one sentence describe biomass heating.

48 In one sentence describe hydro microgeneration.

49 What is a 'smart' electricity meter?

50 What is 'rainwater harvesting'?

51 What is 'grey water recycling'?

Classroom activities

Environmental legislation

Match one of the numbered boxes to each lettered box on the left.

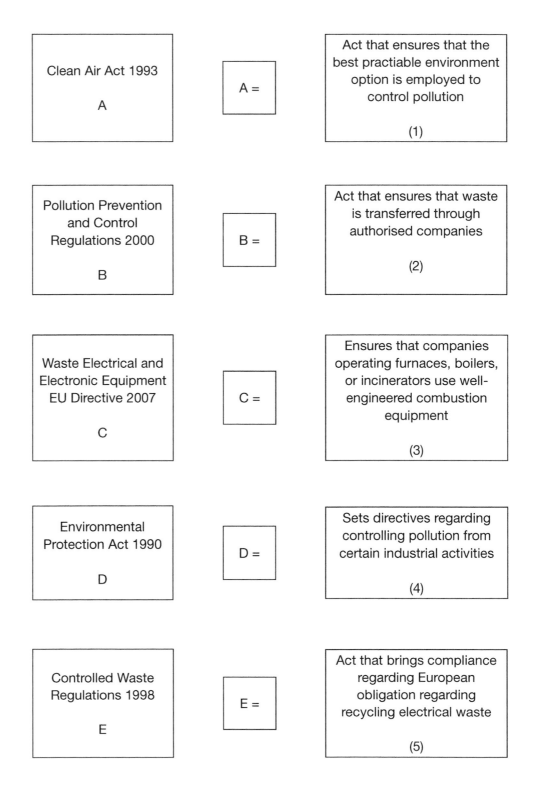

Clean Air Act 1993 A	A =	Act that ensures that the best practiable environment option is employed to control pollution (1)
Pollution Prevention and Control Regulations 2000 B	B =	Act that ensures that waste is transferred through authorised companies (2)
Waste Electrical and Electronic Equipment EU Directive 2007 C	C =	Ensures that companies operating furnaces, boilers, or incinerators use well-engineered combustion equipment (3)
Environmental Protection Act 1990 D	D =	Sets directives regarding controlling pollution from certain industrial activities (4)
Controlled Waste Regulations 1998 E	E =	Act that brings compliance regarding European obligation regarding recycling electrical waste (5)

Environmental terms and roles

Match one of the numbered boxes to each lettered box on the left.

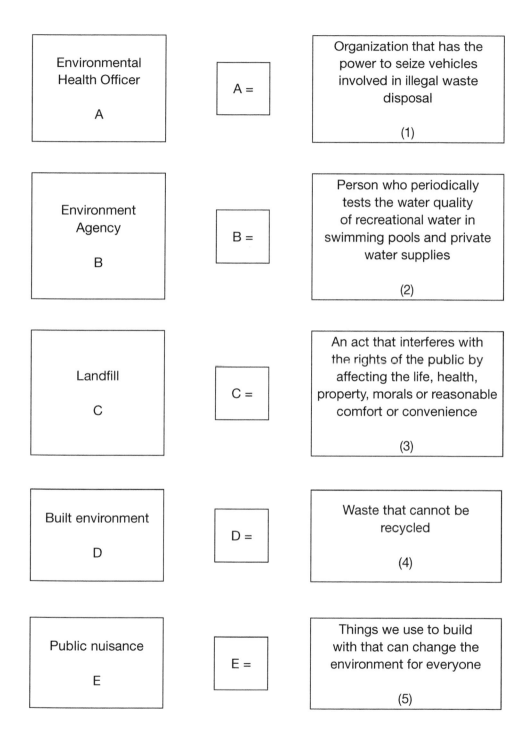

Environmental Health Officer A	A =	Organization that has the power to seize vehicles involved in illegal waste disposal (1)
Environment Agency B	B =	Person who periodically tests the water quality of recreational water in swimming pools and private water supplies (2)
Landfill C	C =	An act that interferes with the rights of the public by affecting the life, health, property, morals or reasonable comfort or convenience (3)
Built environment D	D =	Waste that cannot be recycled (4)
Public nuisance E	E =	Things we use to build with that can change the environment for everyone (5)

QELTK3/002 Understanding environmental legislation, working practices and the principles of environmental technology system

Chapter checklist

Learning outcome	Assessment criteria	Page number
1 Understand the environmental legislation, working practices and principles that are relevant to work activities	The learner can: 1.1 Specify the current, relevant legislation for processing waste, including: • Environmental Protection Act • The Hazardous Waste Regulations • Pollution Prevention and Control Act • Control of Pollution Act • The Control of Noise at Work Regulations • Packaging (Essential Requirements) Regulations • Environment Act • The Waste Electrical and Electronic Equipment Regulations.	2
	1.2 Describe what is meant by the term environment.	2
	1.3 Describe the ways in which the environment may be affected by work activities: • land contamination • air pollution • pollution of water courses.	7
	1.4 Identify and interpret the requirements for electrical installations as outlined in relevant sections of the Building Regulations and the Code for Sustainable Homes.	7
	1.5 State materials and products that are classed as: • hazardous to the environment • recyclable.	11
	1.6 Describe the organizational procedures for processing materials that are classed as: • hazardous to the environment • recyclable.	11
2 Understand how work methods and procedures can reduce material wastage and impact on the environment	2.1 State installation methods that can help to reduce material wastage. 2.2 Explain why it is important to report any hazards to the environment that arise from work procedures. 2.3 Specify environmentally friendly materials, products and procedures that can be used in the installation and maintenance of electrotechnical systems and equipment.	13 15 15
3 Understand how and where environmental technology systems can be applied	3.1 Describe the fundamental operating principles of the following environmental technology systems: • solar photovoltaic • wind energy generation (micro and macro) • micro hydro generation • heat pumps • combined heat and power (CHP) including micro CHP • grey water recycling • rainwater harvesting • biomass heating • solar thermal hot water heating.	15

Learning outcome	Assessment criteria	Page number
	3.2 State the applications and limitations of the following environmental technology systems: • solar photovoltaic • wind energy generation (micro and macro) • micro hydro generation • heat pumps • combined heat and power (CHP) including micro CHP • grey water recycling • rainwater harvesting • biomass heating • solar thermal hot water heating.	15
	3.3 State the Local Authority building control requirements that apply to the installation of environmental technology systems.	16

Understanding the practices and procedures for overseeing and organizing the work environment (electrical installation)

EAL Unit QELTK3/003

Learning outcomes

When you have completed this chapter you should:

1. Understand the types of technical and functional information that are available for the installation of electrotechnical systems and equipment.
2. Understand the procedures for supplying technical and functional information to relevant people.
3. Understand the requirements for overseeing health and safety in the work environment.
4. Understand the requirements for liaising with others when organizing and overseeing work activities.
5. Understand the requirements for organizing and overseeing work programmes.
6. Understand the requirements for organizing the provision and storage of resources that are required for work activities.

EAL Electrical Installation Work – Level 3, 2nd Edition 978 0 367 19564 9
© 2019 T. Linsley. Published by Taylor & Francis. All rights reserved.
https://www.routledge.com/9780367195649

Assessment criteria 1.1

Specify sources of technical and functional information that apply to electrotechnical installations

Assessment criteria 1.2

Interpret technical and functional information and data

To carry out and complete an electrical installation makes use of many forms of information and can be subdivided into two categories namely:

- functional;
- technical.

Functional information is written to explain how an item works or operates, and how things are to be done, assembled or installed. We see this in manufacturer's data sheets. Regulation 643.10 informs us that equipment shall be subjected to functional testing, as appropriate, to verify that it is properly mounted, adjusted, installed and operates correctly.

Technical information is the engineering data contained in, for example, a specification, the IET Regulations, or the On Site Guide or the IET Guidance Notes.

Both types of information are communicated in lots of different ways. For instance technical information in the electrical installation industry is drawn from the following:

- Guidance Notes (1–8);
- GS 38 (Safe Isolation Procedure);
- codes of practice (BS 7671 and *On-Site Guide*);
- manufacturers' data sheets;
- British Standards (BS) and British Standard European Norms (BS EN);
- technical books and drawings;
- the Internet.

Being aware of how to extract information and realizing how credible that information is can therefore act as a safeguard when, for example, confirming that replacement components or equipment are fully compatible with the required specification. It can also help ensure that when changes are made to specific codes of practice such as the wiring regulations (BS 7671) then you can embrace and be vigilant of those changes. For instance, only approved test equipment is designed to comply with the rigorous requirements to carry out the safe isolation procedures listed within GS38 – a document specifically written by the Health and Safety Executive. However, being aware of such a document is not enough, since the actual requirement could be subjected to review and possible amendment, which means it is vital that all practising electricians are pro-active in the pursuit of good practice by attending relevant courses.

Interpreting and recording the testing element of an installation is also indelibly linked with data and information since IET Regulations 641 advise us that

Figure 2.1 Amendment 3 has introduced the term 'electrically skilled' in order to make it clearer who can and cannot work on electrical parts of sites.

every installation shall, during erection and on completion before being put into service, be inspected and tested to verify that the regulations have been met. Regulation 643 also advises that where required, periodic inspection and testing shall be carried out in order to determine if the installation is in a satisfactory condition for continued service. Any such tests therefore must be recorded so that they can prove and maintain safety and also can be used as a comparison during any future re-testing. A schedule of test results is therefore a written record of the results obtained when carrying out the electrical tests required by Part 6 of the IET Regulations.

The schedule brings together all the detailed information about the testing of that particular installation. In the same way, a work schedule brings together the detail of the whole project.

Interpretation of information is therefore very important and can be seen through the following:

- *Specifications* – these are details of the client's requirements, usually drawn up by an architect. For example, the specification may give information about the type of wiring system to be employed or detail the type of luminaires or other equipment to be used.
- *Manufacturer's data* – if certain equipment is specified, let's say a particular type of luminaire or other piece of equipment, then the manufacturer's data sheet will give specific instructions for its assembly and fixing requirements. It is always good practice to read the data sheet before fitting the equipment. A copy of the data sheet should also be placed in the job file for the client to receive when the job is completed.
- *User instructions* – give information about the operation of a piece of equipment. Manufacturers of equipment provide 'user instructions' and a copy should be placed in the job file for the client to receive when the project is handed over.

Key fact

The type of wiring system to be used will be indicated on the specification.

Key fact

When installing equipment always read the manufacturer's data. User instructions will dictate its different modes of operation.

- *Job sheets and time sheets* – give 'on site' information. Job sheets give information about what is to be done and are usually issued by a manager to an electrician. Time sheets are a record of where an individual worker has been spending his time, which job and for how long. This information is used to make up individual wages and to allocate company costs to a particular job. We will look at these again later under the sub-heading 'on-site documentation'.

Let us now look in more detail at these sourses of information.

Drawings and diagrams

Many different types of electrical drawing and diagram can be identified, which give relevant information for the installation, termination and connection of conductors: layout, schematic, block, wiring and circuit diagrams. The type of diagram to be used in any particular application is the one that most clearly communicates the desired information.

Layout drawings or site plan

These are scale drawings based upon the architect's site plan of the building and show the positions of the electrical equipment that is to be installed. The electrical equipment is identified by a graphical symbol.

The standard symbols used by the electrical contracting industry are those recommended by the British Standard BS EN 60617, *Graphical Symbols for Electrical Power, Telecommunications and Electronic Diagrams.* Some of the more common electrical installation symbols are given in Fig. 2.4.

The site plan or layout drawing will be drawn to a scale smaller than the actual size of the building, so to find the actual measurements you must measure the distance on the drawing and then multiply by the scale. For example, if the site plan is drawn to a scale of 1:100, then 10 mm on the site plan represents 1 m measured in the building. A layout drawing is shown in Fig. 2.2 of a small domestic extension. It can be seen that the mains intake position, probably a consumer's unit, is situated in the storeroom, which also contains one light controlled by a switch at the door. The bathroom contains one lighting point controlled by a one-way switch at the door. The kitchen has two doors and a switch is installed at each door to control the fluorescent luminaire. There are also three double sockets situated around the kitchen. The sitting room has a two-way switch at each door controlling the centre lighting point. Two wall lights with built in switches are to be wired, one at each side of the window. Two double sockets and one switched socket are also to be installed in the sitting room. The bedroom has two lighting points controlled independently by two one-way switches at the door.

The wiring diagrams and installation procedures for all these circuits can be found in *EAL Electrical Installation Work: Level 2*.

As-fitted drawings

When the installation is completed a set of drawings should be produced that indicate the final positions of all the electrical equipment. As the building and electrical installation progresses, it is sometimes necessary to modify

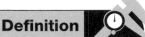

Definition

Layout drawings tend to be drawn to scale so that the positional relationship of the components can be appreciated.

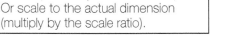

Top tip

When dealing with scale diagrams look at the situation and ask yourself: am I going from the actual to the scale? (Divide by the scale ratio.) Or scale to the actual dimension (multiply by the scale ratio).

Definition

When the installation is completed a set of drawings should be produced that indicate the final positions of all the electrical equipment.

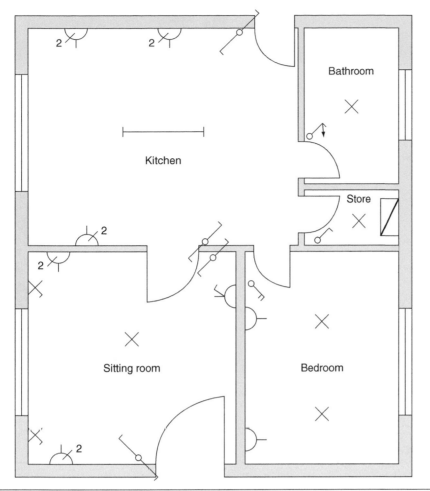

Figure 2.2 Layout drawing or site plan for electrical installation.

the positions of equipment indicated on the layout drawing because, for example, the position of a doorway has been changed. The layout drawings indicate the original intentions for the positions of equipment, while the 'as-fitted' drawing indicates the actual positions of equipment upon completion of the job.

Detail drawings and assembly drawings

These are **additional drawings** produced by the architect to clarify some point of detail. For example, a drawing might be produced to give a fuller description of the suspended ceiling arrangements.

Schematic diagrams

A schematic diagram is a diagram in outline of, for example, a motor starter circuit as shown in Fig 2.3 and 2.4. It uses graphical symbols to indicate the interrelationship of the electrical elements in a circuit. These help us to understand the working operation of the circuit.

An electrical schematic diagram looks very much like a circuit diagram.

> **Definition**
>
> *Detail drawings* are additional drawings produced by the architect to clarify some point of detail.

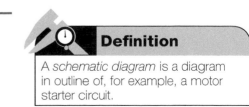

> **Definition**
>
> A *schematic diagram* is a diagram in outline of, for example, a motor starter circuit.

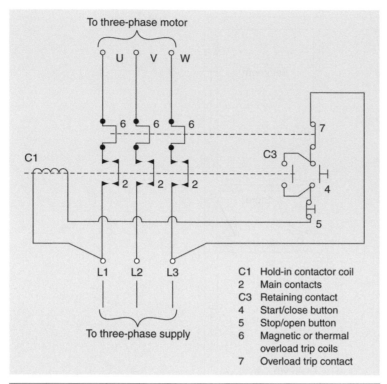

C1   Hold-in contactor coil
2   Main contacts
C3   Retaining contact
4   Start/close button
5   Stop/open button
6   Magnetic or thermal overload trip coils
7   Overload trip contact

Figure 2.3a Schematic diagram – DOL motor starter.

Figure 2.3b Schematic diagrams can be invaluable when dealing with unfamiliar equipment.

Block diagrams

Definition

A *block diagram* is a very simple diagram in which the various items or pieces of equipment are represented by a square or rectangular box.

A block diagram is a very simple diagram in which the various items or pieces of equipment are represented by a square or rectangular box as shown in Fig 2.5. The purpose of the block diagram is to show how the components of the circuit relate to each other and therefore the individual circuit connections are not shown.

Wiring diagrams

Definition

A *wiring diagram* or connection diagram shows the detailed connections between components or items of equipment.

A **wiring diagram** or connection diagram shows the detailed connections between components or items of equipment. They do not indicate how a piece of equipment or circuit works. The purpose of a wiring diagram is to help someone with the actual wiring of the circuit. Figure 2.6 shows the wiring diagram for a two-way lighting circuit.

Try this

The next time you are on site:

- Ask your supervisor to show you the site plans.
- Ask them to show you how the scale works.

Circuit diagrams

Definition

A *circuit diagram* shows most clearly how a circuit works.

A **circuit diagram** shows most clearly how a circuit works. All the essential parts and connections are represented by their graphical symbols. The purpose of a circuit diagram is to help our understanding of the circuit. It will be laid out as clearly as possible, without regard to the physical layout of the actual

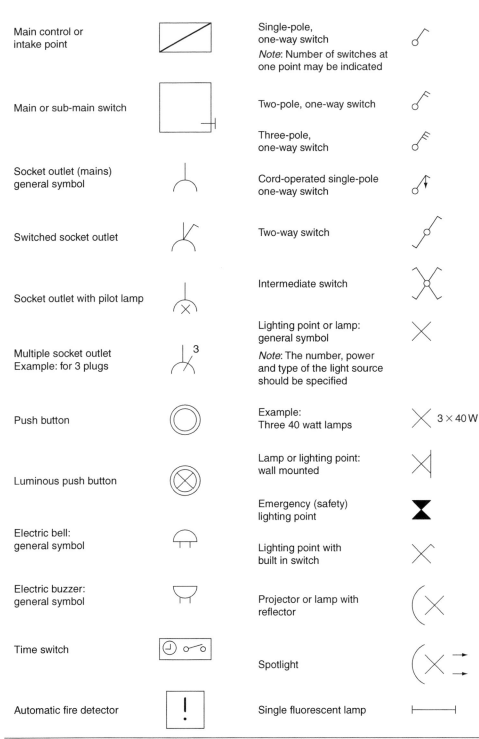

Main control or intake point		Single-pole, one-way switch *Note*: Number of switches at one point may be indicated	
Main or sub-main switch		Two-pole, one-way switch	
		Three-pole, one-way switch	
Socket outlet (mains) general symbol		Cord-operated single-pole one-way switch	
Switched socket outlet		Two-way switch	
Socket outlet with pilot lamp		Intermediate switch	
Multiple socket outlet Example: for 3 plugs		Lighting point or lamp: general symbol *Note*: The number, power and type of the light source should be specified	
Push button		Example: Three 40 watt lamps	$3 \times 40\,W$
Luminous push button		Lamp or lighting point: wall mounted	
Electric bell: general symbol		Emergency (safety) lighting point	
		Lighting point with built in switch	
Electric buzzer: general symbol		Projector or lamp with reflector	
Time switch		Spotlight	
Automatic fire detector		Single fluorescent lamp	

Figure 2.4 Some BS EN 60617 installation symbols.

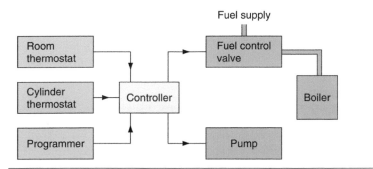

Figure 2.5 Block diagram – space-heating control system (Honeywell Y. Plan).

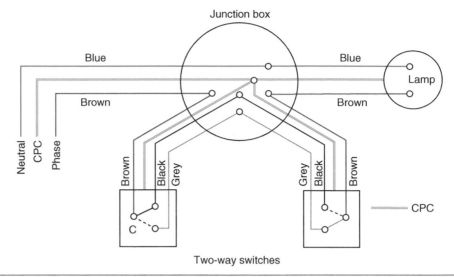

Figure 2.6 Wiring diagram of two-way switch control.

components, and therefore it may not indicate the most convenient way to wire the circuit. The circuit diagram for a domestic space heating system is shown in fig 2.7.

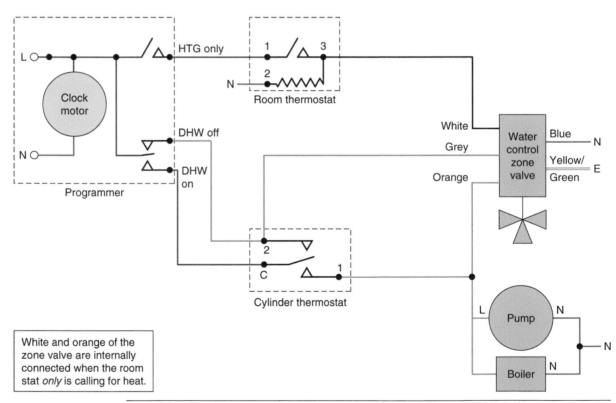

White and orange of the zone valve are internally connected when the room stat *only* is calling for heat.

Figure 2.7 Circuit diagram – space-heating control system (Honeywell Y. plan).

Assessment criteria 1.3

Identify and interpret technical and functional information relating to electrotechnical products or equipment

Detailed product data can be extracted from manufacturers' user manuals, which will describe how to operate equipment through various settings in order to engage various modes of operation and perhaps adjust the equipment for different environments. Examples of such devices include electronic thermostats and air conditioning equipment.

Equally important is to ensure that equipment is rated to the correct values such as working voltage, current or power rating or even breaking capacity, which is a value of current associated with a short circuit fault condition. For instance, a given component might fail because its rating is incorrect, especially if it's under rated in relation to the specification. Replacing such an item without consulting the specification or manufacturer's data will only cause it to fail again.

Assessment criteria 1.4

Describe the work site requirements and procedures in terms of services provision etc.

Assessment criteria 1.5

Identify equipment and systems that are compatible to site operations and requirements

Construction site

The IET Regulations have given us the whole of Section 704 for the regulations which relate to Construction and Demolition site installations.

Construction and demolition sites are potentially dangerous in many ways, tripping, falling, and the constantly changing and temporary nature of the working environment create the hazard. The risk of electric shock is high because of the following factors:

- a construction site or demolition site is, by definition, a temporary state. Upon completion there will be either a building with all the necessary safety features, or a brown field site where the building previously stood.
- there is the possibility of damage to cables and equipment as a consequence of the temporary nature of the site and because the site is not always sealed in the early stages from the weather.
- mobile equipment such as electrical tools and hand lamps with trailing leads will be in use.
- there will be many extraneous conductive parts on site which cannot practically be bonded because of the changing nature of the construction process.

A construction site will have many people working on a building project throughout the construction period. The groundwork people lay the foundations, the steel work is erected, the bricklayers build the walls, the carpenters put on the roof and only when the building is waterproof do the electricians and mechanical service trades begin to install the electrical and mechanical systems. If there was an emergency and the site had to be evacuated, how would you know who, or how many people were on site? You can see that there has to be a

Figure 2.8 Many different skilled jobs are required on site, so it's important that everyone follows a safe system of working.

procedure for logging people in and out, so that the main contractor can identify who is on-site at any one time.

With so many people employed through many different companies and agencies it is little wonder that sites employ both an induction programme and a site visitor procedure. The aim is to bring about a safe system of work which includes the sharing of information about how this is brought about. This means defining who are the key personnel or duty holders, what processes are involved and where the documents in relation to these processes are kept. Furthermore, sign posting where key operations are undertaken will bring about clarity of the specific and potential hazards that exist on site.

Such a programme therefore indicates how the site is complying with the Workplace (Health, Safety and Welfare) Regulations 1992. These regulations ensure that provision is made regarding the following:

- ventilation (is adequate when working inside);
- indoor temperature (16°C if mostly seated/13°C for work involving physical work);
- space (sufficient space to operate freely and therefore safely. An important consideration for an electrician);
- seating (ensuring that workstations are sufficiently comfortable);
- maintenance (ensuring that vital equipment is maintained in a safe and efficient order);
- traffic routes (pedestrians should be given their own walkways).

The induction programme will also share specific information in relation to where materials are to be stored, and procedures and locations involving waste disposal, including the separation of waste and the locations of appropriate containers. Again, this indicates how the site complies with the Environmental Protection Act 1990, the Control of Pollution Act 1989 and the Hazardous Waste Regulations 2005.

Furthermore, on large installations there may be more than one position for the distribution of temporary electrical supplies and cables may radiate from the site of the electrical mains intake position to various sub-mains locations. Although these sort of arrangements would normally have been drafted by an appropriate design engineer, the location, routing and positioning of such services would be indicated on an architect site plan so that all trades are aware of their location.

What employees should never lose sight of is that everyone on site is held responsible to the Health and Safety Act 1974 and this means they have a duty of care towards other workers and visitors, including reporting any suspected safety issue. Witnessing any episodes whereby equipment or service in use is either incompatible to its environment or impinged in its operation must be reported without delay.

Assessment criteria 2.1

State the limits of their responsibility for supplying technical and functional information

Part 2 of The IET Regulations describe people as an ordinary person, an instructed person (electrically) or a skilled person (electrically).

When you begin your apprenticeship you are an 'ordinary person', that is, someone who has no knowledge of the hazards which electricity can create. However, in just a few weeks or months you will become an instructed person electrically, because you will be carrying out electrical work under supervision. When you have completed your apprenticeship you will be a skilled person electrically, able to carry out electrical work safely, and able to perceive risks and avoid the hazards which electricity can create.

There are many different people involved in an electrical installation project and the transfer of information to those persons will largely depend on their roles, experience, position and responsibility within the electrical company.

Members of the construction team and their role within the industry

The construction industry is one of the UK's biggest employers, and carries out contracts to the value of about 10% of the UK's gross national product. In all these various construction projects the electrical industries play an important role, supplying essential electrical services to meet the needs of those who will use the completed building. All the processes involved in the industry, however, are underpinned by the use of information, in essence: a team of professionals working together to produce the desired results. We can call this team of professionals the building team and their interrelationship and how they use and share information is expressed as follows.

The client is the person or group of people with the actual need for the building, such as a new house, office or factory. The client is responsible for financing all the work and, therefore, in effect, employs the entire building team. The client could be an individual or a group of people representing a housing association.

Figure 2.9 Clients often request to be shown around a site.

They would need to specify their requirements to the building contractor regarding the design and specification and would probably engage an architect to do this.

The architect is the client's agent and is considered to be the leader of the building team. The architect must interpret the client's requirements and produce working drawings. During the building process the architect will supervise all aspects of the work until the building is handed over to the client.

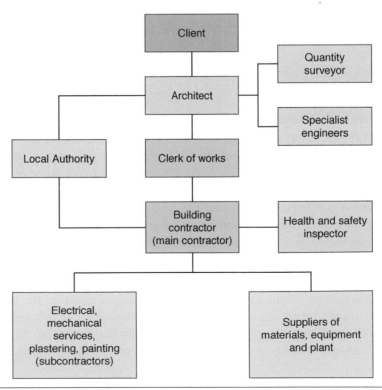

Figure 2.10 The building team.

The quantity surveyor measures the quantities of labour and materials necessary to complete the building work from drawings supplied by the architect. A quantity surveyor must therefore maintain detailed records regarding how the project is progressing in relation to the use of materials and costing.

A site / project manager will have overall responsibility for the successful planning, execution, monitoring and control of a project, usually working for the main construction contractor. They must have a combination of skills, including an ability to ask questions of all the team members and resolve conflicts, as well as more general management skills. The project manager, who is managing the installation, would have access to the latest working drawings and should ensure that all the trades are updated with any changes to the specification that have been drawn from architect's instructions or client requirements.

Building surveyors provide advice on property and construction, including residential, commercial, industrial, leisure and agriculture projects.

They work on the design and development of new buildings as well as the restoration and maintenance of existing ones.

Building surveyors often work to keep buildings in good condition and look for ways to make buildings sustainable. They may be called upon to give evidence in court in cases where building regulations have been breached and as expert witnesses on building defects. Building surveyors have access to detailed plans, including information they would need to share regarding the restrictions that apply due to planning consent, especially if a building is listed.

Contracts managers oversee projects working for the sub-contractor, from the start through to completion, ensuring that work is completed on time and within its budget. Information is often extracted from Gantt charts. For instance, any details of planned or unplanned events that impact on expenditure or project completion would need to be discussed with the client. A contracts manager is sometimes in charge of a single scheme, or may look after several smaller ones.

An estimator in the construction industry is responsible for compiling estimates of how much it will cost to provide a client with products or services. He or she will do this by working out how much a project is likely to cost and quotes accordingly. The *job* involves assessing material, labour and equipment required and comparing quotes from sub-contractors and suppliers.

Construction buyers are responsible for ensuring that the materials required for construction projects are provided to schedule and according to projected budgets. They have a vital part to play in helping to ensure the profitability of contracts since they are responsible for ensuring that the most cost effective and appropriate materials are purchased.

Structural engineers design buildings to withstand stresses and pressures caused by environmental conditions and human use. They ensure buildings and other structures do not deflect, rotate, vibrate excessively or collapse and that they remain stable and secure throughout their use. They work closely with architects and other professional engineers to choose appropriate materials, such as concrete, steel, timber and masonry, to meet design specifications.

Specialist engineers advise the architect during the design stage. They will prepare drawings and calculations on specialist areas of work.

The clerk of works is the architect's 'on-site' representative. He or she will make sure that the contractors carry out the work in accordance with the drawings and other contract documents. They can also agree general matters directly with the building contractor as the architect's representative.

The building control inspector from the local council will ensure that the proposed building conforms to the relevant planning and building legislation.

The health and safety inspectors ensure that the government's legislation concerning health and safety is fully implemented by the building contractor.

The building contractor will enter into a contract with the client to carry out the construction work in accordance with contract documents. The building contractor is usually the main contractor and may engage subcontractors to carry out specialist services such as electrical installation, mechanical services, plastering and painting.

Subcontractors are individuals or companies employed by the main contractor to perform specific, often specialist, tasks as part of the construction process. Specialisms of the subcontractors may include:

- electrician;
- gasfitter;
- plumber;
- heating and ventilation engineer;
- bricklayer;
- carpenter and joiner;
- plasterer;
- tiler;
- decorator;
- groundworker.

The construction of a new building is a complex process that requires a team of professionals working together to produce the desired results. We can call this team of professionals the building team, and their interrelationship is described in Fig. 2.10. Within this is the electrical team.

The electrical team

The electrical contractor is the subcontractor responsible for the installation of electrical equipment within the building.

Electrical installation activities include:

- installing electrical equipment and systems in new sites or locations;
- installing electrical equipment and systems in buildings that are being refurbished because of change of use;
- installing electrical equipment and systems in buildings that are being extended or updated;
- replacement, repairs and maintenance of existing electrical equipment and systems.

Try this

MY TEAM

Sketch a block diagram, similar to those shown in Figs. 2.10 and 2.12, that represents the team in which you work.

Figure 2.11 The electrical contractor is responsible for the installation of electrical equipment.

An electrical contracting firm is made up of a group of individuals with varying duties and responsibilities. There is often no clear distinction between the duties of the individuals, and the responsibilities carried by an employee will vary from one employer to another. If the firm is to be successful, the individuals must work together to meet the requirements of their customers. Good customer relationships are important for the success of the firm and the continuing employment of the employee.

The customer or his or her representatives will probably see more of the electrician and the electrical trainee than the managing director of the firm and, therefore, the image presented by them is very important. They should always be polite and be seen to be capable and in command of the situation. This gives a customer confidence in the firm's ability to meet his or her needs. The electrician and his trainee should be appropriately dressed for the job in hand, which probably means an overall of some kind. Footwear is also important, but is sometimes a difficult consideration for a journeyman electrician. For example, if working in a factory, the safety regulations may insist that protective footwear be worn, but rubber boots with toe protection may be most appropriate for a building site. However, neither of these would be the most suitable footwear for an electrician fixing a new light fitting in the home of the managing director! The electrical installation in a building is often carried out alongside other trades.

It makes sound sense to help other trades where possible and to develop good working relationships with other employees.

The employer has the responsibility of finding sufficient work for his employees, paying government taxes and meeting the requirements of the Health and Safety at Work Act. The rates of pay and conditions for electricians and trainees are determined by negotiation between the Joint Industry Board and the Trades Unions, which will also represent their members in any disputes. Electricians are usually paid at a rate agreed for their grade; movements through the grades are determined by a combination of academic achievement and practical

Key fact

Whilst the detailed drawing and plans will normally be produced by the architect, the clerk of work is the person who monitors that they are being put into place on site.

Top tip

As an employee only answer questions within your full understanding and experience. If unsure, ask your supervisor.

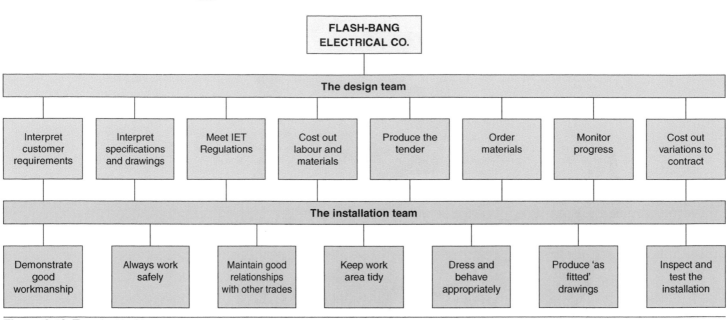

Figure 2.12 The electrical team.

experience. The electrical team will consist of a group of professionals and their interrelationship can be expressed as shown in Fig. 2.12.

Specify organizational policies/procedures for the handover and demonstration of electrotechnical systems, products and equipment, including requirements for confirming and recording handover

The commissioning of the electrical and mechanical systems within a building is a part of the 'handing-over' process of the new building by the architect and main contractor to the client or customer in readiness for its occupation and intended use. To 'commission' means to give authority to someone to check that everything is in working order. If it is out of commission, it is not in working order. Following the completion, inspection and testing of the new electrical installation, the functional operation of all the electrical systems must be tested before they are handed over to the customer. It is during the commissioning period that any design or equipment failures become apparent, and this testing is one of the few quality controls possible on a building services installation.

This is the role of the commissioning engineer, who must assure himself or herself that all the systems are in working order and that they work as they were designed to work. This engineer must also instruct the client's representative, or the staff who will use the equipment, in the correct operation of the systems as part of the handover arrangements.

The commissioning engineer must also test the operation of all the electrical systems, including the motor controls, the fan and air conditioning systems, the fire alarm and emergency lighting systems. However, before testing the emergency systems, it is important to first notify everyone in the building of the intention to test the alarms so that they may be ignored during the period of testing.

Figure 2.13 The commissioning engineer will test the operation of all the electrical systems.

Commissioning has become one of the most important functions within a building project's completion sequence. The commissioning engineer will therefore have access to all relevant contract documents, including the building specifications and the electrical installation certificates as required by the IET Regulations (BS 7671), and have a knowledge of the requirements of the Electricity at Work Regulations and the Health and Safety at Work Act.

In summary, when handing over an installation to the client you are effectively saying to the client, 'here is your installation complete with information or a breakdown of the systems and equipment in place'. The client will inherit an operation manual, as well as technical documents such as: 'fitted drawings', details of the specifications and the electrical installation certificates required by Part 6 of the IET Regulations, warranties and manufacturers guides.

Assessment criteria 2.3

State the appropriateness of different customer relations methods and procedures

Effective communication is not just important in electrical installation, it is true of all industries. When we talk about good communication we are talking about transferring information from one person to another both quickly and accurately. We can do this through various forms, ranging from talking to other people directly and personally, to looking at drawings and plans and entering into a discussion with colleagues or other professionals who have a vested interest in the project. However, interaction also takes place with customers, clients as well as members of the public and therefore it is incredibly important to stress that employees remain respectful at all times – even if irritated. You are representing your company!

Assessment criteria 2.4

Identify methods of providing technical and functional information appropriate to the needs of clients etc.

On-site documentation

A lot of communication between and within larger organizations takes place by completing standard forms or sending internal memos, fax, texting or E.mails. Written messages have the advantage of being 'auditable'. An auditor can follow the paperwork trail to see, for example, who was responsible for ordering certain materials.

When completing standard forms, follow the instructions given and ensure that your writing is legible. Do not leave blank spaces on the form, always specifying 'not applicable' or 'N/A' whenever necessary. Sign or give your name and the date as asked for on the form. Finally, read through the form again to make sure you have answered all the relevant sections correctly. Internal memos are forms of written communication used within an organization; they are not normally used for communicating with customers or suppliers.

Figure 2.14 shows the layout of a typical standard memo form used by Dave Twem to notify John Gall that he has ordered the hammer drill.

Letters and E.mails provide a permanent record of communications between organizations and individuals. They may be handwritten for internal use but formal business letters give a better impression of the organization if they are typewritten. A letter or E.mail should be written using simple concise language, and the tone of the letter should always be polite, even if it is one of complaint. Always include the date of the correspondence. The greeting on a formal letter should be 'Dear Sir/Madam' and conclude with 'Yours faithfully'. A less formal greeting would be 'Dear Mr Smith' and conclude 'Yours sincerely'. Your name and status should be typed below your signature.

FLASH-BANG ELECTRICAL	internal MEMO
From _Dave Twem_	To _John Gall_
Subject _Power Tool_	Date _Thurs 11 Aug. 2018_

Message

Have today ordered Hammer Drill from P.S. Electrical – should be with you end of next week – Hope this is OK. Dave.

Figure 2.14 Typical standard memo form.

Delivery notes

When materials are delivered to site, the person receiving the goods is required to sign the driver's '**delivery note**'. This record is used to confirm that goods have been delivered by the supplier, who will then send out an invoice requesting payment, usually at the end of the month.

The person receiving the goods must carefully check that all the items stated on the delivery note have been delivered in good condition. Any missing or damaged items must be clearly indicated on the delivery note before signing, because by signing the delivery note the person is saying 'yes, these items were delivered to me as my company's representative on that date and in good condition and I am now responsible for these goods'. Suppliers will replace materials damaged in transit provided that they are notified within a set time period, usually three days. The person receiving the goods should try to quickly determine their condition. Has the packaging been damaged, does the container 'sound' like it might contain broken items? It is best to check at the time of delivery if possible, or as soon as possible after delivery and within the notifiable period. Electrical goods delivered to site should be handled carefully and stored securely until they are installed. Copies of delivery notes are sent to head office so that payment can be made for the goods received.

Time sheets

A **time sheet** is a standard form completed by each employee to inform the employer of the actual time spent working on a particular contract or site. This helps the employer to bill the hours of work to an individual job. It is usually a weekly document and includes the number of hours worked, the name of the job and any travelling expenses claimed. Office personnel require time sheets such as that shown in Fig. 2.15 so that wages can be made up.

Job sheets

A **job sheet** or job card such as that shown in Fig. 2.16 carries information about a job that needs to be done, usually a small job. It gives the name and address of the customer, contact telephone numbers, often a job reference number and a brief description of the work to be carried out. A typical job sheet work description might be:

- Job 1 Upstairs lights not working.
- Job 2 Funny fishy smell from kettle socket in kitchen.

An electrician might typically have a 'jobbing day' where he picks up a number of job sheets from the office and carries out the work specified. Job 1, for example, might be the result of a blown fuse, which is easily rectified, but the electrician must search a little further for the fault that caused the fuse to blow in the first place. The actual fault might, for example, be a decayed flex on a pendant drop that has become shorted out, blowing the fuse. The pendant drop should be re-flexed or replaced, along with any others in poor condition. The installation would then be tested for correct operation and the customer given an account of what has been done to correct the fault. General information and assurances about the condition of the installation as a whole might be requested and given before setting off to job 2.

The kettle socket outlet at job 2 is probably getting warm and, therefore, giving off that 'fishy' Bakelite smell because loose connections are causing the Bakelite

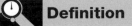

Definition

A *delivery note* is used to confirm that goods have been delivered by the supplier, who will then send out an invoice requesting payment.

Definition

A *time sheet* is a standard form completed by each employee to inform the employer of the actual time spent working on a particular contract or site.

Definition

A *job sheet* or job card such as that shown in Fig. 2.14 carries information about a job which needs to be done, usually a small job.

TIME SHEET

FLASH-BANG ELECTRICAL

Employee's name (Print) --

Week ending --

Day	Job number and/or address	Start time	Finish time	Total hours	Travel time	Expenses
Monday						
Tuesday						
Wednesday						
Thursday						
Friday						
Saturday						
Sunday						

Employee's signature -- Date -------------------

Figure 2.15 Typical time sheet.

socket to burn locally. A visual inspection would confirm the diagnosis. A typical solution would be to replace the socket and repair any damage to the conductors inside the socket box. Check the kettle plug top for damage and loose connections. Make sure all connections are tight before reassuring the customer that all is well; then, off to the next job or back to the office.

The time spent on each job and the materials used are sometimes recorded on the job sheet, but alternatively a daywork sheet can be used. This will

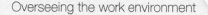

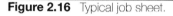

| JOB SHEET | **FLASH-BANG** |
| Job Number .... | **ELECTRICAL** |

Customer name --

Address of job --

Contact telephone No. --

Work to be carried out --

Any special instructions/conditions/materials used

Figure 2.16 Typical job sheet.

depend upon what is normal practice for the particular electrical company. This information can then be used to 'bill' the customer for work carried out.

Daywork sheets or variation order

Daywork is one way of recording variations to a contract, that is, work done that is outside the scope of the original contract. If daywork is to be carried out, the site supervisor must first obtain a signature from the client's representative, for example the architect, to authorize the extra work. A careful record must then be kept on the daywork sheets of all extra time and materials used so that the client can be billed for the extra work. A typical daywork sheet or variation order is shown in Fig. 2.17.

Definition

Daywork is one way of recording variations to a contract.

Reports

On large jobs, the foreman or supervisor is often required to keep a report of the relevant events that happen on the site – for example, how many people from your company are working on site each day, what goods were delivered, whether there were any breakages or accidents, and records of site meetings attended. Some firms have two separate documents, a site diary to record daily events and a weekly report, which is a summary of the week's events extracted from the site diary. The site diary remains on site and the weekly report is sent to head office to keep managers informed of the work's progress.

FLASH-BANG ELECTRICAL

VARIATION ORDER or DAYWORK SHEET

Client name _____

Job number/ref. _____

Date	Labour	Start time	Finish time	Total hours	Office use

Materials quantity	Description	Office use

Site supervisor or F.B. Electrical Representative responsible for carrying out work _____

Signature of person approving work and status e.g.

Client ☐   Architect ☐   Q.S. ☐   Main contractor ☐   Clerk of works ☐

Signature _____

Figure 2.17 Typical daywork sheet or variation order.

Explain the importance of ensuring that information provided is accurate and complete etc.

Aside from reports other methods of transferring information on a modern day site would include texting as well as e-mail; however, a very important point to remember is to think about the person you are communicating with in order to:

- use the correct tone of language;
- ensure information is sufficient for the task in hand;

Top tip

Always write up any verbal agreement so that there is a written record of anything agreed.

Figure 2.18 Always write up verbal agreements to ensure that there is a record of what was agreed.

- ensure information is clear and to the point and is not ambiguous;
- consider the persons in their role/position;
- ensure that it is the correct format.

By following the above advice you will ensure that people in authority are given the relevant information needed for a particular task. If you are not in a position to give that information then you should refer the matter to a more senior person, otherwise misinformation could lead to a loss of reputation, or even a loss of life if vital safety information is not given or understood.

Information including agreements between different trades are also made verbally person to person, therefore it is vital that any such agreements are also recorded on paper so that a written record then exists. This extends to a commissioning engineer, as mentioned previously, who has to hand over an installation and prove that it complies with the actual design specification and certain industry standards such as BS 7671. This is why following the completion of all new electrical work or additional work to an existing installation, the installation must be inspected and tested and an installation certificate issued and signed by a skilled person who must have a sound knowledge of the type of work undertaken, be fully versed in the inspection and testing procedures contained in the IET Regulations (BS 7671) and employ adequate testing equipment. This extends to:

1 Electrical Installation Certificate following the completion of a new installation or for alterations or additions to an existing installation.
2 Minor Electrical Installation Works Certificate: this form is intended to be used for the addition of a socket outlet or lighting point to an existing circuit or for a repair or modification to the installation. The work does not extend to the addition of a new circuit.
3 Electrical Installation Condition Report: all electrical installations must be periodically inspected and tested for the purpose of completing an electrical installation condition report.

Top tip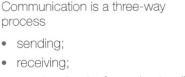

Communication is a three-way process

- sending;
- receiving;
- sender checks for understanding.

Assessment criteria 2.6

Describe methods for checking that relevant persons have an adequate understanding of the technical and non-technical information provided, including appropriate health and safety information

When accidents occur it is almost inevitable that it was caused by a breakdown of communication. The breakdown can occur verbally, written or by the fact that the receiver did not gain full understanding or comprehension of what was required. This is especially true when it comes to handing over information to another person. Consider for a moment you are on the phone and trying to describe and explain the following:

- outstanding work required to complete an installation;
- outstanding work required to complete the testing phase;
- outstanding spares and materials required;
- outstanding tools required;
- list of requested on-site dimensions and measurements.

In trying to recall all this information the possibility for error would be enormous. This is why it is vital that information is: written down, kept up-to-date (if anything changes) and the information is retained and filed for future reference. Other useful strategies when handing something over is to talk and walk through what remains to be done, but allowing time for questioning to ensure that full comprehension has taken place. Often people simply nod their head and say yes but in truth have no real comprehension of what is required.

Another strategy to reduce the possibility of error is to – whenever possible – task the same person who removed certain equipment to re-install it. Equally effective is labelling equipment that has been removed, which will make the re-installation process that much easier. Both are examples of strategies that can underpin best practice.

Figure 2.19 Don't simply nod your head and say yes when you aren't sure of something.

Assessment criteria 3.1

State the applicable health and safety requirements with regard to overseeing the work of others

Assessment criteria 3.2

State the procedures for interpreting risk assessments etc.

Construction site – safe working practice

Many governments have passed laws aimed at improving safety at work, but the most important recent legislation has been the Health and Safety at Work Act 1974. This Act should be thought of as an umbrella Act that other statutory legislation sit under. Following this Act many other statutory Acts have led to employers implementing safe systems of work, including providing safety signs and personal protective equipment (PPE), and provide training in how to recognize and use different types of fire extinguishers.

If your career in the electrical industry is to be a long, happy and safe one, you must always wear appropriate PPE such as footwear and head protection and behave responsibly and sensibly in order to maintain a safe working environment. Before starting work, make a safety assessment: what is going to be hazardous, will you require PPE, do you need any special access equipment? Construction sites can be hazardous because of the temporary nature of the construction process. The surroundings and systems are always changing as the construction process moves to its completion date when everything is finally in place.

Safe methods of working must be demonstrated by everyone at every stage. 'Employees have a duty of care to protect their own health and safety and that of others who might be affected by their work activities.' To make the work area safe before starting work and during work activities, it may be necessary to:

- use barriers or tapes to screen off potential hazards;
- place warning signs as appropriate;
- inform those who may be affected by any potential hazard;
- use a safe isolation procedure before working on live equipment or circuits;
- obtain any necessary 'permits to work' before work begins.

Get into the habit of always working safely and being aware of the potential hazards around you when you are working. Having chosen an appropriate wiring system that meets the intended use and structure of the building and satisfies the environmental conditions of the installation, you must install the system conductors, accessories and equipment in a safe and competent manner.

The structure of the building must be made good if it is damaged during the installation of the wiring system; for example, where conduits and trunking pass

Safety first

Everyone has a responsibility towards health and safety. Failure to act or ignoring procedures that lead to the injury of another is a breach of responsibility and punishable through law.

Figure 2.20 It is important that warning signs are displayed and maintained where appropriate.

through walls and floors. All connections in the wiring system must be both electrically and mechanically sound and the actual conductors must be chosen so that they will carry the full design current under the installed conditions.

If the wiring system is damaged during installation it must be made good to prevent future corrosion. For example, where galvanized conduit trunking or tray is cut or damaged by pipe vices, it must be made good to prevent localized corrosion.

All tools must be used safely and sensibly. Cutting tools should be sharpened and screwdrivers ground to a sharp square end on a grindstone. It is particularly important to check that the plug and cables of hand-held electrically powered tools and extension leads are in good condition. Damaged plugs and cables must be repaired before you use them. All electrical power tools of 110 V and 230 V must be tested with a portable appliance tester (PAT) in accordance with the company's health and safety procedures, but probably at least once each year.

Tools and equipment that are left lying about in the workplace can become damaged or stolen and may also be the cause of people slipping, tripping or falling. Tidy up regularly and put power tools back in their boxes. This is called 'good housekeeping' and is an important part of Appendix F of the *Electrician's Guide to the Building Regulations*. You personally may have no control over the condition of the workplace in general, but keeping your own work area clean and tidy is the mark of a skilled and conscientious craftsman. Finally, when the job is finished, clean up and dispose of all waste material responsibly. This is an important part of your company's 'good customer relationships' with the client. We also know that we have a 'duty of care' for the waste that we produce as an electrical company.

We have also said many times in this book that having a good attitude to health and safety, working conscientiously and neatly, keeping passageways clear and regularly tidying up the workplace are signs of a good and competent craftsman. But what do you do with the rubbish that the working environment produces? Well:

- All the packaging material for electrical fittings and accessories usually goes into either your employer's skip or the skip on site designated for that purpose.
- All the offcuts of conduit, trunking and tray also go into the skip.
- In fact, most of the general site debris will probably go into the skip and the waste disposal company will take the skip contents to a designated local council landfill area for safe disposal.
- The part coils of cable and any other reusable leftover lengths of conduit, trunking or tray will be taken back to your employer's stores area. Here it will be stored for future use and the returned quantities deducted from the costs allocated to that job.
- What goes into the skip for normal disposal into a landfill site is usually a matter of common sense. However, some substances require special consideration and disposal. We will now look at asbestos which is classified as 'special waste' or 'hazardous waste'.

Asbestos is a mineral found in many rock formations. When separated it becomes a fluffy, fibrous material with many uses. It was used extensively in the construction industry during the 1960s and 1970s for roofing material, ceiling and floor tiles, fire resistant board for doors and partitions, for thermal insulation and commercial and industrial pipe lagging.

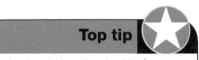

Top tip

Tools should be checked before and after use to ensure that they remain fit for purpose.

Figure 2.21 Asbestos fibres can cause lung disease.

In the buildings where it was installed some 40 years ago, when left alone, it does not represent a health hazard, but those buildings are increasingly becoming in need of renovation and modernization. It is in the dismantling and breaking up of these asbestos materials that the health hazard occurs.

Safe system of work

Understanding how employers implement a safe system of work is fundamental for any operative irrespective of how experienced they are. Formal risk assessments have to be undertaken by an employer in order to highlight any hazards present in the work area. This also includes identifying who is at risk and scrutinizing any existing control measure including its effectiveness. In certain high-risk areas or hazardous environments a permit to work is used, which ensures that an operation is only carried out once full authority has been granted and that the person undertaking the job is fully aware of his or her responsibilities. Finally, used extensively in the electrical installation industry, method statements signpost: exactly how a task is to be implemented; which tools are to be used; and even how waste material is to be disposed of so that no one involved is left in doubt about the job requirements.

Assessment criteria 4.1

Describe techniques for the communication with others

Assessment criteria 4.3

Specify their role in terms of responsibility for other staff etc.

Figure 2.22 Site safety starts before you even enter the site.

An electrician working for an electrical contracting company works as a part of the broader construction industry. This is a multi-million-pound industry carrying out all types of building work, from basic housing to hotels, factories, schools, shops, offices and airports. The construction industry is one of the UK's biggest employers, and carries out contracts to the value of about 10% of the UK's gross national product.

Although a major employer, the construction industry is also very fragmented. Firms vary widely in size, from the local builder employing two or three people to the big national companies employing thousands. Of the total workforce of the construction industry, 92% are employed in small firms of fewer than 25 people. The yearly turnover of the construction industry is about £35 billion. Of this total sum, about 60% is spent on new building projects and the remaining 40% on maintenance, renovation or restoration of mostly housing. In all these various construction projects the electrical industries play an important role, supplying essential electrical services to meet the needs of those who will use the completed building.

Assessment criteria 5.4

Identify within the scope of the work programme and operations their responsibilities

Assessment criteria 4.4

Identify appropriate methods for communicating with and responding to others

The information chain therefore includes a myriad of people as discussed previously when defining how roles are formed within both construction and the electrical team and how senior people within each element tie in with the client.

It is vital therefore that Information is kept current when initial plans or work schedules are changed, and irrespective of the justification, circumstance or reasons behind such change, the details of which must be made available to the relevant people. For instance, there are many examples of those working on site being driven by working drawings, but they have not been made aware that the drawings they are using have been updated, they have been superseded by a later version, since that information has not been passed down to them.

Such matters can cause friction since what has taken place is a waste of materials and labour. Friction can also take place when different trades are at loggerheads since both parties require access to a specific area. During such times a meeting must take place in order to find a way forward.

Assessment criteria 4.5

Specify procedures for re-scheduling work to co-ordinate with changing conditions in the workplace and to coincide with other trades

Assessment criteria 4.6

Clarify organizational procedures for completing the documentation that is required during work operations

One strategy used in building relationships and formulating roles and responsibility is shown below.

Stages of team development

- Forming: Individuals will gather together in order to discuss an ongoing issue or concern and share different ideas or strategies.

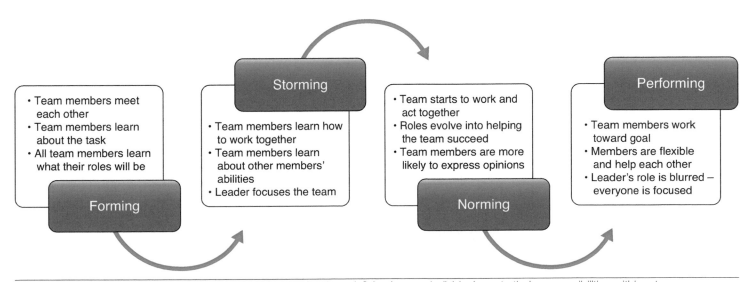

Figure 2.23 Forming, Storming, Norming and Performing involves defining how an individual meets their responsibilities within a team environment.

- Storming: The process when a trade-off of thoughts takes place and also individuals define their role within the team.
- Norming: Moving forward the individuals involved agree certain roles and responsibilities within the team and especially define key processes such as how best to perform a task.
- Performing: Individual responsibilities have now been defined and especially how they fit in with others.

What is important here is that everyone understands their role and this includes cooperation in a wider context. For instance, there is always going to be tension when resources are required to be shared but an element of understanding can disarm such situations. Other examples of site tension can be caused through a lack of support from supervisors or management, and it is well worth reminding those in authority that the Electricity at Work Act 1989 lays down clear requirements in ensuring that through an onus of care, apprentices are given appropriate guidance and supervision based on their skills and experience.

In some cases where disputes cannot be settled in-house then some form of mediation is required through the involvement of a union representative who will negotiate on a member's behalf. Their aim will be to instil the fundamentals of fair working practices. If all relationships are harmonious then it will make for a happier workforce, which in turn will ensure:

- the efficient use of resources;
- customer satisfaction;
- increased efficiency and productivity;
- wastage reduction;
- increased profitability.

There is a lot to learn for any new employee, therefore it is worth stressing that if a problem does occur that is outside of your authority or comprehension then contact your supervisor for professional help and guidance. It is not unreasonable for a young member of the company's team to seek help and guidance from those employees with more experience. This approach would be preferred by most companies rather than having to meet the cost of an expensive blunder.

> **Top tip** ⭐
>
> Although you will be given an induction programme to help you gain familiarity of site processes and procedures there is still a lot of information to remember. Don't be a mushroom if you are being kept in the dark – make it your business to find out!

Assessment criteria 4.2

Describe methods of determining the competence of operatives for whom they are responsible

Employee skills and competencies

Certain skills and understanding of trade related information are maintained through industry driven schemes and are especially important on a construction site, since it is one of the most hazardous workplaces that you can work within.

Consequently, entry on to site is controlled and regulated through the Construction Skills Certification Scheme (CSCS) card. The CSCS is a scheme

Figure 2.24 Instructional communication not only saves time, but it also makes for a safer working environment.

that assesses an individual's knowledge of health and safety policies and procedures in order to ensure that personnel that gain entry recognize and understand what is expected of them, including policies and procedures. The scheme will also differ for a labourer (green card) to a skilled individual (gold card) and therefore allow further involvement and engagement on specific operations and tasks. An electrician is normally afforded an ECS card, which proves recognized and current trade based skills and qualifications as well as having been successful in undertaking a health and safety assessment.

Other industry recognized schemes regarding recognition of trade related skills and experience can be found through JIB membership and being given an electrician's gold card.

That said, all contractors irrespective of their qualifications are still monitored on site to ensure that safety is maintained at all times. In certain conditions written references from previous employers would be required if, for instance, a prospective contractor had worked abroad for a time or had not been active in the industry. The government has approved schemes to be operated by BRE Certification Ltd, British Standards Institution, ELECSA Ltd, NICEIC Certification Services Ltd, certsure.com, and Napit Certification Services Ltd. All the different bodies will operate the scheme to the same criteria and will be monitored by the Department for Communities and Local Government, formerly called the Office of the Deputy Prime Minister.

Installers of environmental technology systems must also be registered under the scheme. The work of members will then be inspected at least once each year. There will be an initial registration and assessment fee and then an annual membership and inspection fee.

Assessment criteria 5.1

Describe how to plan work allocations etc.

Assessment criteria 5.2

Specify procedures for carrying out work activities that will maintain the safety of the work environment etc.

Organizing and overseeing work programmes

Smaller electrical contracting firms will know where their employees are working and what they are doing from day to day because of the level of personal contact between the employer, employee and customer. As a firm expands and becomes engaged on larger contracts, it becomes less likely that there is anyone in the firm with a complete knowledge of the firm's operations, and there arises an urgent need for sensible management and planning skills so that men and materials are on site when they are required and a healthy profit margin is maintained. What is paramount, however, is that work operations also tally with the skills and experience of the workforce so that individuals are not put in a position that exceeds their responsibility or ability.

When the electrical contractor is told that he has been successful in tendering for a particular contract they are committed to carrying out the necessary work within the contract period. They must therefore consider:

- by what date the job must be finished;
- when the job must be started if the completion date is not to be delayed;
- how many men will be required to complete the contract including qualifications, skills, authorizations;
- when certain materials will need to be ordered;
- when the supply authorities must be notified that a supply will be required;
- if it is necessary to obtain authorization from a statutory body for any work to commence.

In thinking ahead and planning the best method of completing the contract, the individual activities or jobs must be identified and consideration given to how the various jobs are interrelated. To help in this process a number of management techniques are available. In this chapter we will consider only two: bar charts and network analysis. The very preparation of a bar chart or network analysis forces the contractor to think deeply, carefully and logically about the particular contract, and it is therefore a very useful aid to the successful completion of the work.

Assessment criteria 5.7

Specify methods of producing and illustrating work programmes such as bar charts, spreadsheets and critical path analysis

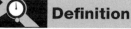

Bar charts

There are many different types of bar chart used by companies but the **objective of any bar chart** is to establish the sequence and timing of the various activities involved in the contract as a whole. They are a visual aid in the process of communication. In order to be useful they must be clearly understood by the people involved in the management of a contract. The chart is constructed on a rectangular basis, as shown in Fig. 2.25.

All the individual jobs or activities which make up the contract are identified and listed separately down the vertical axis on the left-hand side, and time flows from left to right along the horizontal axis. The unit of time can be chosen to suit the length of the particular contract, but for most practical purposes either days or weeks are used.

The simple bar chart shown in Fig. 2.25(a) shows a particular activity A, which is estimated to last two days, while activity B lasts eight days. Activity C lasts four days and should be started on day three. The remaining activities can be interpreted in the same way.

With the aid of colours, codes, symbols and a little imagination, much additional information can be included on this basic chart. For example, the actual work completed can be indicated by shading above the activity line as shown in Fig. 2.25(b) with a vertical line indicating the number of contract days completed; the activities that are on time, ahead of or behind time can easily be identified. Activity B in Fig. 2.25(b) is two days behind schedule, while activity D is two days ahead of schedule. All other activities are on time. Some activities must be completed before others can start. For example, all conduit work must be completely erected before the cables are drawn in. This is shown in Fig. 2.25(b) by activities J and K. The short vertical line between the two activities indicates that activity J must be completed before K can commence.

Useful and informative as the bar chart is, there is one aspect of the contract that it cannot display. It cannot indicate clearly the interdependence of the various activities upon each other, and it is unable to identify those activities that must strictly adhere to the time schedule if the overall contract is to be completed on time, and those activities in which some flexibility is acceptable. To overcome this limitation, in 1959 the Central Electricity Generating Board (CEGB) developed the critical path network diagram, which we will now consider.

Network analysis

In large or complex contracts there are a large number of separate jobs or activities to be performed. Some can be completed at the same time, while others cannot be started until others are completed. A **network diagram** can be used to coordinate all the interrelated activities of the most complex project in such a way that all sequential relationships between the various activities, and the restraints imposed by one job on another, are allowed for. It also provides a method of calculating the time required to complete an individual activity and will identify those activities that are the key to meeting the completion date, called the critical path. Before considering the method of constructing a network diagram, let us define some of the terms and conventions we shall be using.

Definition

There are many different types of bar chart used by companies but the *objective of any bar chart* is to establish the sequence and timing of the various activities involved in the contract as a whole.

Definition

A *network diagram* can be used to coordinate all the interrelated activities of the most complex project.

Activity \ Day number	1	2	3	4	5	6	7	8	9	10	11	12	13	14
A	█	█												
B	█	█	█	█	█	█	█	█						
C			█	█	█	█								
D									█	█	█	█	█	█
E	█	█	█	█										
F	█	█	█	█	█	█								
G		█	█	█	█	█	█							
H						█	█	█	█	█	█	█	█	
I		█	█	█	█	█	█	█						
J	█	█	█	█	█	█	█							
K							█	█	█	█	█	█	█	█
L										█	█	█	█	
M													█	█

(a) A simple bar chart or schedule of work

Activity \ Day number	1	2	3	4	5	6	7	8	9	10	11	12	13	14
A	█	█												
B	█	█	█	█	█	█								
C			█	█	█									
D							█	█	█	█				
E	█	█	█											
F	█	█	█	█	█	█								
G		█	█	█	█	█	█							
H						█	█	█						
I		█	█	█	█	█								
J	█	█	█	█	█									
K								█						
L														
M														

⟶ Number of contract days completed

(b) A modified bar chart indicating actual work completed

Figure 2.25 Bar charts: (a) a simple bar chart or schedule of work; (b) a modified bar chart indicating actual work completed.

Definition

Critical path is the path taken from the start event to the end event which takes the longest time.

Critical path

Critical path is the path taken from the start event to the end event which takes the longest time. This path denotes the time required for completion of the whole contract.

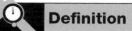

Float time

Float time, slack time or time in hand is the time remaining to complete the contract after completion of a particular activity.

The total float time for any activity is the total leeway available for all activities in the particular path of activities in which it appears. If the float time is used up by one of the early activities in the path, there will be no float left for the remaining activities and they will become critical.

Activities

Activities are represented by an arrow, the tail of which indicates the commencement, and the head the completion of the activity. The length and direction of the arrows have no significance: they are not vectors or phasors. Activities require time, manpower and facilities. They lead up to or emerge from events.

Dummy activities

Dummy activities are represented by an arrow with a dashed line. They signify a logical link only, require no time and denote no specific action or work.

Event

An **event** is a point in time, a milestone or stage in the contract when the preceding activities are finished. Each activity begins and ends in an event. An event has no time duration and is represented by a circle, which sometimes includes an identifying number or letter. Time may be recorded to a horizontal scale or shown on the activity arrows. For example, the activity from event A to B takes nine hours in the network diagram shown in Fig. 2.26.

> **Definition**
>
> *Float time*, slack time or time in hand is the time remaining to complete the contract after completion of a particular activity.
>
> Float time = Critical path time – Activity time

> **Definition**
>
> *Activities* are represented by an arrow, the tail of which indicates the commencement, and the head the completion of the activity.

> **Definition**
>
> *Dummy activities* are represented by an arrow with a dashed line.

> **Definition**
>
> An *event* is a point in time, a milestone or stage in the contract when the preceding activities are finished.

Example 1

Identify the three possible paths from the start event A to the finish event F for the contract shown by the network diagram in Fig. 2.26. Identify the critical path and the float time in each path.

The three possible paths are:

1 event A–B–D–F
2 event A–C–D–F
3 event A–C–E–F

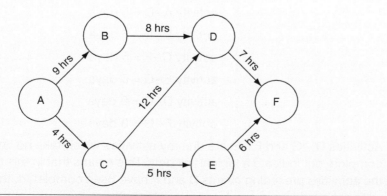

Figure 2.26 A network diagram for Example 1.

(Continued)

Example 1 (Continued)

The times taken to complete these activities are:

1 path A–B–D–F = 9 + 8 + 7 + 24 hours
2 path A–C–D–F = 4 + 12 + 7 + 23 hours
3 path A–C–E–F = 4 + 5 + 6 + 15 hours

The longest time from the start event to the finish event is 24 hours, and therefore the critical path is A – B – D – F.

The float time is given by:

$$\text{Float time} = \text{Critical Path} - \text{Activity time}$$

For path 1, A–B–D–F,

$$\text{Float time} = 24 \text{ hours} - 24 \text{ hours} = 0 \text{ hours}$$

There can be no float time in any of the activities which form a part of the critical path, since a delay on any of these activities would delay completion of the contract. On the other two paths some delay could occur without affecting the overall contract time.

For path 2, A – C – D – F,

$$\text{Float time} = 24 \text{ hours} - 23 \text{ hours} = 1 \text{ hour}$$

For path 3, A – C – E – F,

$$\text{Float time} = 24 \text{ hours} - 15 \text{ hours} = 9 \text{ hours}$$

Example 2

Identify the time taken to complete each activity in the network diagram shown in Fig. 2.27. Identify the three possible paths from the start event A to the final event G and state which path is the critical path.

The time taken to complete each activity using the horizontal scale is:

activity A–B = 2 days

activity A–C = 3 days

activity A–D = 5 days

activity B–E = 5 days

activity C–F = 5 days

activity E–G = 3 days

activity D–G = 0 days

activity F–G = 0 days

Activities D – G and F – G are dummy activities which take no time to complete but indicate a logical link only. This means that in this case, once the activities preceding events D and F have been completed, the contract will not be held up by work associated with these particular paths and they will progress naturally to the finish event.

(Continued)

Example 2 (Continued)

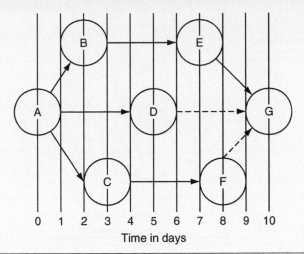

Figure 2.27 A network diagram for Example 2.

The three possible paths are:

1 A–B–E–G
2 A–D–G
3 A–C–F–G

The times taken to complete the activities in each of the three paths are:

$$\text{path 1, A–B–E–G} = 2 + 5 + 3 = 10 \text{ days}$$

$$\text{path 2, A–D–G} = 5 + 0 = 5 \text{ days}$$

$$\text{path 3, A–C–F–G} = 3 + 5 + 0 = 8 \text{ days}$$

The critical path is path 1, A–B–E–G.

Constructing a network

The first step in constructing a network diagram is to identify and draw up a list of all the individual jobs, or activities, that require time for their completion and which must be completed to advance the contract from start to completion. The next step is to build up the arrow network showing schematically the precise relationship of the various activities between the start and end event. The designer of the network must ask these questions:

1 Which activities must be completed before others can commence? These activities are then drawn in a similar way to a series circuit but with event circles instead of resistor symbols.
2 Which activities can proceed at the same time? These can be drawn in a similar way to parallel circuits but with event circles instead of resistor symbols.

Commencing with the start event at the left-hand side of a sheet of paper, the arrows representing the various activities are built up step by step until the final event is reached. A number of attempts may be necessary to achieve a well

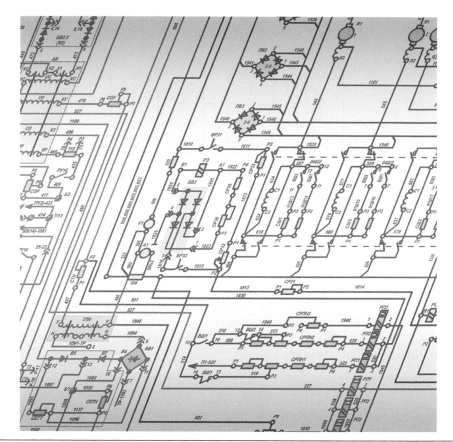

Figure 2.28 Part of an electrical network diagram.

balanced and symmetrical network diagram showing the best possible flow of work and information, but this time is well spent when it produces a diagram that can be easily understood by those involved in the management of the particular contract.

Example 3

A particular electrical contract is made up of activities A – F as described below:

A = an activity taking 2 weeks commencing in week 1

B = an activity taking 3 weeks commencing in week 1

C = an activity taking 3 weeks commencing in week 4

D = an activity taking 4 weeks commencing in week 7

E = an activity taking 6 weeks commencing in week 3

F = an activity taking 4 weeks commencing in week 1

Certain constraints are placed on some activities because of the availability of men and materials and because some work must be completed before other work can commence as follows:

Activity C can only commence when B is completed

Activity D can only commence when C is completed

(Continued)

Example 3 (Continued)

Activity E can only commence when A is completed

Activity F does not restrict any other activity

(a)   Produce a simple bar chart to display the activities of this particular contract.

(b)   Produce a network diagram of the programme and describe each event.

(c)   Identify the critical path and the total contract time.

(d)   State the maximum delay which would be possible on activity E without delaying the completion of the contract.

(e)   State the float time in activity F.

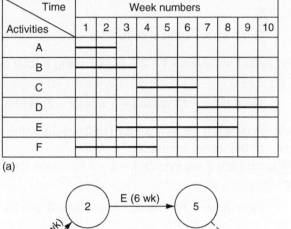

(a)

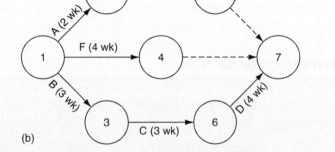

(b)

Figure 2.29 (a) Bar chart and (b) network diagram for Example 3.

(a)   A simple bar chart for this contract is shown in Fig. 2.29(a).

(b)   The network diagram is shown in Fig. 2.29(b). The events may be described as follows:

Event 1 = the commencement of the contract

Event 2 = the completion of activity A and the commencement of activity E

Event 3 = the completion of activity B and the commencement of activity C

Event 4 = the completion of activity F

Event 5 = the completion of activity E

Event 6 = the completion of activity C

Event 7 = the completion of activity D and the whole contract.

(Continued)

Example 3 (Continued)

(c)   There are three possible paths:

   1   via events 1 − 2 − 5 − 7
   2   via events 1 − 4 − 7
   3   via events 1 − 3 − 6 − 7.

The time taken for each path is:

$$\text{path 1} = 2 \text{ weeks} + 6 \text{ weeks} = 8 \text{ weeks}$$

$$\text{path 2} = 4 \text{ weeks} = 4 \text{ weeks}$$

$$\text{path 3} = 3 \text{ weeks} + 3 \text{ weeks} + 4 \text{ weeks} = 10 \text{ weeks}.$$

The critical path is therefore path 3, via events 1 − 3 − 6 − 7, and the total contract time is 10 weeks.

(d)   We have that:

$$\text{Float time} = \text{Critical path time} - \text{Activity time}$$

Activity E is on path 1 via events 1 − 2 − 5 − 7 having a total activity time of 8 weeks

$$\text{Float time} = 10 \text{ weeks} - 8 \text{ weeks} = 2 \text{ weeks}$$

Activity E could be delayed for a maximum of 2 weeks without delaying the completion date of the whole contract.

(e)   Activity F is on path 2 via events 1 − 4 − 7 having a total activity time of 4 weeks

$$\text{Float time} = 10 \text{ weeks} - 4 \text{ weeks} = 6 \text{ weeks}$$

Assessment criteria 5.5

Identify how to determine the estimated time required for the completion of the work required taking into account influential factors such as the deployment and availability of suitable personnel etc.

Assessment criteria 5.6

State the possible consequences of not completing work within the estimated time etc.

Another type of chart is known as a Gantt chart and it enables various stages of an installation to be controlled and tracked against a critical and optimum time frame. Disruptions and delays inevitably happen and can occur due to a variety of reasons such as:

- unavailability of suitable personnel;
- unavailability of suitable materials;
- equipment failure;
- latent (hidden) faults;

- inclement weather;
- scheduling additional requirements or additional specifications.

It is therefore vital that team meetings are held on a regular basis to track progress and discuss any related issues as mentioned above. This will allow certain contingencies to be made available or brought about in order to maintain certain milestones and target dates. Inevitably, certain outdates might have to be changed to accommodate circumstances that are not retrievable, which could lead to a breach of contract and in some circumstances where such matters have been formalized will bring about a heavy fine.

Key fact

Re-scheduling work due to delay must include all the trades involved.

Assessment criteria 5.3

Identify the industry standards that are relevant to activities carried out during the installation of electrotechnical systems and equipment, including the current editions of Management of Health and Safety Regulations etc.

Within the electrical installation industry there are relevant elements and statutory regulations that impact work activities. They include the Health and Safety Act 1974, which is the main umbrella act that other statutory regulations are drawn from such as:

Management of Health and Safety at Work Regulations (MHSWR) 1999

This regulation makes sure that managers ensure that they:

- make arrangements for implementing the health and safety measures to reduce risk;
- appoint competent people to help you implement the arrangements;
- set up emergency procedures;
- provide clear information and training to employees and ensure that policy and procedures are adopted by employees;
- cooperate on health and safety matters with other employers who share the same workplace to coordinate an exchange of information;
- consult their employees;
- review policy.

The Electricity at Work Regulations 1989 (EWR)

The regulations are made under the Health and Safety at Work Act 1974, and enforced by the Health and Safety Executive. The purpose of the regulations is to 'require precautions to be taken against the risk of death or personal injury from electricity in work activities'.

Sex Discrimination Act – for example, discounting a prospective candidate on the basis of their gender.

Employment Relations Act – for example, discounting a prospective candidate because they belong to a trade union.

Employment Rights Act – sets out the statutory rights of employees.

Human Rights Act – failing to treat everyone equally, with fairness, dignity and respect.

Race Relations Act – discounting a prospective candidate on the basis of their race.

Data Protection Act – ensuring that personal information is not given without consent.

Protection from Harassment Act – for example, protecting young apprentices from being subjected to initiation ceremonies.

Equality Act – overlooking a prospective candidate based on gender such as excluding a female employee when engaged in labour intensive tasks.

Disability Discrimination Act – for example, discounting a prospective candidate on the basis of their physical disability despite the fact that it is well within the physical capability of the person.

Key fact

Giving away your boss's personal details even to a prospective customer without consent is in breach of the Data Protection Act.

BS 7671 The IET Regulations

The Institution of Engineering and Technology Requirements for Electrical Installations (the IET Regulations) are non-statutory regulations. They relate principally to the design, selection, erection, inspection and testing of electrical installations, whether permanent or temporary, in and about buildings generally and to agricultural and horticultural premises, construction sites and caravans and their sites. Paragraph 7 of the introduction to the EWR says: 'the IET Wiring Regulations is a code of practice which is widely recognized and accepted in the United Kingdom and compliance with them is likely to achieve compliance with all relevant aspects of the Electricity at Work Regulations.' The IET Wiring Regulations are the national standard in the United Kingdom and apply to installations operating at a voltage up to 1000 V a.c. They do not apply to electrical installations in mines and quarries, where special regulations apply because of the adverse conditions experienced there. The current edition of the IET Wiring Regulations is the 18th Edition 2018.

Construction (Design and Management) Regulations

The Construction (Design and Management) Regulations (CDM) are aimed at improving the overall management of health, safety and welfare throughout all stages of the construction project.

The person requesting that construction work commences, the client, must first of all appoint a 'duty holder', someone who has a duty of care for health, safety and welfare matters on-site. This person will be called a 'planning supervisor'. The planning supervisor must produce a 'pre-tender' health and safety plan and coordinate and manage this plan during the early stages of construction. The client must also appoint a principal contractor who is then required to develop the health and safety plan made by the planning supervisor, and keep it up to date during the construction process to completion. The degree of detail in the health and safety plan should be in proportion to the size of the construction project and recognize the health and safety risks involved on that particular project. Small projects will require simple, straightforward plans; large projects, or those involving significant risk, will require more detail. The CDM Regulations will apply to most large construction projects but they do not apply to the following:

- construction work, other than demolition work, that does not last longer than 30 days and does not involve more than four people;
- construction work carried out inside commercial buildings such as shops and offices, which does not interrupt the normal activities carried out on those premises;
- construction work carried out for a domestic client;
- the maintenance and removal of pipes or lagging which form a part of a heating or water system within the building.

Graphical symbols BS EN

To maintain continuity and consistency, any use of symbols should conform and comply with BS EN 60617.

Assessment criteria 6.1

Interpret the installation specification and work programme to identify resource requirements

Assessment criteria 6.2

Interpret the material schedule to confirm that materials available are the right type etc.

Designing an electrical installation

The designer of an electrical installation must ensure that the design meets the requirements of the IET Wiring Regulations for electrical installations and any other regulations that may be relevant to a particular installation. In doing so the **designer** interprets the electrical requirements of the customer within the regulations, identifies the appropriate types of installation, the most suitable methods of protection and control and the size of cables to be used.

The 18th Edition of the Regulations gives us **a new Appendix 17, Energy Efficiency**. It is intended that this new Appendix will become a new Part 8 in future Amendments to the regulations. The new Appendix provides **recommendations** for the design and erection of electrical installations. All new electrical installation design, and modifications to existing installations must continue to meet the required levels of safety and capacity, but must **NOW ALSO** optimise electrical efficiency, providing the lowest levels of electricity consumption.

The new Appendix considers design and maintenance from the context of improving the efficiency of the electrical installation. There is a change of emphasis in this new section, to incorporate energy efficiency into the electrical installation design as a prerequisite, and not just as an aspiration.

Safety first

The installation of sub-standard materials could lead to a dangerous situation.

The standard makes it clear that any measure taken to make the electrical installation more efficient, **must not compromise the safety of any occupants, property or livestock, and that much of this appendix will not apply to domestic and similar installations**.

A large electrical installation may require many meetings with the customer and his professional representatives in order to identify a specification of what is required. The designer can then identify the general characteristics of the electrical installation and its compatibility with other services and equipment, as indicated in Part 3 of the regulations. The protection and safety of the installation, and of those who will use it, must be considered, with due regard to Part 4 of the regulations. An assessment of the frequency and quality of the maintenance to be expected will give an indication of the type of installation which is most appropriate. The size and quantity of all the materials, cables, control equipment and accessories can then be determined. This is called a '**bill of quantities**'.

It is common practice to ask a number of electrical contractors to tender or submit a price for work specified by the bill of quantities. A tender is a formal offer to supply goods or carry out work at a stated price. The contractor must cost all the materials but also consider such things as vehicles, plant equipment and tools, as well as the labour cost required to install the materials. They then add on profit and overhead costs in order to arrive at a final estimate for the work.

The contractor tendering the lowest cost is usually, but not always, awarded the contract. To complete the contract in the specified time the electrical contractor must use the management skills required by any business to ensure that men and materials are on site as and when they are required. If alterations or modifications are made to the electrical installation as the work proceeds, which are outside the original specification, then a **variation order** must be issued so that the electrical contractor can be paid for the additional work.

The specification for the chosen wiring system will be largely determined by the building construction and the activities to be carried out in the completed building. An industrial building, for example, will require an electrical installation that incorporates flexibility and mechanical protection. This can be achieved by a conduit, tray or trunking installation. In a block of purpose-built flats, all the electrical connections must be accessible from one flat without intruding upon the surrounding flats. A loop-in conduit system, in which the only connections are at the light switch and outlet positions, would meet this requirement.

For a domestic electrical installation an appropriate lighting scheme and multiple socket outlets for the connection of domestic appliances, all at a reasonable cost, are important factors which can usually be met by a PVC-insulated and sheathed wiring system. Whatever the wiring or enclosure system selected it must be inspected and tested during and on completion of the installation.

The final choice of a wiring system must rest with those designing the installation and those ordering the work, but whatever system is employed, good workmanship by a skilled person is required. That said the actual materials selected and then checked on delivery have to be fit for purpose for the job in hand. This is why obtaining materials that are branded with BS, BSEN or the CE mark ensures that they comply with known national and international standards.

Index of Protection (IP) BS EN 60529

The Index of Protection is a code that gives us a means of specifying the suitability of equipment for the environmental conditions in which it will be used and must always be a consideration to ensure that materials and equipment are fit for purpose. Whilst Chapter 3 of this book will cover this element in far greater detail, it is worth remembering that IET Regulation 416.2.1 tells us that where barriers and enclosures have been installed to prevent direct contact with live parts, they must afford a degree of protection not less than IP2X and IPXXB (total protection). This is explained by Fig. 2.30 and on page 124 and Fig 3.5 of this book.

Obtaining information and components

Electricians use electrical wholesalers and suppliers to purchase electrical cable, equipment and accessories. Similar facilities are available in most towns and cities for the purchase of electronic components and equipment. There are also a number of national suppliers who employ representatives who will call at your workshop to offer technical advice and take your order. Some of these national companies also offer a 24-hour telephone order and mail order service, such as RS at www.rswww.com or telephone 08457 201 201. Their full-colour, fully illustrated catalogues also contain an enormous amount of technical information.

For local suppliers you must consult your local phone book and *Yellow Pages*. However, such outlets can also be used to serve as a means of confirming if equipment complies with the standards on the specification.

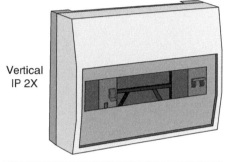

Definition

IP rating is very important to ensure that a piece of equipment is suitable to its environment.

Top horizontal surfaces IP 4X or IPXXB

Vertical IP 2X

Figure 2.30 Horizontal surfaces are afforded higher protection given a higher possibility of moisture ingress.

Assessment criteria 6.3

Specify the storage and transportation requirements for all materials required in the work location

Assessment criteria 6.4

Specify procedures to ensure the safe and effective storage of materials, tools and equipment in the work location

Secure site storage

The ability to resolve a technical practical problem using tools and equipment is a part of what those of us in the electrical industry are about, and so tools are very important to us. As a safety precaution it makes sense not to leave them unattended on site and to keep them in the back of your locked vehicle when not being used. Power tools are expensive, make your working life less difficult and are very easy to be carried away so they too should leave with you at the end of the working day. However, some pieces of equipment are too big to take home such as access equipment, generators and other electrical equipment brought in

Top tip

Failure to lock unused materials and tools could not only make them a target for thieves but also compromise site insurance since the insurance company expects every effort to be made to secure valuable assets.

or hired specifically for a particular job. At the end of the working day these larger pieces of equipment must be made secure.

On large construction sites the main contractors often make secure storage in the shape of metal containers available to their subcontractors and this may be a good solution in some situations. However, the security of a store can be put at risk if there is more than one keyholder. One of the other keyholders might leave the store unlocked and your material and equipment could be stolen through no fault of your own. So what are the alternatives? You could put your materials and equipment inside a large locked box that has been bolted to the wall or floor inside the main contractor's locked store, or you could use motorcycle locks to secure your plant inside the main contractor's locked store. Materials that are to be installed at some later date also require secure storage. Electrical contractors these days often work out of an industrial unit which may also be their Head Office incorporating facilities for 'office work' and will therefore be occupied during the working day. Materials could probably be delivered and stored at this industrial unit and only small quantities taken to site each day by the installing electricians. Alternatively, if you have a good relationship with your local wholesalers, they too will deliver smaller quantities to site so that there is no requirement for storage. You may think that I am painting a rather bleak picture of secure site storage but electrical goods and equipment are very expensive and most desirable to a thief and you must, therefore, give serious consideration to how you will securely store goods on site if you have no other alternative.

Test your knowledge

When you have completed the questions check out the answers at the back of the book.

Note: more than one multiple-choice answer may be correct.

Learning outcome 1

1   Which of the following correctly describes a circuit diagram?
    a.   shows the location of individual parts or components
    b.   shows information about how they are to be connected
    c.   uses symbols to represent all major components
    d.   no information about wiring but explains how the circuit operates.

2   The symbol shown represents:
    a.   wall mounted lamp
    b.   general lamp symbol
    c.   fluorescent lamp
    d.   socket outlet.

3   Which piece of information will you find in 'particular' specifications?
    a.   client requirements
    b.   wiring system
    c.   contract details
    d.   layout drawings.

4   How many 2 way 1 gang light switches are shown in the layout diagram?
    a.   0
    b.   1
    c.   2
    d.   3.

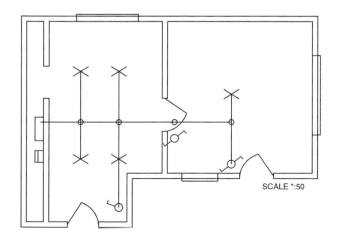

SCALE *:50

5   The symbol shown is:
a.   double switched socket
b.   double socket
c.   fused connection unit
d.   switched fused connection unit.

6   The type of diagram shown is a:
a.   wiring diagram
b.   circuit diagram
c.   block diagram
d.   schematic diagram.

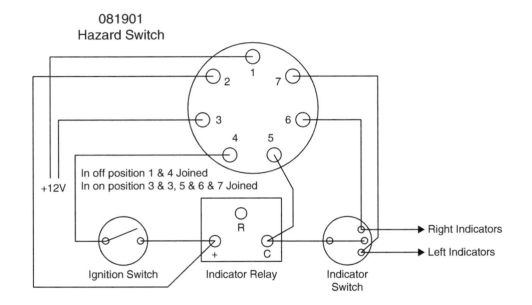

081901
Hazard Switch

In off position 1 & 4 Joined
In on position 3 & 3, 5 & 6 & 7 Joined
+12V

Ignition Switch   Indicator Relay   Indicator Switch

Right Indicators
Left Indicators

7   The operation of test equipment and their different operating modes can be found in:
a.   GS38
b.   manufacturer's user manual
c.   supplier manual
d.   BS 7671.

8   A calibrating laboratory will check equipment accuracy by comparing the test equipment against:
a.   a known accurate source
b.   another piece of test equipment
c.   the results of the last testing schedule
d.   manufacturer's data.

9   For large building sites the scale drawing for distribution of cable routes and the buildings involved will be drawn:
a.   using a ratio between the scale drawing and the actual size
b.   to a very large scale
c.   to a different scale
d.   to a specific local planning scale.

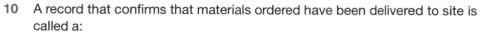

10 A record that confirms that materials ordered have been delivered to site is called a:
   a. job sheet
   b. time sheet
   c. delivery note
   d. daywork sheet.

11 A standard form completed by every employee to inform the employer of the time spent working on a particular site is called a:
   a. job sheet
   b. time sheet
   c. delivery note
   d. daywork sheet.

12 A standard form containing information about work to be done usually distributed by a manager to an electrician is called a:
   a. job sheet
   b. time sheet
   c. delivery note
   d. daywork sheet.

13 A standard form that records any changes or extra work carried out is called a:
   a. job sheet
   b. time sheet
   c. delivery note
   d. daywork sheet.

14 A formal contract to supply goods or carry out work at a stated price is called a:
   a. network analysis
   b. variation order
   c. bar chart
   d. tender.

15 A record of any work done that is outside or in addition to the original electrical contract is called a:
   a. network analysis
   b. variation order
   c. bar chart
   d. tender.

16 A method of showing the separate activities or tasks that make up a large or complex electrical contract is called a:
   a. network analysis
   b. variation order
   c. Gantt chart
   d. tender.

Learning outcome 2

17   Which one of the following is **NOT** part of a handover procedure?
a.   final inspection of the installation by client
b.   copies of testing schedules
c.   demonstration of equipment
d.   copy of BS 7671.

18   Which of the following would be responsible for submitting a tender?
a.   manager
b.   engineer
c.   estimator
d.   inspector.

19   Nominated subcontractors tend to be appointed by:
a.   the main contractor
b.   the other subcontractors
c.   the client/architect
d.   the council.

20   As an apprentice you are asked by the main contractor where the main switch panel is to be located. You are aware of this therefore you should:
a.   tell them that you are unsure just in case you get it wrong
b.   tell them you are only an apprentice and you are not responsible enough
c.   tell them politely where it is to be located but also inform your supervisor
d.   tell them to speak to your supervisor.

21   Whose role is it to manage and control the costs of a building project?
a.   clerk of works
b.   quantity surveyor
c.   estimator
d.   consulting engineer.

22   Trainees who are asked their advice but are unsure of how to respond should:
a.   look it up
b.   make it up
c.   explain that as they are trainees they do not need to know
d.   seek advice from the employer or more experienced colleague.

Learning outcome 3

23   Site organizational procedures are written by management therefore apply to:
a.   management only
b.   all persons
c.   all employees
d.   visitors only.

24 Risk assessments ensure that work places monitor any changing conditions by:
   a. recording all minor accidents
   b. recording all serious accidents
   c. reviewing hazardous situations
   d. reviewing visitor book.

25 Written procedures that inform an employee of how a task is to be undertaken are known as:
   a. risk assessment
   b. permit to work
   c. method statement
   d. COSHH statement.

26 All tools and equipment should be regulated. This means:
   a. secure storage
   b. control and maintained
   c. stored only
   d. maintained only.

27 After a serious injury on site, which of the following applies?
   a. investigation by the HSE
   b. RIDDOR
   c. enter details into the accident book
   d. cancellation of bank holidays.

Learning outcome 4

28 Praising good workmanship is considered to be:
   a. a necessary requirement therefore positive
   b. motivational therefore positive
   c. instructive therefore positive
   d. destructive therefore positive.

29 Which of the following can be thought of as a permanent record?
   a. verbal
   b. telephone
   c. memorandum
   d. letter.

30 A modern-day device that can be used to record various work activity on site which when sent to an assessor can be used as evidence is:
   a. site diary entry
   b. Royal Mail
   c. formal letter
   d. mobile phone.

31 A convenient and quick method of sending draft contracts between all parties concerned is:
   a. video tape
   b. Royal Mail
   c. e-mail
   d. mobile technology.

32 Forming, storming, norming and performing are associated with:

a. sharing of thoughts

b. sharing of ideas

c. sharing of facilities

d. sharing of storage media.

33 Who should check the suitability of an employee?

a. the employer

b. the trade regulator

c. the awarding body

d. the union.

34 Serious issues that affect how various trades cooperate on site should be dealt with:

a. within 24 hours

b. as soon as possible

c. at the next site meeting

d. the following morning.

Learning outcome 5

35 If a company fails to complete an installation within the contracted time specified the consequence could include:

a. breach of contract and penalty charges if stated within the contract

b. extra time being allocated alongside further payment

c. breach of contract and imprisonment

d. breach of contract and penalty charges even if not stated within contract.

36 Which task starts on day 1 and lasts for 1 day?

a. task A

b. task B

c. tasks A and B

d. No task.

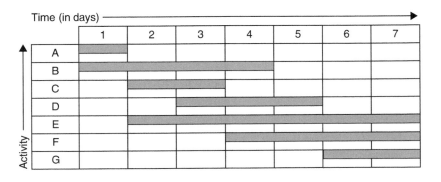

37 Identify the critical path.

a. A to E

b. F

c. B to C to D

d. A to E plus F.

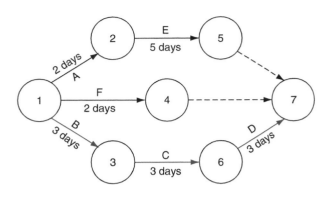

38 Positive discrimination is when:
   a. someone is given a job not based on their ability to do the job
   b. someone is given a job because they are the best candidate
   c. only white middle class people are given the job
   d. an extra position is created.

39 Equal opportunity is when:
   a. someone is given a job not based on their ability to do the job
   b. someone is given a job because they are the best candidate
   c. only white middle class people are given the job
   d. everyone is given a job.

40 Any individual subjected to violence and abuse through an initiation ceremony has been subject to:
   a. victimization
   b. discrimination
   c. a rite of passage
   d. equal opportunity.

41 When work has to be re-scheduled who needs to be involved?
   a. all trades on site
   b. all trades affected
   c. all subcontractors involved
   d. all nominated subcontractors involved.

42 Which of the following could affect the completion of an installation?
   a. weather
   b. motivation of workers
   c. surplus of materials
   d. suitable personnel.

43 A graph that shows the sequence or time to be taken on various electrical activities within the contract as a whole is called a:
   a. network analysis
   b. variation order
   c. Gantt chart
   d. tender.

2

Learning outcome 6

44 Using materials that do not include a CE mark could lead to:
   a. a dangerous situation
   b. reliable materials
   c. a pay rise
   d. damaging reputation.

45 Any materials received and not for immediate use:
   a. should be stored securely
   b. should be sent back
   c. should be archived
   d. should be sealed up.

46 The suitability of equipment in relation to its environment is associated with:
   a. shock protection
   b. basic protection
   c. index protection
   d. external influences.

47 The horizontal surface of an enclosure should be given an IP protection of:
   a. IP 2X
   b. IP 4X
   c. IP67
   d. IP66.

48 The vertical surface of an enclosure should be given an IP protection of:
   a. IP 2X
   b. IP 4X
   c. IP67
   d. IP66.

49 Which of the following is dangerous?
   a. Installing larger rated cables than specified
   b. Installing a smaller rated protective device
   c. installing sub-standard materials
   d. installing metal conduit without RCD protection.

50 A low resistance ohm meter is used to check:
   a. continuity of CPC
   b. earth loop impedance
   c. insulation resistance
   d. transient voltages.

51 Write down the difference between block, circuit, layout and site positional diagrams.

52 Write down the difference between: a memo, formal letter and a report.

53 Write down the roles NICEIC, ECA, JIB perform within the electrical installation industry.

54 How does a bar chart help with the organization of a work programme?

55 State five methods of making your work area safe on a construction site.

56   Why are good relationships important between yourself and the customer and other trades on site when carrying out work activities?

57   Briefly state why time sheets, fully and accurately completed, are important to:
    a.   an employer
    b.   an employee.

58   State the reasons why you should always present the right image to a client, customer or his representative.

59   Briefly describe what we mean by a schedule of work. Who would use a bar chart or schedule of work in your company and why?

Classroom activities

Site paperwork

Match one of the numbered boxes to each lettered box on the left.

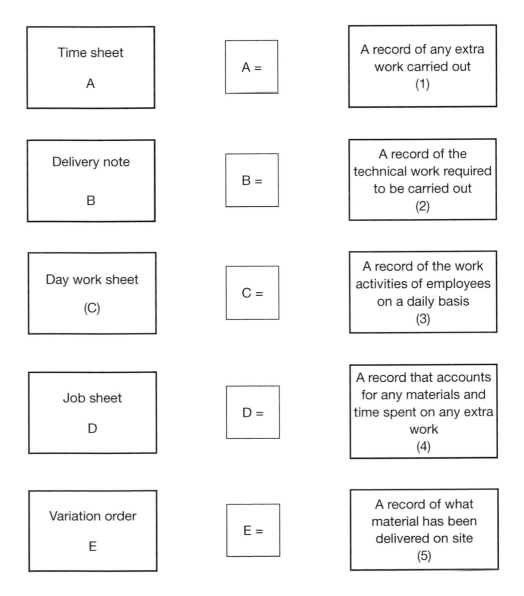

Time sheet	A =	A record of any extra work carried out
A		(1)

Delivery note	B =	A record of the technical work required to be carried out
B		(2)

Day work sheet	C =	A record of the work activities of employees on a daily basis
(C)		(3)

Job sheet	D =	A record that accounts for any materials and time spent on any extra work
D		(4)

Variation order	E =	A record of what material has been delivered on site
E		(5)

On-Site Guide symbols

Match one of the numbered boxes to each lettered box on the left.

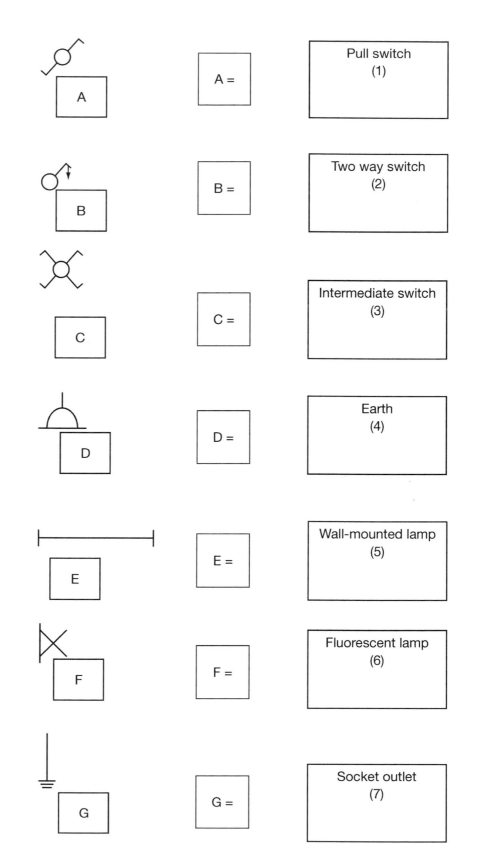

A	A =	Pull switch (1)
B	B =	Two way switch (2)
C	C =	Intermediate switch (3)
D	D =	Earth (4)
E	E =	Wall-mounted lamp (5)
F	F =	Fluorescent lamp (6)
G	G =	Socket outlet (7)

QELTK3/003 Understanding the practices and procedures for overseeing and organizing the work environment (electrical installation)

Chapter checklist

Learning outcome	Assessment criteria	Page number
1. Understand the types of technical and functional information that are available for the installation of electrotechnical systems and equipment.	1.1 Specify sources of technical and functional information that apply to electrotechnical installations, including: • manufacturer information and data • supplier information and data • information from their employing organization • installation • specifications • client/customer specifications • specifications, drawings and diagrams.	52

	1.2 Interpret technical and functional information and data from:
	• manufacturer information and data
	  • materials
	  • components
	  • equipment
	  • measuring and test instruments
	• supplier information and data
	  • materials
	  • components
	  • equipment
	  • measuring and test instruments
	• information from their employing organization
	• installation specifications
	• client/customer specifications
	• specifications, drawings and diagrams
	• records and certificates for –
	  • inspection
	  • testing
	  • installation completion.
		52

	1.3 Identify and interpret technical and functional information relating to electrotechnical products or equipment: • operation • controls • settings • adjustments.	58
	1.4 Describe the work site requirements and procedures in terms of: • services provision • ventilation provision • waste disposal procedures • equipment and material storage • health and safety requirements • access by personnel.	59
	1.5 Identify equipment and systems that are compatible to site operations and requirements.	59

Learning outcome	Assessment criteria		Page number
2. Understand the procedures for supplying technical and functional information to relevant people.	2.1	State the limits of their responsibility for supplying technical and functional information to: • clients • customers • major contractors • other services • site managers.	61
	2.2	Specify organizational policies/procedures for the handover and demonstration of electrotechnical systems, products and equipment, including requirements for confirming and recording handover.	66
	2.3	State the appropriateness of different customer relations methods and procedures.	67
	2.4	Identify methods of providing technical and functional information appropriate to the needs of: • clients • customers • major contractors • other services • site managers.	68
	2.5	Explain the importance of ensuring that: • information provided is accurate and complete • information is provided clearly, courteously and professionally • copies of information provided are retained • the installation, on completion, functions in accordance with the specification, is safe and complies with industry standards.	72
	2.6	Describe methods for checking that relevant persons have an adequate understanding of the technical and non-technical information provided, including appropriate health and safety information.	74
3. Understand the requirements for overseeing health and safety in the work environment.	3.1	State the applicable health and safety requirements with regard to overseeing the work of others.	74
	3.2	State the procedures for: • interpreting risk assessments • applying method statements • monitoring changing conditions in the workplace • complying with site organizational procedures • managing health and safety on site • organizing the safe and secure storage of tools and materials.	74
4. Understand the requirements for liaising with others when organizing and overseeing work activities.	4.1	Describe techniques for the communication with others for the purpose of: • motivation • instruction • monitoring • cooperation.	77
	4.2	Describe methods of determining the competence of operatives for whom they are responsible, such as: • checking competency cards (e.g. CSCS cards, JIB cards)	80

Learning outcome	Assessment criteria		Page number
		• checking technical qualifications • written references from previous employers • informal monitoring of performance on site • Competent Person Scheme Registration.	

	4.3	Specify their role in terms of:
		• responsibility for other staff
		• liaison with their employer
		• communication with –
		  • customers
		  • clients
		  • site managers
		  • major contractors (where appropriate)
		  • subcontractors (where appropriate)
		  • other services
		  • the public.
			77

	4.4	Identify appropriate methods for communicating with and responding to others, including: • customers • clients • site managers • major contractors (where appropriate) • subcontractors (where appropriate) • other services • the public.	78
	4.5	Specify procedures for re-scheduling work to coordinate with changing conditions in the workplace and to coincide with other trades.	79
	4.6	Clarify organizational procedures for completing the documentation that is required during work operations.	79
5. Understand the requirements for organizing and overseeing work programmes.	5.1	Describe how to plan: • work allocations • duties of operatives for whom they are responsible • coordination with other services and personnel.	82
	5.2	Specify procedures for carrying out work activities that will: • maintain the safety of the work environment • maintain cost effectiveness • ensure compliance with the programmes of work.	82
	5.3	Identify the industry standards that are relevant to activities carried out during the installation of electrotechnical systems and equipment, including the current editions of: • Management of Health and Safety Regulations • Health and Safety at Work Act • Electricity at Work Regulations • Construction Design and Management • BS 7671 Requirements for Electrical Installations • BS EN Graphical Symbols • Employment Rights Act • Data Protection Act • Disability Discrimination Act • Race Relations Act	91

Learning outcome	Assessment criteria		Page number
		• Sex Discrimination Act	
		• Human Rights Act.	
	5.4	Identify within the scope of the work programme and operations their responsibilities.	78
	5.5	Identify how to determine the estimated time required for the completion of the work required taking into account influential factors such as: • the deployment and availability of suitable personnel • the delivery and availability of equipment, components and materials • weather conditions • work to be completed by other services • specification variations.	90
	5.6	State the possible consequences of not: • completing work within the estimated time • meeting the requirements of the programme of work • using the specified materials • installing materials and equipment as specified.	90
	5.7	Specify methods of producing and illustrating work programmes such as: • bar charts • spreadsheets • critical path analysis.	83
6. Understand the requirements for organizing the provision and storage of resources that are required for work activities.	6.1	Interpret the installation specification and work programme to identify resource requirements for the following: • materials • components • plant • vehicles • equipment • labour • tools • measuring and test instruments.	93
	6.2	Interpret the material schedule to confirm that materials available are: • the right type • fit for purpose • in the correct quantity • suitable for work to be completed cost efficiently.	93
	6.3	Specify the storage and transportation requirements for all materials required in the work location.	95
	6.4	Specify procedures to ensure the safe and effective storage of materials, tools and equipment in the work location.	95

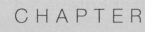

EAL Unit ELEC3/04a

Electrical installation planning, preparing and designing

Learning outcomes

When you have completed this chapter you should:

1. Understand how to plan for the installation of wiring systems and equipment.
2. Understand protection against overcurrent.
3. Understand earthing and protection.
4. Understand the electrical design procedure.
5. Understand how to prepare the worksite.

EAL Electrical Installation Work – Level 3, 2nd Edition 978 0 367 19564 9
© 2019 T. Linsley. Published by Taylor & Francis. All rights reserved.
https://www.routledge.com/9780367195649

Statutory laws

Acts of Parliament are made up of Statutes. Statutory Regulations have been passed by Parliament and have, therefore, become laws. Non-compliance with the laws of this land may lead to prosecution by the Courts and possible imprisonment for offenders.

We shall now look at some of the Statutory Regulations as they apply not only to the employers, employees and contractors within the electrical industry, but equally to visitors on site.

The Health and Safety at Work Act 1974

Many governments have passed laws aimed at improving safety at work, but the most important recent legislation has been the Health and Safety at Work Act 1974. This Act should be thought of as an umbrella Act that other statutory legislation sit under. The purpose of the Act is to provide the legal framework for stimulating and encouraging high standards of health and safety at work; the Act puts the responsibility for safety at work on both workers and managers.

Duty of care

The employer has a duty to care for the health and safety of employees (Section 2 of the Act). To do this he or she must ensure that:

- the working conditions and standard of hygiene are appropriate;
- the plant, tools and equipment are properly maintained;
- the necessary safety equipment – such as personal protective equipment (PPE), dust and fume extractors and machine guards – is available and properly used;
- the workers are trained to use equipment and plant safely.

Failure to comply with the Health and Safety at Work Act is a criminal offence and any infringement of the law can result in heavy fines, a prison sentence or both. This would apply to an employer who could be prosecuted if they knowingly allow an employee to work and that employee places other people at risk of possible injury.

Employees have a duty to care for their own health and safety and that of others who may be affected by their actions, including fellow employees and members of the public (Section 7 of the Act). To do this they must:

- take reasonable care to avoid injury to themselves or others as a result of their work activity;
- cooperate with their employer, helping him or her to comply with the requirements of the Act;
- not interfere with or misuse anything provided to protect their health and safety.

Definition

Statutory Regulations are acts of parliament and must be met. Non-statutory regulations are technical documents and should be thought of as codes of practice.

Figure 3.1 Both workers and managers are responsible for health and safety on site.

Key fact

There may be regional differences in some Statutory Laws such as The Health and Safety at Work Act as they apply to Northern Ireland and Wales. Where this is relevant, colleges should seek the advice of the examination board.

The Electricity at Work Regulations 1989 (EWR)

This legislation came into force in 1990 and replaced earlier regulations such as the Electricity (Factories Act) Special Regulations 1944. The regulations are made under the Health and Safety at Work Act 1974, and enforced by the Health and Safety Executive. The purpose of the regulations is to 'require precautions to be taken against the risk of death or personal injury from electricity in work activities'.

Section 4 of the EWR tells us that 'all systems must be constructed so as to prevent danger …, and be properly maintained … Every work activity shall be carried out in a manner which does not give rise to danger … In the case of work of an electrical nature, it is preferable that the conductors be made dead before work commences.'

The EWR do not tell us specifically how to carry out our work activities but they can be used in a court of law as evidence to claim compliance of other statutory requirements. If proceedings were brought against an individual for breaking the EWR, the only acceptable defence would be 'to prove that all reasonable steps were taken and all diligence exercised to avoid the offence' (Regulation 29).

An electrical contractor could reasonably be expected to have 'exercised all diligence' if the installation was wired according to the IET Wiring Regulations (see below). However, electrical contractors must become more 'legally aware' following the conviction of an electrician for manslaughter at Maidstone Crown Court in 1989. The court accepted that an electrician had caused the death of another man as a result of his shoddy work in wiring up a central heating system. He received a nine-month suspended prison sentence. This case has set an important legal precedent, and in future any tradesman or professional who causes death through negligence or poor workmanship risks prosecution and possible imprisonment.

The EWR is split into 16 regulations of its own. These regulations apply to any person who is engaged with electrical work: employers, the self-employed and employees, including certain classes of trainees.

Regulation 1 Citation and commencement

The first regulation puts the EWR into its context and cites that the EWR came into force on 1 April 1990.

Regulation 2 Interpretation

Brings about certain terms used in the EWR such as how we define terms such as: system, conductor and even what we mean by danger.

A system, for instance, is defined as:

> an electrical system in which all the electrical equipment is, or may be, electrically connected to a common source of electrical energy, and includes such source and such equipment.

This means that the term 'system' includes all the constituent parts of a system, including the conductors and all the electrical equipment that fits within it.

'Electrical equipment' as defined in the regulations includes every type of electrical equipment from, for example, a 400 kV overhead line to a battery-powered hand lamp. The reason that the EWR apply to even low powered equipment is that although the risk of electric shock might be low, there might still be a risk of explosion for example.

A very important distinction is made regarding the terms 'charged' and 'live'. This is because when electricians carry out safe isolation procedures they must ensure that all forms of energy are removed from a circuit including any batteries or other devices such as capacitors that can store charge.

Consequently, the term 'dead' means: a conductor that is not 'live' nor 'charged'.

Regulation 3 Persons on whom duties are imposed by these regulations

This regulation gives a clear statement of who the EWR applies to and makes a statement:

> It shall be the duty of every employee while at work to comply with the provisions of these regulations in so far as they relate to matters which are within his/her control.

This means that a trainee electrician, although not fully qualified, must adhere to the EWR as well as always cooperating with their employer. Moreover, any office worker for instance must realize where their expertise and authority lie and cannot interfere with electrical equipment.

The EWR also defines the distinction between the terms 'Absolute' and 'Reasonably Practicable'. Absolute means that something must be met irrespective of time or cost, whilst reasonably practicable means that a duty holder must decide the extent of the risks involved with the job in question against the costs involved as well as the actual difficulty in implementing safeguards.

Regulation 4 Systems, work activities and protective equipment

The definition of a system has already been defined above, but this regulation ensures that systems are designed so that as far as is reasonably practicable they do not pose a danger to anybody. This includes scheduling maintenance activities so that the equipment selected for that system must be fit for purpose.

Regulation 5 Strength and capability of electrical equipment

This regulation ensures that the system and all related equipment can withstand certain electromechanical/chemical stresses and temperature rises during normal operation, overload conditions and even fault current.

Regulation 6 Adverse or hazardous environments

The regulation draws attention to the kinds of adverse conditions where danger could arise if equipment is not designed properly or fit for purpose. This consideration includes impact damage, weather conditions, temperature, and even explosive conditions such as dust rich environments.

Regulation 7 Insulation, protection and placing of conductors

This regulation looks at the danger surrounding electric shock, and looks to insulate conductors or place/shield them so that people cannot directly touch any live parts and therefore receive either an electric shock or burn.

Regulation 8 Earthing or other suitable precautions

This regulation looks at how systems are protected to ensure that danger and specifically electric shock is minimized if faults occur. Both basic and fault protection measures apply, which includes earthing, bonding, separation and insulation of live parts.

Regulation 9 Integrity of referenced conductors

The objective of this regulation is to prevent certain conductors that are designed for electrical safety from being altered, which then brings about danger of electric shock. One of the most efficient earthing systems is called PME, which links the line and neutral conductors. However, any interference or fault in combined conductors can bring about possibly high dangerous voltages being developed across parts not normally live and therefore bringing about danger of electric shock.

Regulation 10 Connections

The objective of this regulation is to define the requirement regarding joints and connection. It specifies that all electrical connections need to be both mechanically and electrically strong as well as being suitable for use. In essence all electrical connections need to be low in resistance, but a problem occurs when joints are not formed properly because they create high resistance joints, which in turn creates areas where power is not normally dissipated and electrical fires are created.

Regulation 11 Means for protecting from excess of current

The objective of this regulation is the requirement to include protective devices to interrupt the supply when excess current is drawn. This is normally provided through fuses and circuit-breakers as well as additional protection through RCDs.

Regulation 12 Means for cutting off the supply and for isolation

The objective of this regulation is to install where necessary a suitable means of electrical isolation. For instance, if the control equipment of an electrical motor is in a different room to the motor, then the control equipment must be encased in a lockable enclosure.

It is worth reminding readers that 'isolation' means that all forms of energy are removed from a circuit, including any batteries or other devices such as capacitors that can store charge.

Regulation 13 Precautions for work on equipment made dead

This regulation ensures that adequate precautions shall be taken to prevent electrical equipment which has been made dead from becoming live when work is carried out on or near that equipment. In essence what this regulation is proposing is that electricians always carry out a safe isolation procedure. The regulation also states that where reasonable a written procedure known as a permit to work is used to authorize and control electrical maintenance.

Regulation 14 Work on or near live conductors

This regulation ensures that no person shall be engaged in any work activity on or so near any live conductor (other than one suitably covered with insulating material so as to prevent danger) where danger may arise unless:

a. it is unreasonable in all the circumstances for it to be dead; and
b. it is reasonable in all the circumstances for him to be at work on or near it while it is live; and
c. suitable precautions (including where necessary the provision of suitable protective equipment) are taken to prevent injury.

In other words, there is an expectation that electricians only work on dead supplies unless there is a reason against it and you can justify that reason.

The regulation also specifies that the test instrument used to establish that a circuit is dead has to be fit for purpose or 'approved'. Also included is mention of procedures regarding working in and around overhead power lines as well as quoting the Health and Safety at Work Act 1974 with regard to the provision of suitably trained first aiders at places of work.

Regulation 15 Working space, access and lighting

This regulation ensures that for the purposes of stopping injury, adequate working space, adequate means of access, and adequate lighting shall be provided when working with electrical equipment. For instance, when live conductors are in the immediate vicinity adequate space would allow electricians to pull back away from the conductors without hazard as well as allowing space for people to pass one another safely without hazard.

Regulation 16 Persons to be competent to prevent danger and injury

This regulation ensures that people practising with electrical work are: technically knowledgeable, experienced and competent to carry out that work activity. Electrical apprentices can engage in electrical work activity but must be supervised accordingly.

The main objective of the regulation is to ensure that people are not placed at risk due to a lack of skills on the part of themselves or others in dealing with electrical equipment.

Remember, the EWR 1989 regulations if defied can be used in any proceedings for an offence under this regulation.

6 Pack Regulations

As was previously highlighted the Health and Safety at Work Act 1974 is the main umbrella act that other statutory acts are drawn from. Alongside it five other acts form what is known as the 6 Pack Regulations and are shown below:

The Management of Health and Safety at Work Regulations 1999

The Health and Safety at Work Act 1974 places responsibilities on employers to have robust health and safety systems and procedures in the workplace. Directors and managers of any company that employs more than five employees can be held personally responsible for failures to control health and safety. The Management of Health and Safety at Work Regulations 1999 tell us that employers must systematically examine the workplace, the work activity and the management of safety in the establishment through a process of 'risk assessments'. A record of all significant risk assessment findings must be kept in a safe place and be available to an HSE inspector if required. Information based on these findings must be communicated to relevant staff and, if changes in work behaviour patterns are recommended in the interests of safety, they must be put in place. The process of risk assessment is considered in detail later in this chapter.

Risks that may require a formal assessment in the electrical industry might be:

- working at heights;
- using electrical power tools;
- falling objects;
- working in confined places;

- electrocution and personal injury;
- working with 'live' equipment;
- using hire equipment;
- manual handling – pushing – pulling – lifting;
- site conditions – falling objects – dust – weather – water – accidents and injuries.

And any other risks that are particular to a specific type of workplace or work activity.

Personal Protective Equipment (PPE) at Work Regulations 1998

PPE is defined as all equipment designed to be worn, or held, to protect against a risk to health and safety. This includes most types of protective clothing, and equipment such as eye, foot and head protection, safety harnesses, lifejackets and high-visibility clothing. Under the Health and Safety at Work Act, employers must provide free of charge any PPE and employees must make full and proper use of it.

Safety first

Always wear or use the PPE (personal protective equipment) provided by your employer for your safety.

Figure 3.2 Always wear appropriate personal protective equipment for the task at hand.

Provision and Use of Work Equipment Regulations 1998

These regulations tidy up a number of existing requirements already in place under other regulations such as the Health and Safety at Work Act 1974, the Factories Act 1961 and the Offices, Shops and Railway Premises Act 1963. The Provision and Use of Work Equipment Regulations 1998 place a general duty on employers to ensure minimum requirements of plant and equipment. If an employer has purchased good-quality plant and equipment which is well maintained, there is little else to do. Some older equipment may require modifications to bring it into line with modern standards of dust extraction, fume extraction or noise, but no assessments are required by the regulations other than those generally required by the Management Regulations 1999 discussed previously.

Workplace Health, Safety and Welfare Regulations 1992

This regulation specifies the general requirements and expectation of accommodation standards for nearly all workplaces. A breach of this regulation would be seen as a crime, punishable following any successful conviction.

The Control of Substances Hazardous to Health Regulations 2002 (COSSH)

Figure 3.3 Asbestos is an extremely toxic substance when it breaks down and causes lung disease.

The original COSHH Regulations were published in 1988 and came into force in October 1989. They were re-enacted in 1994 with modifications and improvements, and the latest modifications and additions came into force in 2002.

The COSHH Regulations control people's exposure to hazardous substances in the workplace. Regulation 6 requires employers to assess the risks to health from working with hazardous substances, to train employees in techniques that will reduce the risk and provide personal protective equipment (PPE) so that employees will not endanger themselves or others through exposure to hazardous substances. Employees should also know what cleaning, storage and disposal procedures are required and what emergency procedures to follow. The necessary information must be available to anyone using hazardous substances as well as to visiting HSE Inspectors. Hazardous substances include:

1 any substance that gives off fumes causing headaches or respiratory irritation;
2 man-made fibres that might cause skin or eye irritation (e.g. loft insulation);
3 acids causing skin burns and breathing irritation (e.g. car batteries, which contain dilute sulphuric acid);
4 solvents causing skin and respiratory irritation (strong solvents are used to cement together PVC conduit fittings and tube);
5 fumes and gases causing asphyxiation (burning PVC gives off toxic fumes);
6 cement and wood dust causing breathing problems and eye irritation;
7 exposure to asbestos – although the supply and use of the most hazardous asbestos material is now prohibited, huge amounts were installed between 1950 and 1980 in the construction industry and much of it is still in place today.

In their latest amendments, the COSHH Regulations focus on giving advice and guidance to builders and contractors on the safe use and control of asbestos products. These can be found in Guidance Notes EH 71 or visit www.hse.uk/hiddenkiller.

Remember: where PPE is provided by an employer, employees have a duty to use it to safeguard themselves.

Working at Height Regulations

Figure 3.4 Working at height is a risk that requires a formal assessment in the electrical industry.

Working above ground level creates added dangers and slows down the work rate of the electrician. New Work at Height Regulations came into force on 6 April 2005. Every precaution should be taken to ensure that the working platform is appropriate for the purpose and in good condition. This is especially important since the main cause of industrial deaths comes from working at height.

Manual Handling Operations Regulations 1992 (as amended)

In effect, any activity that requires an individual to lift, move or support a load will be classified as a manual handling task. More than a third of all reportable injuries are believed to involve incorrect lifting techniques or carrying out manual handling operations without using mechanical lifting devices. Companies will also train their personnel through specific persons being appointed as manual handling advisors or safety representatives who will carry out both induction and refresher training in order to educate their workforce on correct lifting techniques and manual handling procedures.

Health and Safety (display screen equipment) Regulations 1992

These regulations are concerned with providing specific parameters on the expected safety and health requirements and implications for those personnel who work with display screen equipment. The equipment must be scrutinized so that it is not a source of risk for operators and should include the:

- display screen;
- keyboard;
- user space;
- chair;
- lighting (glare);
- software.

The Construction (Design and Management) Regulations 1994

The Construction (Design and Management) Regulations (CDM) are aimed at improving the overall management of health, safety and welfare throughout all stages of the construction project.

The person requesting that construction work commence, the client, must first of all appoint a 'duty holder', someone who has a duty of care for health, safety and welfare matters on-site. This person will be called a 'planning supervisor'. The planning supervisor must produce a 'pre-tender' health and safety plan and coordinate and manage this plan during the early stages of construction. The client must also appoint a principal contractor who is then required to develop the health and safety plan made by the planning supervisor, and keep it up to date during the construction process to completion. The degree of detail in the health and safety plan should be in proportion to the size of the construction project and recognize the health and safety risks involved on that particular project. Small projects will require simple, straightforward plans; large projects, or those involving significant risk, will require more detail. The CDM Regulations will apply to most large construction projects but they do not apply to the following:

- construction work, other than demolition work, that does not last longer than 30 days and does not involve more than four people;
- construction work carried out inside commercial buildings such as shops and offices, which does not interrupt the normal activities carried out on those premises;
- construction work carried out for a domestic client;
- the maintenance and removal of pipes or lagging that form a part of a heating or water system within the building.

The Electricity Safety, Quality and Continuity Regulations 2002 (formerly Electricity Supply Regulations 1989)

The Electricity Safety, Quality and Continuity Regulations 2002 are issued by the Department of Trade and Industry. They are statutory regulations that are enforceable by the laws of the land. They are designed to ensure a proper and safe supply of electrical energy up to the consumer's terminals.

These regulations impose requirements upon the regional electricity companies regarding the installation and use of electric lines and equipment. The regulations are administered by the Engineering Inspectorate of the Electricity Division of the Department of Energy and will not normally concern the electrical contractor, except that it is these regulations that lay down the earthing requirement of the electrical supply at the meter position.

Definition

The 'duty holder' is someone who has a duty of care for health, safety and welfare matters on-site. This phrase recognizes the level of responsibility that electricians are expected to take on as part of their job in order to control electrical safety in the work environment.

The regional electricity companies must declare the supply voltage and maintain its value between prescribed limits or tolerances.

The government agreed on 1 January 1995 that the electricity supplies in the United Kingdom would be harmonized with those of the rest of Europe. Thus the voltages used previously in low-voltage supply systems of 415 V and 240 V have become 400 V for three-phase supplies and 230 V for single-phase supplies. The permitted tolerances to the nominal voltage have also been changed from 6% to +10% and +6%. This gives a voltage range of 216–253 V for a nominal voltage of 230 V and 376–440 V for a nominal supply voltage of 400 V.

The next proposed change is for the tolerance levels to be adjusted to +10% of the declared nominal voltage (IET Regulation, Appendix 2:14).The frequency is maintained at an average value of 50 Hz over 24 hours so that electric clocks remain accurate.

Regulation 29 gives the area boards the power to refuse to connect a supply to an installation that in their opinion is not constructed, installed and protected to an appropriately high standard. This regulation would only be enforced if the installation did not meet the requirements of the IET Regulations for Electrical Installations.

Control of asbestos at work regulations

In October 2010 the HSE launched a national campaign to raise awareness among electricians and other trades of the risk to their health of coming into contact with asbestos. It is called the 'Hidden Killer Campaign' because approximately six electricians will die each week from asbestos-related diseases. For more information about asbestos hazards, visit www.hse.uk/hiddenkiller.

Non-statutory regulations

Statutory laws and regulations are written in a legal framework; some don't actually tell us how to comply with the laws at an everyday level.

Non-statutory regulations and codes of practice interpret the statutory regulations, telling us how we can comply with the law.

They have been written for every specific section of industry, commerce and situation, to enable everyone to comply with or obey the written laws. When the Electricity at Work Regulations (EWR) tell us to 'ensure that all systems are constructed so as to prevent danger' they do not tell us how to actually do this in a specific situation. However, the IET Regulations tell us precisely how to carry out our electrical work safely in order to meet the statutory requirements of the EWR. In Part 1 of the IET Regulations, at 114, it states: 'the Regulations are non-statutory. They may, however, be used in a court of law in evidence to claim compliance with a statutory requirement.' If your electrical installation work meets the requirements of the IET Regulations, you will also meet the requirements of EWR.

Over the years, non-statutory regulations and codes of practice have built upon previous good practice and responded to changes by bringing out new editions of the various regulations and codes of practice to meet the changing needs of industry and commerce.

We will now look at one non-statutory regulation, what is sometimes called 'the electrician's bible', the most important set of regulations for anyone

Definition

Statutory laws and regulations are written in a legal framework; some don't actually tell us how to comply with the laws at an everyday level.

Definition

Non-statutory regulations and codes of practice interpret the statutory regulations, telling us how we can comply with the law.

working in the electrical industry, the BS 7671: 2008 Requirements for Electrical Installations, IET Wiring Regulations 18th Edition.

The IET Wiring Regulations 18th Edition requirements for electrical installations to BS 7671: 2018

The Institution of Engineering and Technology Requirements for Electrical Installations (the IET Regulations) are non-statutory regulations. They relate principally to the design, selection, erection, inspection and testing of electrical installations, whether permanent or temporary, in and about buildings generally and to agricultural and horticultural premises, construction sites and caravans and their sites.

Paragraph 7 of the introduction to the EWR says: 'the IET Wiring Regulations is a code of practice that is widely recognized and accepted in the United Kingdom and compliance with them is likely to achieve compliance with all relevant aspects of the Electricity at Work Regulations.' The IET Wiring Regulations are the national standard in the United Kingdom and apply to installations operating at a voltage up to 1000 V a.c. They do not apply to electrical installations in mines and quarries, where special regulations apply because of the adverse conditions experienced there. The current edition of the IET Wiring Regulations is the 18th Edition 2018. The main reason for incorporating the IET Wiring Regulations into British Standard BS 7671: 2008 was to create harmonization with European Standards.

The IET Regulations take account of the technical intent of the CENELEC European Standards, which in turn are based on the IEC International Standards.

The purpose in harmonizing British and European Standards is to help develop a single European market economy so that there are no trade barriers to electrical goods and services across the European Economic Area.

To assist electricians in their understanding of the regulations a number of guidance notes have been published. The guidance notes that I will frequently make reference to in this book are those contained in the *On-Site Guide*. Eight other guidance notes booklets are also currently available. These are:

- *Selection and Erection;*
- *Isolation and Switching;*
- *Inspection and Testing;*
- *Protection against Fire;*
- *Protection against Electric Shock;*
- *Protection against Overcurrent;*
- *Special Locations;*
- *Earthing and Bonding.*

These guidance notes are intended to be read in conjunction with the regulations.

The IET Wiring Regulations are the electrician's bible and provide the authoritative framework of information for anyone working in the electrical industry.

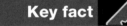

Key fact

IET regulations
- They are the UK National Standard for all electrical work.
- They are the 'electrician's bible'.
- Comply with the IET Regulations and you also comply with Statutory Regulations (IET Regulations 1.14)

Assessment criteria 1.2

Specify the criteria that affect the selection of wiring systems and equipment

Whilst the designer of any electrical installation is the person who interprets the electrical requirements of the customer and matches it to the regulations, an installer is also closely linked to this process and indeed on some projects the designer and senior installer are one and the same. The aim is to select a wiring system and equipment that is suitable for the actual environment.

A large electrical installation project may require quite a few scheduled meetings with the customer in order to identify a precise specification to meet their needs. This will also give an opportunity to discuss any ergonomic and aesthetic requirements since being pleasing to the eye is often an important aspect of customer expectation. Other important considerations include scrutinizing the general characteristics of the electrical installation, including what earthing systems are available and the nature of the supply. Furthermore, compatibility of installed equipment with other services, as indicated in Part 3 of the regulations, is equally important in order to ensure that one system or even a single component does not affect an additional element.

Other considerations will include external influences such as ingress of moisture or other solid objects. Looking at maintainability on the other hand is the provision of maintenance and therefore an assessment of the frequency and quality required or expected will give an indication of the type of installation which is most appropriate. Another consideration is safety services such as the requirement to install such things as emergency lighting and fire detection. The protection and safety of the installation users is a paramount consideration, with due regard to Part 4 of BS 7671 the wiring regulations.

From this the size and quantity of all the materials, cables, control equipment and accessories can then be determined. This is called a 'bill of quantities'. The suitability specification for the chosen wiring system will be largely determined by the building construction and the activities to be carried out in the completed building.

An industrial building, for example, will require an electrical installation that incorporates flexibility and mechanical protection, especially if moving heavy plant or vehicles are likely to or may possibly impact the installation. The possibility of vibration would also be a necessary consideration, for example the final connection to motors tends to be through a flexible conduit to offset any inherent juddering/vibration.

In a block of purpose-built flats, all the electrical connections must be accessible from one flat without intruding upon the surrounding flats. A loop-in conduit system, in which the only connections are at the light switch and outlet positions, would meet this requirement. For a domestic electrical installation an appropriate lighting scheme and multiple socket outlets for the connection of domestic appliances, all at a reasonable cost, are important factors, which can usually be met by a PVC-insulated and sheathed wiring system.

Temperature must also be a consideration either through the temperature of the ambient air, or through the harmful effects of heat and thermal radiation developed by electrical equipment. The impact of solar radiation for, example, is such that a PVC conduit for instance has to be fitted with expansion couplers every 5 m to make good any expansion.

Further contemplation may be made in relation to potential explosive atmospheres such as petrol stations or even such locations such as a flour mill, for example; given the rich content of the air they must not be exposed to an ignition source including the contacts of normal switches. This is why

Definition

The *designer* of any electrical installation is the person that interprets the electrical requirements of the customer within the regulations.

the possibility of dust, moisture content, chemical fumes and gas is part of the external influence assessment found in appendix 5 of the wiring regulations in order to reflect upon the correct index protection rating for equipment, which will determine what size body part can enter the accessory or if indeed extraction processes are required.

Index of Protection (IP) BS EN 60529

IET Regulation 416.2.1 tells us that where barriers and enclosures have been installed to prevent direct contact with live parts, they must afford a degree of protection not less than IP2X and IPXXB, total protection, but what does this mean? The Index of Protection is a code that gives us a means of specifying the suitability of equipment for the environmental conditions in which it will be used. The tests to be carried out for the various degrees of protection are given in the British and European Standard BS EN 60529.

The code is written as IP (Index of Protection) followed by two numbers XX. The first number gives the degree of protection against the penetration of solid objects into the enclosure. The second number gives the degree of protection against water penetration. For example, a piece of equipment classified as IP45 will have barriers installed that prevent a 1 mm diameter rigid steel bar from making contact with live parts and be protected against the ingress of water from jets of water applied from any direction. Where a degree of protection is not specified, the number is replaced by an 'X', which simply means that the degree of protection is not specified, although some protection may be afforded. The 'X' is used instead of '0' since '0' would indicate that no protection was given. The index of protection codes are shown in Fig. 3.5.

Appendix 5 of the IET Regulations identifies the required IP classification for electrical equipment being used in hazardous conditions and requiring water protection as follows:

- IPX1 or IPX2 where water vapour occasionally condenses on electrical equipment;
- IPX3 where sprayed water forms a continuous film on the floor;
- IPX4 where equipment may be subjected to splashed water, e.g. construction sites;
- IPX5 where hosed water is regularly used, e.g. car washing;
- IPX6 for seashore locations, e.g. marinas and piers;
- IPX7 for locations which may become flooded, immersing equipment in water;
- IPX8 where electrical equipment is permanently immersed in water, e.g. swimming pools.

Appendix 5 of the IET Regulations also identifies the required IP classification to prevent dust and objects penetrating electrical equipment as follows:

- IP2X to prevent penetration by solid objects as thick as a finger, approximately 12 mm;
- IP3X to prevent penetration by small objects of which the smallest is 2.5 mm;
- IP4X to prevent penetration by very small objects of which the smallest is 1.0 mm;
- IP5X where light dust penetration would not harm the electrical equipment;
- IP6X where dust must not penetrate the equipment.
- IPXXB means total protection.

First number		Second number	
(DEGREE OF PROTECTION AGAINST SOLID OBJECT PENETRATION)		**(DEGREE OF PROTECTION AGAINST WATER PENETRATION)**	
0	Non-protected.	0	Non-protected.
1	Protected against a solid object greater than 50mm, such as a hand.	1	Protected against water dripping vertically, such as condensation.
2	Protected against a solid object greater than 12mm, such as a finger.	2	Protected against dripping water when tilted up to 15∞.
3	Protected against a solid object greater than 2.5mm, such as a tool or wire.	3	Protected against water spraying at an angle of up to 60∞.
4	Protected against a solid object greater than 1.0mm, such as thin wire or strips.	4	Protected against water splashing from any direction.
5	Dust protected. Prevents ingress of dust sufficient to cause harm.	5	Protected against jets of water from any direction.
6	Dust tight. No dust ingress.	6	Protected against heavy seas or powerful jets of water. Prevents ingress sufficient to cause harm.
		7	Protected against harmful ingress of water when immersed to a depth of between 150mm and 1m.
		8	Protected against submersion. Suitable for continuous immersion in water.

Figure 3.5 Index of protection codes.

External influences can also encourage corrosion. which means it has to be a major consideration. It can be improved by using galvanized steel conduit for instance, or encasing MICC conductors with a PVC oversheath. That said, electrolytic corrosion can occur when combining copper and aluminium with an aluminium conductor and copper crimp for instance. Galvanic action then occurs through a process known as dissimilar metal corrosion, therefore the designer must be mindful of such aspects. Certain locations would also be exposed to damaging winds and even seismic activity.

Examining how the chosen installation method will affect the building structure is also an important consideration and we will examine that next.

Assessment criteria 1.3

Select wiring systems and equipment appropriate to the situation and use

An electrical installation is made up of many different electrical circuits:

* lighting circuits;
* power circuits;
* single-phase domestic circuits;
* three-phase industrial or commercial circuits environments;
* emergency systems;
* security systems.

Part 5 of the IET Regulations tells us that electrical equipment and materials must be chosen so that they are suitable for the actual environment in question. The condition of the location will dictate what kind of system is required taking into account: ambient temperature; the presence of moisture especially if that can lead to corrosion; potential impact damage; inherent vibration and exposure to solar radiation. PVC insulated and sheathed cables are manufactured cheaply for the domestic market, but are not necessary long lasting and lack any depth regarding mechanical protection. They would not necessarily be a suitable wiring type for a sub-main since they are mostly buried underground and therefore a PVC/SWA cable would be preferable. These two types of cable are shown in Figs. 3.50 and 3.51.

MI cables are waterproof, heatproof and corrosion-resistant with some mechanical protection. They are designed so that if the cable is flattened then the distance between the inner conductors remains the same. Often, given its qualities, it is the cable choice for hazardous or high-temperature installations such as oil refineries, chemical works, boiler houses and petrol pump installations. MI is also non-ageing, which is why it is often used in listed buildings reducing a need to replace the wiring, which would befall other wiring types. An MI cable with terminating gland and seal is shown in Fig. 3.52.

One-way lighting circuit

The simplest form of lighting circuit is a one way controlled light. This is where only one switch is used to operate the lights such as a bedroom light. In this circuit, the line conductor has a single pole overcurrent protective device (fuse or circuit-breaker) fitted in line with a switch which provides functional switching (on and off). A one way switch simply opens or closes the circuit as shown in Fig 3.6.

In a domestic installation or when using PVC/PVC cable running the line conductor to the switch and the neutral to the lamp is not possible so we use the three plate method and 'loop' the line through the ceiling rose as shown in Fig 3.8 or, we can use the joint box method shown in Fig 3.9.

The switch line, returning from the switch to the lamp, is the blue insulated conductor in the PVC/PVC cable (it is possible to buy cables with two brown

1 Way Switch

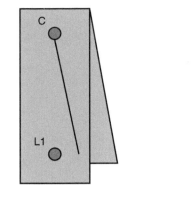
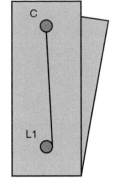

Switch position off, contacts open   Switch position on, contacts closed

Figure 3.6 Shows how circuit is complete when common links to L1.

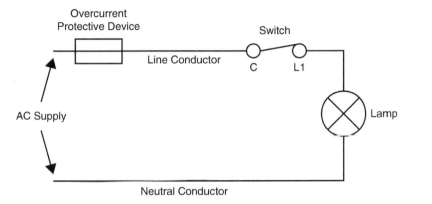

Figure 3.7 A circuit diagram for a 1 way lighting circuit.

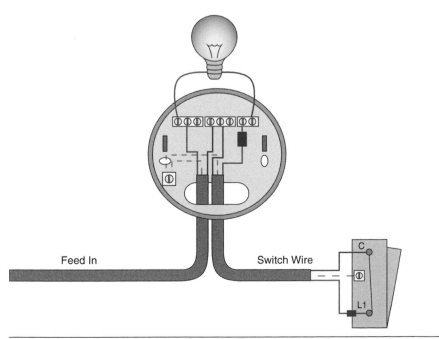

Figure 3.8 Wiring diagram of 1 way lighting circuit using the Loop in Method.

insulated conductors). As this wire is a switched line, the regulations state we must indicate this can be live and therefore we add a brown sleeve to show the blue is not acting as the neutral. The wiring diagram of a 1 way lighting circuit using the joint box method is shown in Fig 3.9.

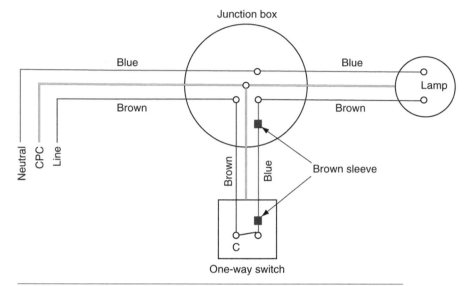

Figure 3.9 Wiring diagram of one-way switch control.

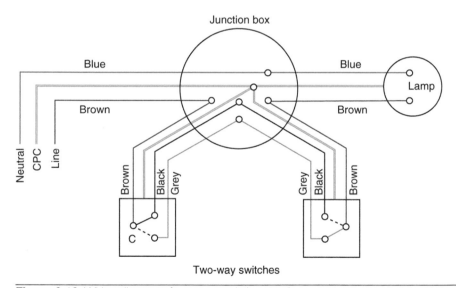

Figure 3.10 Wiring diagram of two-way switch control.

Two-way lighting circuit

A two-way lighting circuit is used where two separate switching positions are required such as a staircase with a switch at either end. The overcurrent protective device is fitted in the line conductor before it feeds the switches, controlling the supply to the light. A two-way switch differs from the one-way switch because the feed in is connected to either one or the other of the two outlet terminals. The switch acts as a changeover, switching the connections between the common and position 1 or the common and position 2 dependent on the switch position. Fig 3.10 shows the joint box method of wiring a two way lighting circuit.

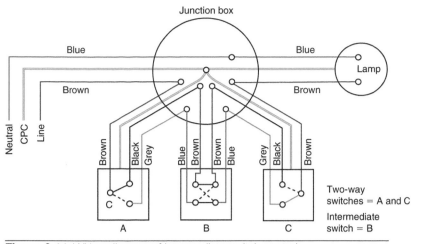

Figure 3.11 Wiring diagram of intermediate switch control.

Intermediate switching

On a lighting circuit where the light requires three or more switches, an intermediate switch is used. This may be a staircase with a landing part way or a long corridor for example. The wiring is carried out in the same way as a two-way installation, with the intermediate switch or switches added between the two, two-way switches. Fig 3.11 shows the joint box method of wiring intermediate switch control.

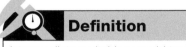

Definition

Intermediate switching would typically be used in long corridors.

Junction boxes

When wiring a lighting circuit, the designer will look at the fittings and decide if it is possible or suitable to terminate multiple cables at the accessory. It may be decided that making the joints in a joint box is preferable and to run a cable to the switches or fittings for ease of terminations. Junction boxes with fixed terminals or adaptable boxes housing connector blocks, often referred to as RB4s, could be used. BS 7671 Regulation 526.3 requires all screw terminals to be accessible for maintenance once installed and this must be considered when designing such a circuit.

SELV lighting

Many spotlight type fittings work, use or operate using SELV (Separated Extra Low Voltage). For the system to be classified as SELV, the voltage must not exceed 50 V ac or 120V dc and the circuit protective conductor may not be connected to the load. This is achieved by use of a transformer to BS EN 61558-2-6, where the low voltage (230 V) circuit is connected to the primary side and the ELV fitting is connected to the secondary side. Plugs and sockets used for the connections must not be interchangeable so as to avoid the fittings being connected to a higher voltage than it is designed for.

Many SELV systems use LED lighting due to its high efficiency, however not all SELV transformers are dimmable.

Key fact

The limitation regarding an A1 ring circuit is not the number of sockets but the maximum area it can supply (100 m²).

Socket outlet Circuits

Whereas most equipment is wired on a radial circuit, it is common practice in the UK to wire BS1363 13A socket outlets on a ring final circuit. BS 7671 describes a ring final circuit as 'starting and finishing at the distribution board, where it is connected to a 30A or 32A overcurrent protective device'.

The *On-Site Guide* describes this arrangement as an A1 circuit and while it does not limit the number of sockets that can be connected to this arrangement, it does give a maximum floor area that the circuit may feed as 100m^2.

The minimum cable size for the live conductors of a ring final circuit is given as 2.5 mm^2 when using thermoplastic cables (1.5 mm for mineral insulated cable as it has a higher current carrying capacity) and at first glance this looks like the cable is rated lower than the protective device and will not comply to other regulations. Due to the ring arrangement, the current will split within the circuit and this design is acceptable to British and European Standards. However, care should be taken when designing the circuit to provide reasonable sharing of the load in each leg of the ring.

The number of socket outlets on a ring final circuit is unlimited but the load will determine if more than one circuit is required. In this case, the designer should look to distribute permanently connected equipment across the circuits to avoid overloading any one circuit and reduce the inconvenience in the case of a fault or when maintenance is being carried out.

Number of socket outlets

When installing socket outlets they should be numerous enough and positioned such that all equipment can be connected conveniently without the need for extension leads. The *On-Site Guide* provides Table H7 in Appendix H to give guidance on the minimum numbers required in assorted locations.

Spurs

A socket that is not connected within the ring but fed via a single cable is known as a spur. The total number of un-fused spurs connected to a ring should not exceed the total number of sockets and stationary equipment connected directly to the ring. A non-fused spur should only feed one single or one twin socket outlet or a piece of permanently connected equipment to avoid overloading the cable feeding it. The connection to the ring final circuit should be made at the terminals of an existing socket outlet, the origin of the circuit or in a junction box, and the size of the live conductors should be 2.5 mm^2 minimum. When a supply to more than one outlet is required and it is not possible to incorporate the addition into the ring, a fused connection unit (FCU) may be used.

The FCU may be connected directly into the ring or as a spur and incorporates a BS EN 1362 fuse. As the largest BS EN 1362 fuse is 13 A, this is the maximum load that can be drawn and the number of sockets fed by the FCU is only limited by the load being connected not exceeding 13 A. The total number of FCUs connected to the ring final circuit is unlimited.

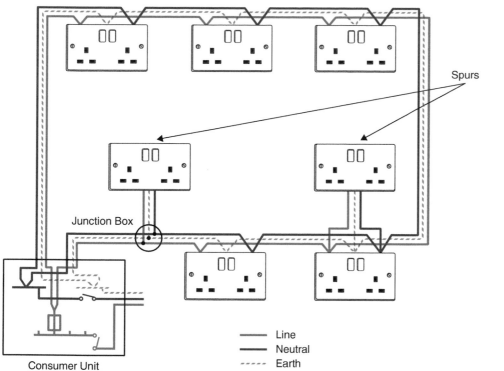

Figure 3.13 Switched Fused Connection Unit

Line
Neutral
Earth

Consumer Unit

Junction Box

Spurs

Figure 3.12 Only one unfused spur is allowed per socket outlet.

As the wiring beyond the FCU is protected by a 13 A fuse, the cross sectional area of the live conductors can be reduced and a 1.5 mm^2 cable is allowed (1 mm^2 for mineral insulated cable).

Define the requirements of standard radial final socket circuits

The *On-Site Guide* also recognizes that socket outlets may be wired on radial final circuits. Table H2.1 classifies the circuits as A2 and A3 where an A2 circuit is protected by a 30 A or 32 A overcurrent protective device, wired in 4 mm^2 thermoplastic cable (2.5 mm^2 for MI cable) and a maximum floor area of 75 m^2.

The A3 circuit uses 2.5 mm^2 cable (1.5 mm^2 for MI) and therefore the overcurrent protective device is reduced to 20 A. This circuit has a maximum floor area of 50 m^2.

Table 3.1 General requirements for standard socket outlet circuits

	Type of circuit Overcurrent protective device		Minimum conductor size (mm^2)	Maximum floor area Square Meters
A1	Ring 30 or 32	any type of device	2.5	100
A2	Radial 30 or 32	cartridge fuse or circuit breaker	4	75
A3	Radial 20	any type of device	2.5	50

Standard circuit arrangements for loads and equipment

Common domestic circuits include cookers, showers and immersion heaters. All of these appliances are wired on radial final circuits and the rating of the circuit is determined by an assessment of the current demand of the appliance.

Water-heating circuits

A small, single-point over-sink type water heater may be considered as a permanently connected appliance and so may be connected to a ring circuit through a fused connection unit. A water heater of the immersion type is usually rated at a maximum of 3 kW, and could be considered as a permanently connected appliance, fed from a fused connection unit. However, many immersion heating systems are connected into storage vessels of about 150 litres in domestic installations, and the *On-Site Guide* states that immersion heaters fitted to vessels in excess of 15 litres should be supplied by their own dedicated circuit (*On-Site Guide* Appendix H5).

Therefore, immersion heaters must be wired on a separate radial circuit when they are connected to water vessels that hold more than 15 litres. Figure 3.20 shows the wiring arrangements for an immersion heater. Every switch must be a double-pole (DP) switch and out of reach of anyone using a fixed bath or shower when the immersion heater is fitted to a vessel in a bathroom.

> **Key fact**
>
> Certain high current load equipment such as large water heaters, cookers and showers should be fitted with a double pole switch in order to isolate both the line and neutral conductors.

Figure 3.14 Immersion heaters in excess of 15 litres should be supplied by their own dedicated circuit and controlled through a double pole switch.

Supplementary equipotential bonding to pipework will only be required as an addition to fault protection (IET Regulation 415.2) if the immersion heater vessel is in a bathroom that does not have:

- all circuits protected by a 30 mA RCD; and also
- protective equipotential bonding (IET Regulation 701.415.2).

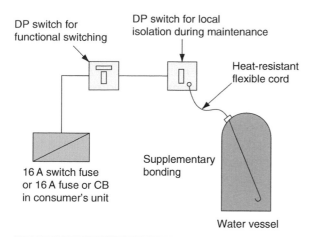

Figure 3.15 Immersion heater wiring.

Electric space-heating circuits

Electrical heating systems can be broadly divided into two categories: unrestricted local heating and off-peak heating.

Unrestricted local heating may be provided by portable electric radiators that plug into the socket outlets of the installation. Fixed heaters that are wall mounted or inset must be connected through a fused connection and incorporate a local switch, either on the heater itself or as a part of the fuse connecting unit. Heating appliances where the heating element can be touched must have a DP switch that disconnects all conductors. This requirement includes radiators that have an element inside a silica-glass sheath.

Figure 3.16 A radiator can be controlled through remote sensors and electronic wireless control.

Off-peak heating systems may provide central heating from storage radiators, ducted warm air or underfloor heating elements. All three systems use the thermal storage principle, whereby a large mass of heat-retaining material is heated during the off-peak period and allowed to emit the stored heat throughout the day. The final circuits of all off-peak heating installations must be fed from a separate supply controlled by an electricity board time clock.

When calculating the size of cable required to supply a single-storage radiator, it is good practice to assume a current demand equal to 3.4 kW at each point. This will allow the radiator to be changed at a future time with the minimum disturbance to the installation. Each storage heater must have a 20 A DP means of isolation adjacent to the heater and the final connection should be via a flex outlet. See Fig. 3.17 for wiring arrangements.

Ducted warm air systems have a centrally sited thermal storage heater with a high storage capacity. The unit is charged during the off-peak period, and a fan drives the stored heat in the form of warm air through large air ducts to outlet grilles in the various rooms. The wiring arrangements for this type of heating are shown in Fig. 3.18.

The single-storage heater is heated by an electric element embedded in bricks and rated between 6 and 15 kW depending upon its thermal capacity. A radiator of this capacity must be supplied on its own circuit, in cable capable of carrying the maximum current demand and protected by a fuse or circuit-breaker (CB) of 30, 45 or 60 A as appropriate. At the heater position, a DP switch must be installed to terminate the fixed heater wiring. The flexible cables used for the final connection to the heaters must be of the heat-resistant type.

Floor-warming installations use the thermal storage properties of concrete. Special cables are embedded in the concrete floor screed during construction. When current is passed through the cables they become heated, the concrete absorbs this heat and radiates it into the room. The wiring arrangements are

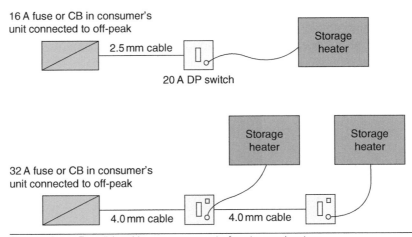

Figure 3.17 Possible wiring arrangements for storage heaters.

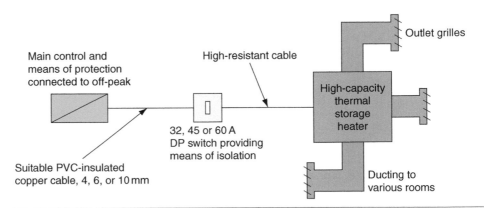

Figure 3.18 Ducted warm air heating system.

shown in Fig. 3.19. Once heated, the concrete will give off heat for a long time after the supply is switched off and is, therefore, suitable for connection to an off-peak supply.

Underfloor heating cables installed in bathrooms or shower rooms must incorporate an earthed metallic sheath or be covered by an earthed metallic grid connected to the protective conductor of the supply circuit (IET Regulation 701.753).

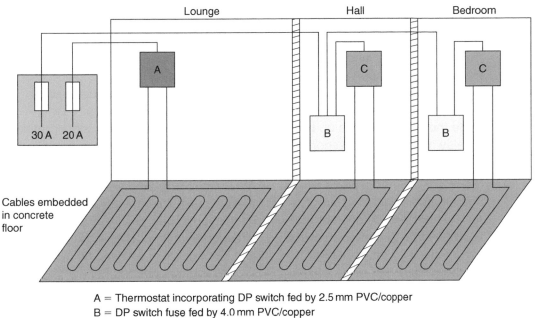

A = Thermostat incorporating DP switch fed by 2.5 mm PVC/copper
B = DP switch fuse fed by 4.0 mm PVC/copper
C = Thermostat fed by 2.5 mm PVC/copper

Figure 3.19 Floor-warming installations.

Emergency management systems

Certain systems are designed to respond during emergency situations, especially if the primary source of power fails.

Emergency lighting (BS 5266 and BS EN 1838)

Emergency lighting should be planned, installed and maintained to the highest standards of reliability and integrity, so that it will operate satisfactorily when called into action, no matter how infrequently this may be. Emergency lighting is not required in private homes because the occupants are familiar with their surroundings, but in public buildings people are in unfamiliar surroundings. In an emergency people do not always act rationally, but well illuminated and easily identified exit routes can help to reduce panic.

Emergency lighting is provided for two reasons: to illuminate escape routes, called 'escape' lighting; and to enable a process or activity to continue after a normal lights failure, called 'standby' lighting. Escape lighting is usually required by local and national statutory authorities under legislative powers. The escape lighting scheme should be planned so that identifiable features and obstructions are visible in the lower levels of illumination which may prevail during an

Figure 3.20 Escape routes should remain illuminated in the event of a powercut.

emergency. Exit routes should be clearly indicated by signs and illuminated to a uniform level, avoiding bright and dark areas.

Standby lighting is required in hospital operating theatres and in industry, where an operation or process once started must continue, even if the mains lighting fails. Standby lighting may also be required for security reasons. The cash points in public buildings may need to be illuminated at all times to discourage acts of theft occurring during a mains lighting failure.

Emergency supplies

Since an emergency occurring in a building may cause the mains supply to fail, the emergency lighting should be supplied from a source that is independent from the mains supply. In most premises the alternative power supply would be from batteries, but generators may also be used. Generators can have a large capacity and duration, but a major disadvantage is the delay of time while the generator runs up to speed and takes over the load. In some premises a delay of more than 5 seconds is considered unacceptable, and in these cases a battery supply is required to supply the load until the generator can take over.

The emergency lighting supply must have an adequate capacity and rating for the specified duration of time (IET Regulation 313.2). BS 5266 and BS EN1838 state that after a battery is discharged by being called into operation for its specified duration of time, it should be capable of once again operating for the specified duration of time following a recharge period of no longer than 24 hours. The duration of time for which the emergency lighting should operate will be specified by a statutory authority but is normally 1–3 hours. The British Standard states that escape lighting should operate for a minimum of 1 hour.

Standby lighting operation time will depend upon financial considerations and the importance of continuing the process or activity.

There are two possible modes of operation for emergency lighting installations: maintained and non-maintained.

Maintained emergency lighting

In a maintained system the emergency lamps are continuously lit using the normal supply when this is available, and change over to an alternative supply when the mains supply fails. The advantage of this system is that the lamps are continuously proven healthy and any failure is immediately obvious. It is a wise precaution to fit a supervisory buzzer or LED indicator in the emergency supply to prevent accidental discharge of the batteries, since it is not otherwise obvious which supply is being used.

Maintained emergency lighting is normally installed in theatres, cinemas, discotheques and places of entertainment where the normal lighting may be dimmed or extinguished while the building is occupied. The emergency supply for this type of installation can also be supplied from a central battery, the emergency lamps being wired in parallel from the low-voltage supply as shown in Fig. 3.21. Escape sign lighting units used in commercial facilities should be wired in the maintained mode.

Definition

Emergency lighting is not required in private homes because the occupants are familiar with their surroundings, but in public buildings people are in unfamiliar surroundings. In an emergency people do not always act rationally, but well-illuminated and easily identified exit routes can help to reduce panic.

Definition

Emergency lighting is provided for two reasons: to illuminate escape routes, called *'escape' lighting*; and to enable a process or activity to continue after a normal lights failure, called *'standby' lighting.*

Definition

In a *maintained system* the emergency lamps are continuously lit using the normal supply when this is available, and change over to an alternative supply when the mains supply fails.

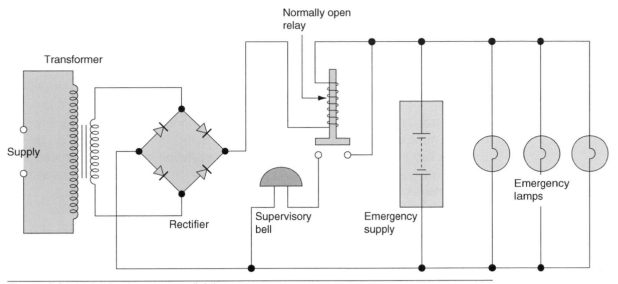

Figure 3.21 Maintained emergency lighting.

Non-maintained emergency lighting

In a **non-maintained system** the emergency lamps are only illuminated if the normal mains supply fails. Failure of the main supply de-energizes a solenoid and a relay connects the emergency lamps to a battery supply, which is maintained in a state of readiness by a trickle charge from the normal mains supply. When the normal supply is restored, the relay solenoid is energized, breaking the relay contacts, which disconnects the emergency lamps, and the charger recharges the battery. Figure 3.22 illustrates this arrangement.

The disadvantage with this type of installation is that broken lamps are not detected until they are called into operation in an emergency, unless regularly maintained. The emergency supply is usually provided by a battery contained within the luminaire, together with the charger and relay, making the unit self-contained.

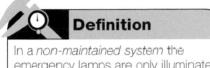

Definition

In a *non-maintained system* the emergency lamps are only illuminated if the normal mains supply fails.

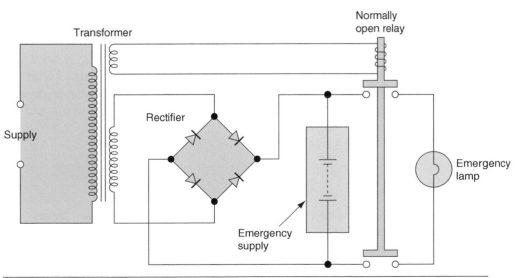

Figure 3.22 Non-maintained emergency lighting.

Self-contained units are cheaper and easier to install than a central battery system, but the central battery can have a greater capacity and duration, and permit a range of emergency lighting luminaires to be installed.

Maintenance

The contractor installing the emergency lighting should provide a test facility that is simple to operate and secure against unauthorized interference, usually a simple key switch. The emergency lighting installation must be segregated completely from any other wiring, so that a fault on the main electrical installation cannot damage the emergency lighting installation (IET Regulation 528.1). Figure 3.23 shows a trunking that provides for segregation of circuits. The batteries used for the emergency supply should be suitable for this purpose. Motor vehicle batteries are not suitable for emergency lighting applications, except in the starter system of motor-driven generators. The fuel supply to a motor-driven generator should be checked. The battery room of a central battery system must be well ventilated and, in the case of a motor-driven generator, adequately heated to ensure rapid starting in cold weather.

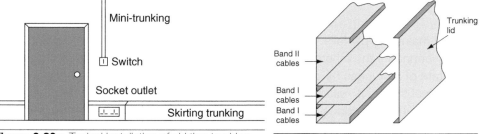

Figure 3.23a Typical installation of skirting trunking and mini-trunking.

Figure 3.23b Segregation of cables in trunking.

The British Standard recommends that the full load should be carried by the emergency supply for at least one hour in every six months. After testing, the emergency system must be carefully restored to its normal operative state. A record should be kept of each item of equipment and the date of each test by a qualified or responsible person. It may be necessary to produce the record as evidence of satisfactory compliance with statutory legislation to a duly authorized person.

Self-contained units are suitable for small installations of up to about 12 units. The batteries contained within these units should be replaced about every five years, or as recommended by the manufacturer and be connected to the a.c. mains supply through a 'test' switch.

Avoiding shutdown of IT equipment

Every modern office now contains computers, and many systems are linked together or networked. Most computer systems are sensitive to variations or distortions in the mains supply and many computers incorporate filters that produce high-protective conductor currents of around 2 or 3 mA. This is clearly not a fault current, but is typical of the current that flows in the circuit protective conductor of IT equipment under normal operating conditions. IET Regulations 543.7.1 and 4 deal with the earthing requirements for the installation

Key fact

IET regulations
Because of filters incorporated within computers, the cumulative effects with a bank of computers are that they can cause sufficiently **high-protective conductor currents** to cause nuisance tripping of RCDs.

of equipment having high-protective conductor currents. IET Guidance Note 7 recommends that IT equipment should be connected to double sockets as shown in Fig. 5.13.

Surge protection

A transient overvoltage or surge is a voltage spike of very short duration. It may be caused by a lightning strike or a switching action on the system. It sends a large voltage spike for a few microseconds down the mains supply, which is sufficient to damage sensitive electronic equipment. Supplies to computer circuits must be 'clean' and 'secure'. Mainframe computers and computer networks are sensitive to mains distortion or interference, which is referred to as 'noise'. Noise is mostly caused by switching an inductive circuit, which causes a transient spike, or by brush gear making contact with the commutator segments of an electric motor. These distortions in the mains supply can cause computers to 'crash' or provoke errors and are shown in Fig. 3.24.

To avoid this, a 'clean' supply is required for the computer network. This can be provided by taking the ring or radial circuits for the computer supplies from a point as close as possible to the intake position of the electrical supply to the building. A clean earth can also be taken from this point, which is usually one core of the cable and not the armour of an SWA cable, and distributed around the final wiring circuit. Alternatively, the computer supply can be cleaned by means of a filter such as that shown in Fig. 3.25 or by installing surge protection devices.

Secure supplies

The mains electrical supply in the United Kingdom is extremely reliable and secure. However, the loss of supply to a mainframe computer or computer network for even a second can cause the system to 'crash', and hours, or even days, of work can be lost.

One solution to this problem is to protect 'precious' software systems with an uninterruptible power supply (UPS). A **UPS** is essentially a battery supply electronically modified to provide a clean and secure a.c. supply. The UPS is plugged into the mains supply and the computer systems are plugged into the UPS.

A UPS to protect a small network of, say, six PCs is physically about the size of one PC hard drive and is usually placed under or at the side of an operator's desk. It is best to dedicate a ring or radial circuit to the UPS and either to connect the computer equipment permanently or to use non-standard outlets to discourage the unauthorized use and overloading of these special supplies by, for example, other workers or cleaning staff kettles.

Finally, remember that most premises these days contain some computer equipment and systems. Electricians intending to isolate supplies for testing or modification should *first check and then check again* before they finally isolate the supply in order to avoid loss or damage to computer systems.

A surge protection device is a device intended to limit transient over voltages, and to divert damaging surge currents away from sensitive equipment. Section 534 of the IET Regulations contain the requirements for the installation of surge protective devices (SPDs) to limit transient over voltages where required by section 443 of BS 7671:2018 or where specified by the designer.

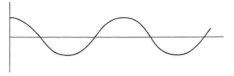

Clean supply

Spikes, caused by an over-voltage transient surging through the mains

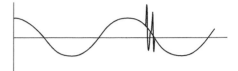

'Noise': unwanted electrical signals picked up by power lines or supply cords

Figure 3.24 Distortion in the a.c. mains supply.

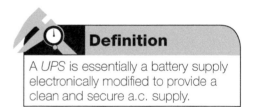

$R = 100\,\Omega$
$C - 0.1\,\mu F$

Figure 3.25 A simple noise suppressor.

Definition

A *UPS* is essentially a battery supply electronically modified to provide a clean and secure a.c. supply.

Figure 3.26 Fire alarm circuits are wired as either normally open or normally closed.

Fire alarm circuits (BS 5839 and BS EN 54-2: 1998)

Through one or more of the various statutory Acts, all public buildings are required to provide an effective means of giving a warning of fire so that life and property may be protected. An effective system is one that gives a warning of fire while sufficient time remains for the fire to be put out and any occupants to leave the building.

Fire alarm circuits are wired as either normally open or normally closed. In a **normally open circuit**, the alarm call points are connected in parallel with each other so that when any alarm point is initiated the circuit is completed and the sounder gives a warning of fire. The arrangement is shown in Fig. 3.27. It is essential for some parts of the wiring system to continue operating even when attacked by fire. For this reason the master control and sounders should be wired in MI or FP 200 cable. The alarm call points of a normally open system must also be wired in MI or FP 200 cable, unless a monitored system is used. In its simplest form this system requires a high-value resistor to be connected across the call-point contacts, which permits a small current to circulate and operate an indicator, declaring the circuit healthy. With a monitored system, PVC insulated cables may be used to wire the alarm call points.

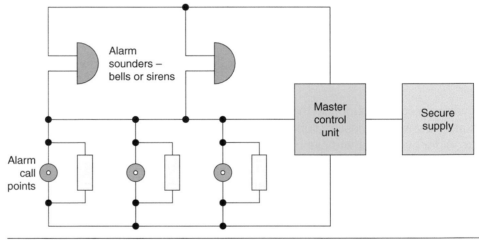

Figure 3.27 A simple normally open fire alarm circuit.

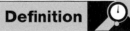

Definition

In a *normally closed circuit*, the alarm call points are connected in series to normally closed contacts as shown in Fig. 3.33. When the alarm is initiated, or if a break occurs in the wiring, the alarm is activated.

In a **normally closed circuit**, the alarm call points are connected in series to normally closed contacts as shown in Fig. 3.28. When the alarm is initiated, or if a break occurs in the wiring, the alarm is activated. The sounders and master control unit must be wired in MI or FP 200 cable, but the call points may be wired in PVC insulated cable since this circuit will always 'fail safe'.

Alarm call points

Manually operated alarm call points should be provided in all parts of a building where people may be present, and should be located so that no one need walk for more than 30 m from any position within the premises in order to give an alarm. A break glass manual call point is shown in Fig. 3.29. They should be located on exit routes and, in particular, on the floor landings of staircases and exits to the street. They should be fixed at a height of 1.4 m above the floor at easily accessible, well illuminated and conspicuous positions. Automatic

Definition

Manually operated alarm call points should be provided in all parts of a building where people may be present, and should be located so that no one need walk for more than 30 m from any position within the premises in order to give an alarm.

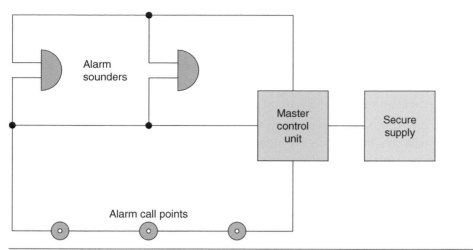

Figure 3.28 A simple normally closed fire alarm circuit.

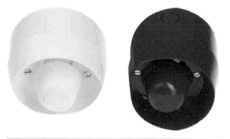

Figure 3.29 Breakglass manual call point.

detection of fire is possible with heat and smoke detectors. These are usually installed on the ceilings and at the top of stairwells of buildings because heat and smoke rise. Smoke detectors tend to give a faster response than heat detectors, but whether manual or automatic call points are used should be determined by their suitability for the particular installation. They should be able to discriminate between a fire and the normal environment in which they are to be installed.

Sounders

The positions and numbers of sounders should be such that the alarm can be distinctly heard above the background noise in every part of the premises. The sounders should produce a minimum of 65 dB, or 5 dB above any ambient sound which might persist for more than 30 seconds. If the sounders are to arouse sleeping persons then the minimum sound level should be increased to 75 dB at the bedhead. Bells, hooters or sirens may be used but in any one installation they must all be of the same type. Examples of sounders are shown in Fig. 3.30. Normal speech is about 5 dB.

> **Definition**
>
> The positions and numbers of *sounders* should be such that the alarm can be distinctly heard above the background noise in every part of the premises.

Fire alarm design considerations

Since all fire alarm installations must comply with the relevant statutory regulations, good practice recommends that contact be made with the local fire prevention officer at the design stage in order to identify any particular local regulations and obtain the necessary certification. Larger buildings must be divided into zones so that the location of the fire can be quickly identified by the emergency services. The zones can be indicated on an indicator board situated in, for example, a supervisor's office or the main reception area. In selecting the zones, the following rules must be considered:

1 Each zone should not have a floor area in excess of 2000 m².
2 Each zone should be confined to one storey, except where the total floor area of the building does not exceed 300 m².
3 Staircases and very small buildings should be treated as one zone.
4 Each zone should be a single fire compartment. This means that the walls, ceilings and floors are capable of containing the smoke and fire.

At least one fire alarm sounder will be required in each zone, but all sounders in the building must operate when the alarm is activated. The main sounders

Figure 3.30 Typical fire alarm sounders.

may be silenced by an authorized person, once the general public have been evacuated from the building, but the current must be diverted to a supervisory buzzer, which cannot be silenced until the system has been restored to its normal operational state. A fire alarm installation may be linked to the local fire brigade's control room by the telecommunication network, if the permission of the fire authority and local telecommunication office is obtained.

The electricity supply to the fire alarm installation must be secure in the most serious conditions. In practice the most reliable supply is the mains supply, backed up by a 'standby' battery supply in case of mains failure. The supply should be exclusive to the fire alarm installation, fed from a separate switch fuse, painted red and labelled, 'Fire Alarm – Do Not Switch Off'. Standby battery supplies should be capable of maintaining the system in full normal operation for at least 24 hours and, at the end of that time, be capable of sounding the alarm for at least 30 minutes.

Fire alarm circuits are Band 1 circuits and consequently cables forming part of a fire alarm installation must be physically segregated from all Band II circuits unless they are insulated for the highest voltage (IET Regulations 528.1 and 560.7.1).

Intruder alarms

Definition

PIR detector units allow a householder to switch lighting units on automatically wherever the area covered is approached by a moving body whose thermal radiation differs from the background.

The installation of security alarm systems in the United Kingdom is already a multi-million-pound business and yet it is also a relatively new industry. As society becomes increasingly aware of crime prevention, it is evident that the market for security systems will expand.

Not all homes are equally at risk, but all homes have something of value to a thief. Properties in cities are at highest risk, followed by homes in towns and villages, and at least risk are homes in rural areas. A nearby motorway junction can, however, greatly increase the risk factor. Flats and maisonettes are the most vulnerable, with other types of property at roughly equal risk. Most intruders are young, fit and foolhardy opportunists. They ideally want to get in and away quickly but, if they can work unseen, they may take a lot of trouble to gain access to a property by, for example, removing the glass from a window.

Most intruders are looking for portable and easily saleable items such as mobile phones, television sets, home computers, jewellery, cameras, silverware, money, cheque books or credit cards. The Home Office has stated that only 7% of homes are sufficiently protected against intruders, although 75% of householders believe they are secure. Taking the simplest precautions will reduce the risk, while installing a security system can greatly reduce the risk of a successful burglary.

Security lighting

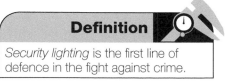

Definition

Security lighting is the first line of defence in the fight against crime.

Security lighting is the first line of defence in the fight against crime. A recent study carried out by Middlesex University has shown that in two London boroughs the crime figures were reduced by improving the lighting levels. Police forces agree that homes and public buildings that are externally well illuminated are a much less attractive target for the thief.

Security lighting installed on the outside of the home may be activated by external detectors. These detectors sense the presence of a person outside the protected property and additional lighting is switched on. This will deter most potential intruders while also acting as courtesy lighting for visitors (Fig. 3.31).

Passive infra-red detectors

Passive infra-red (PIR) detector units allow a householder to switch on lighting units automatically whenever the area covered is approached by a moving body whose thermal radiation differs from the background. This type of detector is ideal for driveways or dark areas around the protected property. It also saves energy because the lamps are only switched on when someone approaches the protected area. The major contribution to security lighting comes from the 'unexpected' high-level illumination of an area when an intruder least expects it. This surprise factor often encourages the potential intruder to 'try next door'.

PIR detectors are designed to sense heat changes in the field of view dictated by the lens system. The field of view can be as wide as 180°, as shown by the diagram in Fig. 3.32. Many of the 'better' detectors use a split lens system so that a number of beams have to be broken before the detector switches on the security lighting. This capability overcomes the problem of false alarms, and a typical PIR is shown in Fig. 3.33.

PIR detectors are often used to switch LED or tungsten halogen floodlights because, of all available luminaires, tungsten halogen and LEDs offer instant high-level illumination. Light fittings must be installed out of reach of an intruder in order to prevent sabotage of the security lighting system.

Figure 3.31 Security lighting reduces crime.

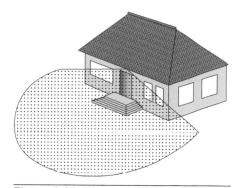

Figure 3.32 PIR detector and field of detection.

Figure 3.33 A typical PIR detector.

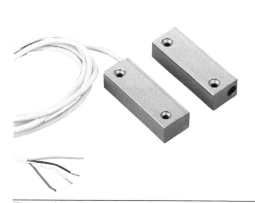

Figure 3.34 Proximity switches for perimeter protection.

Intruder alarm systems

Alarm systems are now increasingly considered to be an essential feature of home security for all types of homes and not just property in high-risk areas. An intruder alarm system serves as a deterrent to a potential thief and often reduces home insurance premiums. In the event of a burglary they alert the occupants, neighbours and officials to a possible criminal act and generate fear and uncertainty in the mind of the intruder, which encourages a more rapid departure. Intruder alarm systems can be broadly divided into three categories – those that give perimeter protection, space protection or trap protection. A system can comprise one or a mixture of all three categories.

Definition

An *intruder alarm system* serves as a deterrent to a potential thief and often reduces home insurance premiums.

A **perimeter protection system** places alarm sensors on all external doors and windows so that an intruder can be detected as he or she attempts to gain access to the protected property. This involves fitting proximity switches to all external doors and windows.

A **movement or heat detector** placed in a room will detect the presence of anyone entering or leaving that room. PIR detectors and ultrasonic detectors give space protection. Space protection does have the disadvantage of being triggered by domestic pets but it is simpler and, therefore, cheaper to install.

Perimeter protection involves a much more extensive and, therefore, expensive installation, but is easier to live with.

Trap protection places alarm sensors on internal doors and pressure pad switches under carpets on through routes between, for example, the main living area and the master bedroom. If an intruder gains access to one room he cannot move from it without triggering the alarm.

Proximity switches

These are designed for the discreet protection of doors and windows. They are made from moulded plastic and are about the size of a chewing-gum packet, as shown in Fig. 3.34. One moulding contains a reed switch, the other a magnet, and when they are placed close together the magnet maintains the contacts of the reed switch in either an open or closed position. Opening the door or window separates the two mouldings and the switch is activated, triggering the alarm.

PIR detectors

These are activated by a moving body that is warmer than the surroundings.

The PIR shown in Fig. 3.33 has a range of 12 m and a detection zone of 110° when mounted between 1.8 and 2 m high.

Intruder alarm sounders

Alarm sounders give an audible warning of a possible criminal act. Bells or sirens enclosed in a waterproof enclosure, such as shown in Fig. 3.35, are suitable.

It is usual to connect two sounders on an intruder alarm installation, one inside to make the intruder apprehensive and anxious, hopefully encouraging a rapid departure from the premises, and one outside. The outside sounder should be displayed prominently since the installation of an alarm system is thought to deter the casual intruder and a ringing alarm encourages neighbours and officials to investigate a possible criminal act.

Control panel

The control panel, such as that shown in Fig. 3.36, is at the centre of the intruder alarm system. All external sensors and warning devices radiate from the control panel. The system is switched on or off at the control panel using a switch or coded buttons. To avoid triggering the alarm as you enter or leave the premises, there are exit and entry delay times to allow movement between the control panel and the door.

Supply

The supply to the intruder alarm system must be secure and this is usually achieved by an a.c. mains supply and battery back-up. Nickel–cadmium rechargeable cells are usually mounted in the sounder housing box.

Figure 3.35 Intruder alarm sounder.

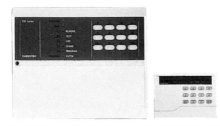
Figure 3.36 Intruder alarm control panel with remote keypad.

Optical fibre cables

The introduction of fibre-optic cable systems and digital transmissions will undoubtedly affect future cabling arrangements and the work of the electrician.

Networks based on the digital technology currently being used so successfully by the telecommunications industry are very likely to become the long-term standard for computer systems. Fibre-optic systems dramatically reduce the number of cables required for control and communications systems, and this will in turn reduce the physical room required for these systems. Fibre-optic cables are also immune to electrical noise when run parallel to mains cables and, therefore, the present rules of segregation and screening may change in the future. There is no spark risk if the cable is accidentally cut and, therefore, such circuits are intrinsically safe.

Optical fibre cables are communication cables made from optical-quality plastic,the same material from which spectacle lenses are manufactured. The energy is transferred down the cable as digital pulses of laser light as against current flowing down a copper conductor in electrical installation terms. The light pulses stay within the fibre-optic cable because of a scientific principle known as 'total internal refraction' which means that the laser light bounces down the cable and when it strikes the outer wall it is always deflected inwards and, therefore, does not escape out of the cable, as shown in Fig. 3.38.

The cables are very small because the optical quality of the conductor is very high and signals can be transmitted over great distances. They are cheap to produce

Figure 3.37 Optical fibre cables.

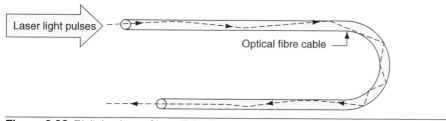

Figure 3.38 Digital pulses of laser light down an optical fibre cable.

and lightweight because these new cables are made from high-quality plastic and not high-quality copper. Single-sheathed cables are often called 'simplex' cables and twin-sheathed cables 'duplex', that is, two simplex cables together in one sheath. Multicore cables are available containing up to 24 single fibres.

Fibre-optic cables look like steel wire armour (SWA) cables (but of course are lighter) and should be installed in the same way and given the same level of protection as SWA cables. Avoid tight-radius bends if possible and kinks at all costs. Cables are terminated in special joint boxes that guarantee cable ends are cleanly cut and butted together, which ensure the continuity of the light pulses. Fibre-optic cables are Band I circuits when used for data transmission and must therefore be segregated from other mains cables to satisfy the IET Regulations. The testing of fibre-optic cables requires that special instruments be used to measure the light attenuation (i.e. light loss) down the cable. Finally, when working with fibre-optic cables, electricians should avoid direct eye contact with the low-energy laser light transmitted down the conductors.

Telephone socket outlets

The installation of telecommunications equipment could, for many years, only be undertaken by British Telecom engineers, but today an electrical contractor may now supply and install telecommunications equipment.

On new premises the electrical contractor may install sockets and the associated wiring to the point of intended line entry, but the connection of the incoming line to the installed master socket must only be made by the telephone company's engineer.

On existing installations, additional secondary sockets may be installed to provide an extended plug-in facility, as shown in Fig. 3.39. Any number of secondary sockets may be connected in parallel, but the number of telephones which may be connected at any one time is restricted.

Each telephone or extension bell is marked with a ringing equivalence number (REN) on the underside. Each exchange line has a maximum capacity of REN 4 and, therefore, the total REN values of all the connected telephones must not exceed four if they are to work correctly.

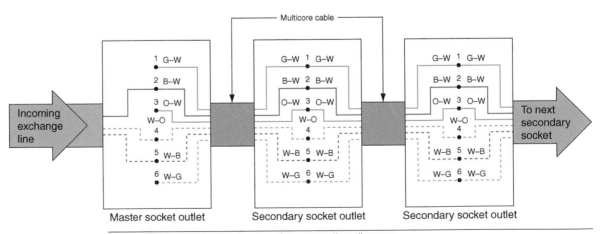

Figure 3.39 Telephone circuit outlet connection diagram.

An extension bell may be connected to the installation by connecting the two bell wires to terminals 3 and 5 of a telephone socket. The extension bell must be of the high impedance type having an REN rating. All equipment connected to a BT exchange line must display the green circle of approval.

The multi-core cable used for wiring extension socket outlets should be of a type intended for use with telephone circuits, which will normally be between 0.4 and 0.68 mm in cross-section. Telephone cable conductors are identified in Table 3.2 and the individual terminals in Table 3.3. The conductors should be connected as shown in Fig. 3.44. Telecommunications cables are Band I circuits and must be segregated from Band II circuits containing mains cables (IET Regulation 528.1).

Table 3.2 Telephone cable identification

Code	Base colour	Stripe
G–W	Green	White
B–W	Blue	White
O–W	Orange	White
W–O	White	Orange
W–B	White	Blue
W–G	White	Green

Table 3.3 Telephone socket terminal identification. Terminals 1 and 6 are frequently unused, and therefore four-core cable may normally be installed. Terminal 4, on the incoming exchange line, is only used on a PBX line for earth recall

Socket terminal	Circuit
1	Spare
2	Speech circuit
3	Bell circuit
4	Earth recall
5	Speech circuit
6	Spare

Building management system

A building management system is a computer system that adjusts certain services within a building to meet the needs of the users. This means certain sensors are utilized in order to monitor certain key parameters such as temperature and thereafter the system programmer will use this information to vary and effect the amount of control required. Modern systems can control a number of buildings and therefore can be aligned to operate as efficiently as possible by being active during building occupancy.

Assessment criteria 1.4

Interpret sources of information relevant to planning of the installation of wiring systems and equipment (covered in Chapter 2 on pages 68 to 71)

Assessment criteria 1.5

Plan wiring system routes

Assessment criteria 1.6

Specify the positional requirements of wiring systems and equipment

Support and fixing methods for electrical equipment

It is vital that any wiring system must be planned properly in order to identify the optimum routes available. This also includes avoiding thermal insulation since it causes a de-rating effect – it reduces its current carrying capacity since due to the property of thermal insulation the wiring system will be exposed to higher temperatures. Other considerations involved in planning would be the distances required when drilling holes into wooden beams, for instance, so that load bearing structures are not weakened. The requirement is that there should be 300 mm laterally between holes and 50 mm from the hole to both the top and bottom of the joist. If those dimensions cannot be achieved then a different route must be taken.

Individual conductors may be installed in trunking or conduit and individual cables may be clipped directly to a surface or laid on a tray using the wiring system that is most appropriate for the particular installation. The installation method chosen will depend upon the contract specification, the fabric of the building and the type of installation – domestic, commercial or industrial. It is important that the wiring systems and fixing methods are appropriate for the particular type of installation and compatible with the structural materials used in the building construction. The support system must not be liable to premature collapse in the event of a fire as required by the 18th Edition of the Regulations in section 521.10. The electrical installation must be compatible with the installed conditions, and as has already been discussed when drilling into the fabric of the building, so as not to weaken load bearing girders or joists.

The installation designer must consider the following questions:

* Does this wiring system meet the contract specification?
* Is the wiring system compatible with this particular installation?
* Do I need to consider any special regulations such as those required by agricultural and horticultural installations, swimming pools or flameproof installations?
* Will this type of electrical installation be aesthetically acceptable and compatible with the other structural materials?

The installation electrician must consider the following questions:

* Am I using materials and equipment that meet the relevant British Standards and the contract specification?

Key fact

Materials marked 'CE' are compliant with European Standards. BASEC approved cables comply with rigid manufacturing standards and specifications

- Am I using an appropriate fixing method for this wiring system or piece of equipment?
- Will the structural material carry the extra load that my conduits and cables will place upon it?
- Will my fixings and fittings weaken the existing fabric of the building?
- Will the electrical installation interfere with other supplies and services?
- Will all terminations and joints be accessible upon completion of the erection period? (IET Regulations 513.1 and 526.3.)
- Will the materials being used for the electrical installation be compatible with the intended use of the building?
- Am I working safely and efficiently and in accordance with the IET Regulations (BS 7671)?

A domestic installation usually calls for a PVC insulated and sheathed wiring system. These cables are generally fixed using plastic clips incorporating a masonry nail, which means that the cables can be fixed to wood, plaster or brick with almost equal ease. However, the 3rd Amendment to the Regulations introduced a new Regulation 521.11.201 which requires wiring systems in escape routes be supported in such a manner that they will not prematurely collapse in the event of a fire.

The 18th Edition of the Regulations at 521.10.202 now requires the same fire proof support for all systems throughout the installation, not just escape routes.

However, it is only those wiring systems that might collapse before or during the period when the rescue services would enter a building, that require a fixing which will not collapse prematurely. So, cables installed in metal trunking, or conduit, or cables laid on top of cable tray or ladder rack will require no extra fixing considerations because they are adequately supported. Also, wiring systems fixed within the fabric of the building which is not liable to premature collapse in the event of a fire is safe in these circumstances.

It is the cable systems fixed with plastic clips or inside plastic trunking which will now require our consideration. These systems can fail when subject to either direct flame or the hot products of combustion leading to wiring systems hanging down and causing an entanglement risk as a result of the fire.

This makes it impossible for us to use non metallic cable clips, cable ties or plastic trunking as the only means of support for PVC wiring system. The regulation tells us that where non-metallic cable systems are used, a suitable means of fire resistant support and retention must be used to prevent cables falling down in the event of a fire. Note 4 to this Regulation advises that suitably spaced steel or copper clips, saddles or ties are examples which will meet this requirement. Cables must be run straight and neatly between clips fixed at equal distances which provide adequate support for the cable so that it does not become damaged by its own weight (IET Regulation 522.8.4), shown in Table 3.4 (page 162). A commercial or industrial installation might call for a conduit or trunking wiring system. A conduit is a tube, channel or pipe in which insulated conductors are contained. The conduit, in effect, replaces the PVC outer sheath of a cable, providing mechanical protection for the insulated conductors. A conduit installation can be rewired easily or altered at any time and this flexibility, coupled with mechanical protection, makes conduit installations popular for commercial and industrial applications. Steel conduits and trunking are, however, much heavier than single cables and, therefore, need substantial and firm fixings and supports. A wide range of support brackets is available for fixing conduit, trunking and tray installations to the fabric of a commercial or industrial installation. Some of these are shown in Fig. 3.40 and 3.48.

When a heavier or more robust fixing is required to support cabling or equipment, a nut and bolt or screw fixing is called for. Wood screws may be screwed directly into wood but when fixing to stone, brick or concrete it is first necessary to drill a hole in the masonry material which is then plugged with a material (usually plastic) to which a screw can be secured. For the most robust fixing to masonry materials, an expansion bolt such as that made by Rawlbolt should be used. For lightweight fixings to hollow partitions or plasterboard, a spring toggle can be used.

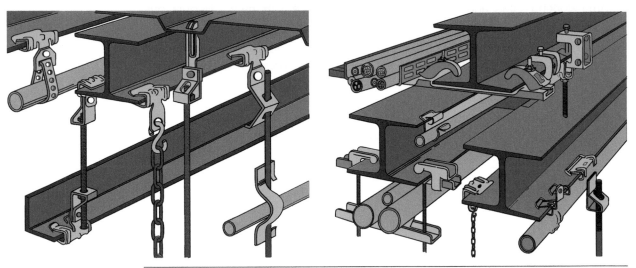

Figure 3.40 Some manufactured girder supports for electrical equipment.

Plasterboard cannot support a screw fixing directly into itself but the spring toggle spreads the load over a larger area, making the fixing suitable for light loads.

Let us look in a little more detail at individual joining, support and fixing methods.

Joining materials

Plastic can be joined with an appropriate solvent. Metal may be welded, brazed or soldered, but the most popular method of on-site joining of metal on electrical installations is by nuts and bolts or rivets.

A nut and bolt joint may be considered a temporary fastening since the parts can easily be separated if required by unscrewing the nut and removing the bolt. A rivet is a permanent fastening since the parts riveted together cannot be easily separated.

Two pieces of metal joined by a bolt and nut and by a machine screw and nut are shown in Fig. 3.41. The nut is tightened to secure the joint. When joining trunking or cable trays, a round-head machine screw should be used with the head inside to reduce the risk of damage to cables being drawn into the trunking or tray.

Thin sheet material such as trunking is often joined using a pop riveter. Special rivets are used with a hand tool, as shown in Fig. 3.42. Where possible, the parts to be riveted should be clamped and drilled together with a clearance hole for the rivet. The stem of the rivet is pushed into the nose bush of the riveter until the alloy sleeve of the rivet is flush with the nose bush (a). The rivet is then placed in the hole and the handles squeezed together (b). The alloy sleeve is compressed and the rivet stem will break off when the rivet is set and the joint complete (c). To release the broken-off stem piece, the nose bush is turned upward and the handles opened sharply. The stem will fall out and is discarded (d).

Bracket supports

Conduit and trunking may be fixed directly to a surface such as a brick wall or concrete ceiling, but where cable runs are across girders or other steel framework, spring steel clips may be used but support brackets or clips may require manufacturing.

The brackets are usually made from flat iron, which is painted after manufacturing to prevent corrosion. They may be made on-site by the electrician or, if many brackets are required, the electrical contractor may make a working sketch with dimensions and have the items manufactured by a blacksmith or metal fabricator.

Bolt and nut   Machine screw and nut

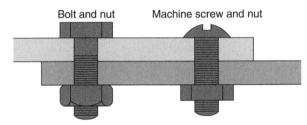

Figure 3.41 Joining of metal.

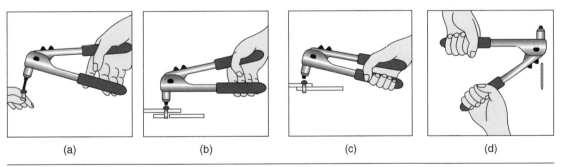

    (a)   (b)   (c)   (d)

Figure 3.42 Metal joining with pop rivets.

The type of bracket required will be determined by the installation, but Fig. 3.43 gives some examples of brackets which may be modified to suit particular circumstances.

Fixing methods

PVC insulated and sheathed wiring systems are usually fixed with PVC clips in order to comply with IET Regulations 522.8.3 and 4. The clips are supplied in various sizes to hold the cable firmly, and the fixing nail is a hardened masonry nail. Figure 3.44 shows a cable clip of this type. The use of a masonry nail means that fixings to wood, plaster, brick or stone can be made with equal ease.

When heavier cables, trunking, conduit or luminaires have to be fixed, a screw fixing is often needed. Wood screws may be screwed directly into wood but when fixing to brick, stone, plaster or concrete it is necessary to drill a hole in the

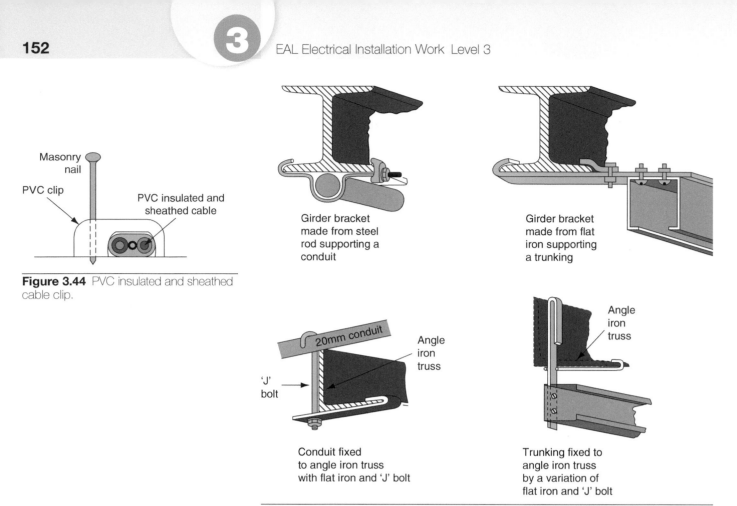

Figure 3.44 PVC insulated and sheathed cable clip.

Figure 3.43 Bracket supports for conduits and trunking.

masonry material, which is then plugged with a material to which the screw can be secured.

Plastic plugs

A **plastic plug** is made of a hollow plastic tube split up to half its length to allow for expansion. Each size of plastic plug is colour coded to match a wood screw size.

A hole is drilled into the masonry, using a masonry drill of the same diameter, to the length of the plastic plug (see Fig. 3.45). The plastic plug is inserted into the hole and tapped home until it is level with the surface of the masonry. Finally, the fixing screw is driven into the plastic plug until it becomes tight and the fixture is secure.

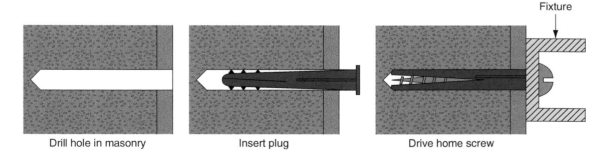

Drill hole in masonry   Insert plug   Drive home screw

Figure 3.45 Screw fixing to plastic plug.

Expansion bolts

The most well-known **expansion bolt** is made by Rawlbolt and consists of a split iron shell held together by a steel ferrule at one end and a spring wire clip at the other end. Tightening the bolt draws up an expanding bolt inside the split iron shell, forcing the iron to expand and grip the masonry. Rawlbolts are for heavy-duty masonry fixings (see Fig. 3.46).

A hole is drilled in the masonry to take the iron shell and ferrule. The iron shell is inserted with the spring wire clip end first so that the ferrule is at the outer surface. The bolt is passed through the fixture, located in the expanding nut and tightened until the fixing becomes secure.

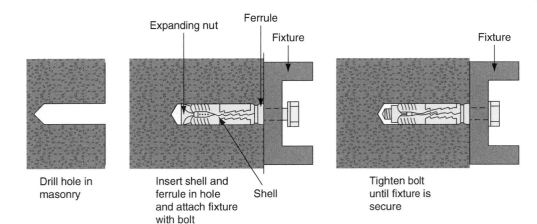

Expanding nut   Ferrule

Fixture

Fixture

Drill hole in masonry

Insert shell and ferrule in hole and attach fixture with bolt   Shell

Tighten bolt until fixture is secure

Figure 3.46 Expansion bolt fixing.

Spring toggle bolts

A **spring toggle bolt** provides one method of fixing to hollow partition walls which are usually faced with plasterboard and a plaster skimming. Plasterboard and plaster wall or ceiling surfaces are not strong enough to support a load fixed directly into the plasterboard, but the spring toggle spreads the load over a larger area, making the fixing suitable for light loads.

A hole is drilled through the plasterboard and into the cavity. The toggle bolt is passed through the fixture and the toggle wings screwed into the bolt. The toggle wings are compressed and passed through the hole in the plasterboard and into the cavity where they spring apart and rest on the cavity side of the plasterboard. The bolt is tightened until the fixing becomes firm. The bolt of the spring toggle cannot be removed after fixing without the loss of the toggle wings. If it becomes necessary to remove and refix the fixture a new toggle bolt will have to be used.

Given all that has been listed it is not surprising that the necessary skills can only be acquired through demonstration of the correct attitude and dedication to hone their craft through a NVQ Level 3 qualification in order to demonstrate the required understanding and proficiency to become a skilled person.

Definition

The most well-known *expansion bolt* is made by Rawlbolt and consists of a split iron shell held together by a steel ferrule at one end and a spring wire clip at the other end. Tightening the bolt draws up an expanding bolt inside the split iron shell, forcing the iron to expand and grip the masonry. Rawlbolts are for heavy-duty masonry fixings.

Top tip

Inserting rawl plugs

When placing rawl plugs into masonry or tile surfaces, ensure the plug is pushed several millimetres past the surface.

Definition

A *spring toggle bolt* provides one method of fixing to hollow partition walls which are usually faced with plasterboard and a plaster skimming.

Definition

A skilled *person* is one who has the ability to perform a particular task properly.

Assessment criteria 1.7

Interpret the requirements of BS 7671 for the planned installation

Wiring systems and enclosures

The final choice of a wiring system must rest with those designing the installation in order to meet the installation criteria against the actual environment and the client's wishes. But whatever system is employed, good workmanship by skilled or instructed persons and the use of proper materials is essential for compliance with the regulations (IET Regulation 134.1.1).

Most cables can be considered to be constructed in three parts: the conductor, which must be of a suitable cross-section to carry the load current; the insulation, which has a colour or number code for identification; and the outer sheath, which may contain some means of providing mechanical protection from impact damage. Some conductors are also given a PVC over-sheath in order to protect from corrosion when used outside.

The conductors of a cable are made of either copper or aluminium but aluminium only has approximately 61% of the conductivity of copper, which means that to carry the same current the aluminium conductor would have to be bigger. Given that it is lighter aluminium is used in overhead power transmission cables, which accounts for the fact that aluminium tends to be manufactured at sizes >16 mm.

Conductors can be stranded or solid. Solid conductors are only used in fixed wiring installations and tend to be shaped in larger cables such as 25 mm². Stranded conductors are more flexible and conductor sizes from 4.0 to 25 mm² contain seven strands. A 10 mm² conductor, for example, has seven 1.35 mm diameter strands, which collectively make up the 10 mm² cross-sectional area of the cable. Conductors above 25 mm² have more than seven strands, depending upon the size of the cable. Flexible cords have multiple strands of very fine wire conductors, as fine as one strand of human hair. This gives the cable its very flexible quality.

Figure 3.47 Brown is live or phase conductor, blue is neutral and green/yellow is earth.

New wiring colours

Over twenty-five years ago the United Kingdom agreed to adopt the European colour code for flexible cords, that is, brown for live or phase conductor, blue for the neutral conductor and green combined with yellow for earth conductors.

However, no similar harmonization was proposed for non-flexible cables used for fixed wiring. These were to remain as red for live or phase conductor, black for the neutral conductor and green combined with yellow for earth conductors.

On 31 March 2004, the IET published Amendment No. 2 to BS 7671: 2001, which specified new cable core colours for all fixed wiring in UK electrical installations.

These new core colours 'harmonize' the United Kingdom with the practice in mainland Europe.

Fixed cable core colours up to 2006

- Single-phase supplies red line conductors, black neutral conductors, and green combined with yellow for earth conductors.
- Three-phase supplies red, yellow and blue line conductors, black neutral conductors and green combined with yellow for earth conductors.

These core colours could not be used after 31 March 2006.

Figure 3.48 Electrical wires connected within a plug.

New (harmonized) fixed cable core colours

- Single-phase supplies brown line conductors, blue neutral conductors and green combined with yellow are used for earth conductors (just like flexible cords).
- Three-phase supplies brown, black and grey line conductors, blue neutral conductors and green combined with yellow are used for earth conductors.

Cable core colours used from 31 March 2004 onwards

Extensions or alterations to existing single-phase installations do not require marking at the interface between the old and new fixed wiring colours. However, a warning notice must be fixed at the consumer unit or distribution fuse board which states:

> Caution – this installation has wiring colours to two versions of BS 7671. Great care should be taken before undertaking extensions, alterations or repair that all conductors are correctly identified.

Alterations to three-phase installations must be marked at the interface L1, L2, L3 for the lines and N for the neutral. Both new and old cables must be marked.

These markings are preferred to coloured tape and a caution notice is again required at the distribution board. Appendix 7 of BS 7671: 2008 deals with harmonized cable core colours.

PVC insulated and sheathed cables

PVC which stands for polyvinylchloride is the common type of cable found in domestic and commercial installations. Because its inner insulation and outer sheath are made from thermoplastic it is known as PVC/PVC and this type of cable may be clipped direct to a surface or sunk in plaster. This type of cable is also used without a sheath and therefore becomes single-core thermoplastic or singles, when installed in conduit or trunking. It is the simplest and least expensive cable. Figure 3.50 shows a sketch of a twin and earth cable.

It is referred to in BS 7671 as '70°C thermoplastic insulated and sheathed flat cable with protective conductor'. This is in contrast to thermosetting, which can operate to an upper temperature range of 90°.

Top tip

Appendix K of the *On-Site Guide* identifies changes to cable core colour identification.

Figure 3.49 PVC, or polyvinylchloride, is commonly used in domestic and commercial installations.

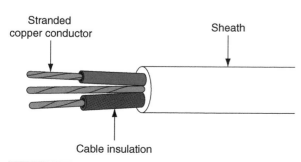

Figure 3.50 A twin and earth PVC insulated sheathed cable

Key fact

Single-core PVC cables must always be installed in protective enclosures and never clipped direct to a surface.

PVC/SWA cable

PVC insulated steel wire armour cables are used for wiring underground between buildings, for main supplies to dwellings, rising sub-mains and industrial installations. They are used where some mechanical protection of the cable conductors is required.

The conductors are covered with colour-coded PVC insulation and then contained either singly or with others in a PVC sheath (see Fig. 3.51). Around this sheath is placed an armour protection of steel wires twisted along the length of the cable, and a final PVC sheath covering the steel wires protects them from corrosion. The armour sheath also provides the circuit protective conductor (CPC) and the cable is simply terminated using a compression gland.

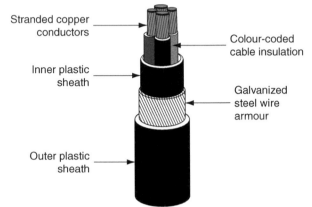

Figure 3.51 A four-core PVC/SWA cable.

Fire performance cables

There are two different types of cable on the market that have improved fire performance compared with standard cables. Those that react in a fire, give low smoke emission and reduced flame propagation, and those that are fire resistant, which will continue to operate during a fire when subjected to a specified flame source for a period of time.

Cables for fire alarm applications are now split into 'standard', where survival time of the cable is 30 minutes, and 'enhanced', where the survival time increases to 120 minutes.

Fixings, terminals and accessories must offer the same level of protection against fire as the cable you use.

Mineral insulated cable also known as Pyro and FP200 cable are the most commonly used examples of these cables.

MI cable

A mineral insulated (MI) cable has a seamless copper sheath, which makes it waterproof and fire- and corrosion-resistant. These characteristics often make it the only cable choice for hazardous or high-temperature installations such as oil refineries and chemical works, boiler houses and furnaces, petrol pump and fire alarm installations.

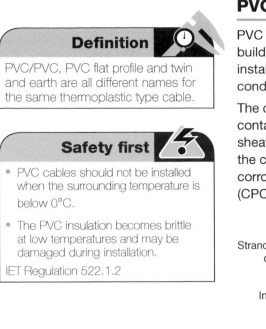

Definition

PVC/PVC, PVC flat profile and twin and earth are all different names for the same thermoplastic type cable.

Safety first

- PVC cables should not be installed when the surrounding temperature is below 0°C.

- The PVC insulation becomes brittle at low temperatures and may be damaged during installation.

IET Regulation 522.1.2

Definition

BS 7671 specifies that the sounding element of a fire detection system must be installed using fire resistant cables.

Definition

Mineral insulated cable does not **emit** any smoke or **toxic** substances and does not allow any flame propagation.

The cable has a small overall diameter when compared with alternative cables and may be supplied as bare copper or with a PVC over sheath. It is colour coded orange for general electrical wiring, white for emergency lighting or red for fire alarm wiring. The copper outer sheath provides the CPC, and the cable is terminated with a pot and sealed with compound and a compression gland (see Fig. 3.52).

The copper conductors are embedded in a white powder, magnesium oxide, which is non-ageing and non-combustible, but which is hygroscopic, which means that it readily absorbs moisture from the surrounding air, unless

> **Top tip**
>
> Mineral insulated cable must be properly sealed to avoid ingress of water especially since magnesium oxide is hygroscopic (absorbs water). Once assembled it is therefore good practice to check the glanding by testing its insulation resistance.

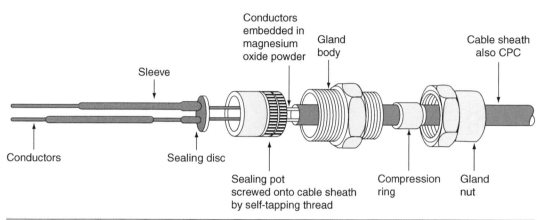

Figure 3.52 MI cable with terminating seal and gland.

adequately terminated. The termination of an MI cable is a complicated process requiring the electrician to demonstrate a high level of practical skill and expertise for the termination to be successful.

The marking on mineral insulated cables and accessories indicates the size and number of the conductors and its rating of light or heavy gauge. As an example, a mineral insulated cable sealing pot marked 3L1.5 would be constructed of 3 cores each of 1.5 mm² c.s.a. and light gauge (up to 500 V). Heavy gauge cable is rated up to 750 V.

> **Key fact**
>
> Because of its non-ageing property mineral insulated cable is used in listed buildings.

FP 200 cable

FP 200 cable is similar in appearance to an MI cable in that it is a circular tube, or the shape of a pencil, and is available with a red or white sheath. However, it is much simpler to use and terminate than an MI cable.

The cable is available with either solid or stranded conductors that are insulated with 'insudite', a fire-resistant insulation material. The conductors are then screened by wrapping an aluminium tape around the insulated conductors, that is, between the insulated conductors and the outer sheath. This aluminium tape screen is applied metal side down and in contact with the bare CPC.

The sheath is circular and made of a robust thermoplastic low-smoke, zero halogen material.

FP 200 is available in 2, 3, 4, 7, 12 and 19 cores with a conductor size range from 1.0 to 4.0 mm².

The cable is as easy to use as a PVC-insulated and sheathed cable. No special terminations are required; the cable may be terminated through a grommet into a knock-out box or terminated through a simple compression gland.

> **Definition**
>
> The difference between *light and heavy gauge MI cable* is that light gauge has a thinner outer copper sheath than heavy gauge.

The cable is a fire-resistant cable, primarily intended for use in fire alarms and emergency lighting installations, or it may be embedded in plaster.

High-voltage overhead cables

Suspended from cable towers or pylons, overhead cables must be light, flexible and strong. The cable is constructed of stranded aluminium conductors formed around a core of steel-stranded conductors (see Fig. 3.53). The aluminium conductors carry the current and the steel core provides the tensile strength required to suspend the cable between pylons. The cable is not insulated since it is placed out of reach and insulation would only add to the weight of the cable.

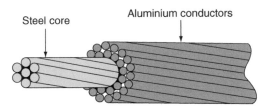
Steel core   Aluminium conductors

Figure 3.53 132 kV overhead cable construction.

Optical fibre cables

The introduction of fibre-optic cable systems and digital transmissions will undoubtedly affect future cabling arrangements and the work of the electrician.

Networks based on the digital technology currently being used so successfully by the telecommunications industry are very likely to become the long-term standard for computer systems. Fibre-optic systems dramatically reduce the number of cables required for control and communications systems, and this will in turn reduce the physical room required for these systems. Fibre-optic cables are also immune to electrical noise when run parallel to mains cables and, therefore, the present rules of segregation and screening may change in the future. There is no spark risk if the cable is accidentally cut and, therefore, such circuits are intrinsically safe. Intrinsic safety is described later under the heading 'Working in hazardous areas', Chapter 5.

Optical fibre cables are communication cables made from optical-quality plastic, the same material from which spectacle lenses are manufactured. The energy is transferred down the cable as digital pulses of laser light, as against current flowing down a copper conductor in electrical installation terms. The light pulses stay within the fibre-optic cable because of a scientific principle known as 'total internal refraction', which means that the laser light is actually bouncing along the cable by striking and deflecting off the inner wall as shown in Figure 3.43.

The biggest advantage of fibre-optic cables is the speed with which they are able to transfer data. In 2014, researchers developed a new fibre optic technology capable of transferring data at a rate of 255 terabits per second (Tbps) – more data than the total traffic flowing across the internet at peak time. At such speeds, it is possible to transfer a 1 GB movie in 0.03 milliseconds, or a one terabyte file – 1000 GB of data – in just 0.03 seconds.

Commercial communication providers offer packages ranging from 300 MB/sec to 1 GB/sec so while this speed is not yet available to the public, the potential for huge development is there.

Definition

Optical fibre cables are communication cables made from optical-quality plastic, the same material from which spectacle lenses are manufactured. The energy is transferred down the cable as digital pulses of laser light, as against current flowing down a copper conductor in electrical installation terms.

Definition

Fibre-optics operate on a process known as *total internal reflection*.

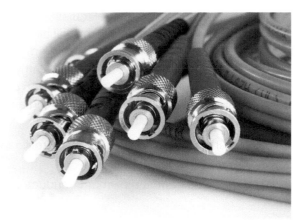

Figure 3.54 Optical fibre cable.

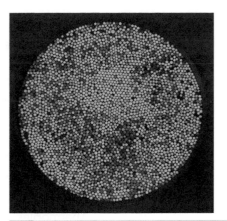

Figure 3.55 Cross-section of fibre-optic cable.

The cables are very small because the optical quality of the conductor is very high and signals can be transmitted over great distances. They are cheap to produce and lightweight because these new cables are made from high-quality plastic and not high-quality copper. Single-sheathed cables are often called 'simplex' cables and twin-sheathed cables 'duplex', that is, two simplex cables together in one sheath. Multi-core cables are available containing up to 24 single fibres.

Fibre-optic cables look like steel wire armour cables (but of course they are lighter) and should be installed in the same way, and given the same level of protection, as SWA cables. Avoid tight-radius bends if possible and kinks at all costs. Cables are terminated in special joint boxes that ensure cable ends are cleanly cut and butted together to ensure the continuity of the light pulses.

Fibre-optic cables are Band I circuits when used for data transmission and must therefore be segregated from other mains cables to satisfy the IET Regulations.

The testing of fibre-optic cables requires special instruments to measure the light attenuation (i.e. light loss) down the cable. Finally, when working with fibre-optic cables, electricians should avoid direct eye contact with the low-energy laser light transmitted down the conductors.

Data cables

Category 5 cable (cat 5), cat 5e and cat 6 cables are twisted pair cables used for computer networks and carrying telephony and video signals. The cable consists of four pairs of conductors, twisted in pairs to minimize interference. It is usually terminated in punch down, insulation displacement connections to the RJ45 specification. The length of the data circuit is restricted to 100 m and is usually restricted to 90 m to allow for the length of the patch leads to be added.

Temperature limits of cables

While there are many cable insulation materials, they are often divided into two types, thermoplastic or thermosetting. The current capacity of a cable is determined by the type of insulation as the build up of heat will have a detrimental effect on the insulation.

⭐ **Top tip**

Fibre optic cables still need to be handled with care and respect. Mishandling or installing an overly tight bending radius can damage its ability to transmit data.

Figure 3.56 Data cables.

Thermoplastic materials are composed of chains of molecules (polyethylene for example). When heat is applied the energy will allow the bonds to separate and the material can flow (melt) and be reformed.

Thermosetting materials are formed when materials such as polyethylene undergo specific heating or chemical processes. During this process the individual chains become cross linked by smaller molecules making a rigid structure. Thermosetting materials cannot be melted or remoulded.

Due to the structure, thermosetting materials are generally more heat resistance and have greater strength.

BS 7671 gives current carrying capacities based on a maximum conductor temperature of:

- Thermoplastic 70°C
- Thermosetting 90°C

Definition

To offset any heat damage *Butyl* is a type of heat resistant cable that is used as the final connection to a water heater.

Choosing an appropriate wiring system

An electrical installation is made up of many different electrical circuits, lighting circuits, power circuits, single-phase domestic circuits and three-phase industrial or commercial circuits.

Whatever the type of circuit, the circuit conductors are contained within cables or enclosures.

Part 5 of the IET Regulations tells us that electrical equipment and materials must be chosen so that they are suitable for the installed conditions, taking into account temperature, the presence of water, corrosion, mechanical damage, vibration or exposure to solar radiation. Therefore, PVC insulated and sheathed cables are suitable for domestic installations but, for a cable requiring mechanical protection and suitable for burying underground, a PVC/SWA cable would be preferable. Category 5 cable (cat 5), cat 5e and cat 6 cables are twisted pair cables used for computer networks and carrying telephony and video signals as seen in Figure 3.56.

MI cables are waterproof, heatproof and corrosion-resistant with some mechanical protection. These qualities often make it the only cable choice for hazardous or high-temperature installations such as oil refineries, chemical works, boiler houses and petrol pump installations. An MI cable with terminating gland and seal is shown in Fig. 3.52.

The final choice of a wiring system must rest with those designing the installation and those ordering the work, but whatever system is employed, good workmanship by skilled or instructed persons and the use of proper materials is essential for compliance with the IET Regulations (IET Regulation 134.1.1). The necessary skills can be acquired by an electrical trainee who has the correct attitude and dedication to his craft.

PVC insulated and sheathed cable installations are used extensively for lighting and socket installations in domestic dwellings. Mechanical damage to the cable caused by impact, abrasion, penetration, compression or tension must be minimized during installation (IET Regulation 522.6.1). The cables are generally fixed using plastic clips incorporating a masonry nail, which means the cables can be fixed to wood, plaster or brick with almost equal ease. Cables should be run horizontally or vertically, not diagonally, down a wall. All kinks should be removed so that the cable is run straight and neatly between clips fixed at equal distances providing adequate support for the cable so that it does not become damaged by its own weight (IET Regulation 522.8.4 and Table D1 of the *On-Site Guide*). A Table of cable supports is shown in Table 3.4. Where cables are bent, the radius of the bend should not cause the conductors to be damaged (IET Regulation 522.8.3 and Table D5 of the *On-Site Guide*).

Cable supports to prevent the premature collapse of cables in the event of fire was introduced by the 3rd Amendment to the IET Regulations at 521.11.201 which requires wiring systems in escape routes to be supported in a manner that they will not prematurely collapse in the event of a fire.

Wiring systems which drop and hang across escape routes due to a failure of the means of support in fire conditions have the potential to entangle persons using the escape route including fire fighters entering the building to put out the fire.

The 18th Edition of the Regulations at 521.10.202 now requires the same fire proof support for all systems throughout the installation, not just escape routes.

It is the cable systems fixed with plastic clips or inside plastic trunking which will now require our consideration. These systems can fail when subject to either direct flame or the hot products of combustion leading to wiring systems hanging down and causing an entanglement risk as a result of the fire.

This makes it impossible for us to use non metallic cable clips, cable ties or plastic trunking as the only means of support for PVC wiring system. The regulation tells us that where non-metallic cable systems are used, a suitable means of fire resistant support and retention must be used to prevent cables falling down in the event of a fire. Note 4 of the Regulation advises that suitably spaced steel or copper clips, saddles or ties are examples which will met this requirement.

In the future, if an electrical contractor carrying out an Electrical Installation Condition Report observes PVC cable systems supported only by non-metallic cable supports, they are to record a code C3 (Improvement Recommended) on the report.

Terminations or joints in the cable may be made in ceiling roses, junction boxes, or behind sockets or switches, provided that they are enclosed in a non-ignitable material, are properly insulated and are mechanically and electrically secure (IET Regulation 526). All joints must be accessible for inspection, testing and maintenance when the installation is completed (IET Regulation 526.3).

Where PVC insulated and sheathed cables are concealed in walls, floors or partitions, they must be provided with a box incorporating an earth terminal at each outlet position. PVC cables do not react chemically with plaster, as do some cables, and consequently PVC cables may be buried under plaster. Further protection by channel or conduit is only necessary if mechanical protection from nails or screws is required or to protect them from the plasterer's trowel.

However, IET Regulation 522.6.202 now tells us that where PVC cables are to be embedded in a wall or partition at a depth of less than 50 mm they should be run along one of the permitted routes shown in Fig. 3.58. Figure 3.57 shows a typical PVC installation. To identify the most probable cable routes, IET Regulation 522.6.202 tells us that outside a zone formed by a 150 mm border all around a wall edge, cables can only be run horizontally or vertically to a point or accessory if they are contained in a substantial earthed enclosure, such as a conduit, which can withstand nail penetration, as shown in Fig. 3.58.

Where the accessory or cable is fixed to a wall that is less than 100 mm thick, protection must also be extended to the reverse side of the wall if a position can be determined.

Top tip

Electrical connections and terminations need to be both electrically and mechanically sound

Table 3.4 Spacing of cable supports

Overall diameter of cable (mm)	Maximum spacings of clips								
	PVC sheathed cables				Armoured cables		Mineral insulated copper sheathed cables	
	Generally		In caravans						
	Horizontal (mm)	Vertical (mm)	Horizontal (mm)	Vertical (mm)	Horizontal (mm)	Vertical (mm)	Horizontal (mm)	Vertical (mm)
1	2	3	4	5	6	7	8	9
Not exceeding 9	250	400	250 (for all sizes)	400 (for all sizes)	–	–	600	800
Exceeding 9 and not exceeding 15	300	400			350	450	900	1200
Exceeding 15 and not exceeding 20	350	450			400	550	1500	2000
Exceeding 20 and not exceeding 40	400	550			450	600	–	–

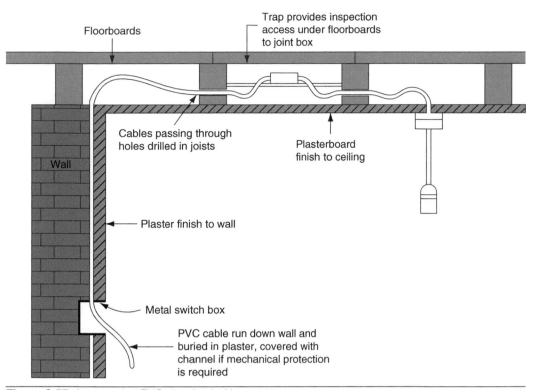

Figure 3.57 A concealed PVC sheathed wiring system.

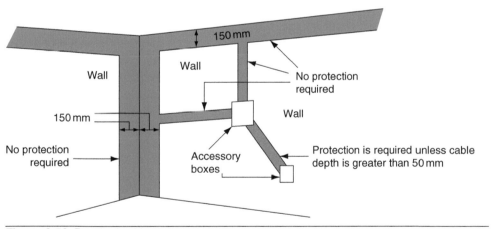

Figure 3.58 Permited cable routes.

Where none of this protection can be complied with the cable must be given additional protection with a 30 mA RCD (IET Regulation 522.6.202).

Where cables pass through walls, floors and ceilings the hole should be made good with incombustible material such as mortar or plaster to prevent the spread of fire (IET Regulations 527.1.2 and 527.2.1). Cables passing through metal boxes should be bushed with a rubber grommet to prevent abrasion of the cable.

Holes drilled in floor joists through which cables are run should be 50 mm below the top or 50 mm above the bottom of the joist to prevent damage to the cable by nail penetration (IET Regulation 522.6.201), as shown in Fig. 3.59. PVC cables should not be installed when the surrounding temperature is below 0°C or when the cable temperature has been below 0°C for the previous 24 hours because the insulation becomes brittle at low temperatures and may be damaged during installation.

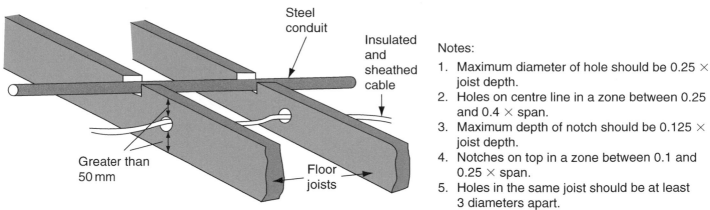

Notes:
1. Maximum diameter of hole should be 0.25 × joist depth.
2. Holes on centre line in a zone between 0.25 and 0.4 × span.
3. Maximum depth of notch should be 0.125 × joist depth.
4. Notches on top in a zone between 0.1 and 0.25 × span.
5. Holes in the same joist should be at least 3 diameters apart.

Figure 3.59 Correct installation of cables in floor joists.

! Try this

Definitions

In the margin write down a short definition of a 'skilled or instructed person'.

Conduit installations

A conduit is a tube, channel or pipe in which insulated conductors are contained. The conduit, in effect, replaces the PVC outer sheath of a cable, providing mechanical protection for the insulated conductors. A conduit installation can be rewired easily or altered at any time, and this flexibility, coupled with mechanical protection, makes conduit installations popular for commercial and industrial applications. There are three types of conduit used in electrical installation work: steel, PVC and flexible.

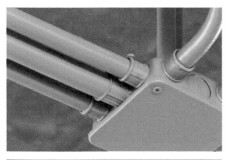

Figure 3.60 A conduit is a channel or pipe in which insulated conductors are contained.

Steel conduit

Steel conduits are made to a specification defined by BS 4568 and are either heavy gauge welded or solid drawn. Heavy gauge is made from a sheet of steel welded along the seam to form a tube and is used for most electrical installation work. Solid drawn conduit is a seamless tube, which is much more expensive and only used for special gas-tight, explosion-proof or flame proof installations.

Conduit is supplied in 3.75 m lengths and typical sizes are 16, 20, 25 and 32 mm. Conduit tubing and fittings are supplied in a black enamel finish for internal use or hot galvanized finish for use on external or damp installations. A wide range of fittings is available and the conduit is fixed using saddles or pipe hooks, as shown in Fig. 3.61.

Metal conduits are threaded with stocks and dies and bent using special bending machines. The metal conduit is also utilized as the CPC and, therefore, all connections must be screwed up tightly and all burrs removed so that cables will not be damaged as they are drawn into the conduit. Metal conduits containing a.c. circuits must contain phase and neutral conductors in the same conduit to prevent eddy currents from flowing, which would result in the metal conduit becoming hot (IET Regulations 521.5.1, 522.8.1 and 522.8.11).

Back outlet box	Terminal box	Through box	'T' or three-way box

Saddle	Space bar saddle	Distance saddle	Pipe hook, or crampet not used for surface work

Figure 3.61 Conduit fitting and saddles.

PVC conduit

PVC conduit used on typical electrical installations is heavy gauge standard impact tube manufactured to BS 4607. The conduit size and range of fittings are the same as those available for metal conduit. PVC conduit is most often joined by placing the end of the conduit into the appropriate fitting and fixing with a PVC solvent adhesive. PVC conduit can be bent by hand using a bending spring of the same diameter as the inside of the conduit. The spring is pushed into the conduit to the point of the intended bend and the conduit then bent over the knee. The spring ensures that the conduit keeps its circular shape. In cold weather, a little warmth applied to the point of the intended bend often helps to achieve a more successful bend.

The advantages of a PVC conduit system are that it may be installed much more quickly than steel conduit and is non-corrosive, but it does not have the mechanical strength of steel conduit. Since PVC conduit is an insulator it cannot be used as the CPC and a separate earth conductor must be run to every outlet.

It is not suitable for installations subjected to temperatures below 25°C or above 60°C. Where luminaires are suspended from PVC conduit boxes, precautions must be taken to ensure that the lamp does not raise the box temperature or that the mass of the luminaire supported by each box does not exceed the maximum recommended by the manufacturer (IET Regulations 522.1 and 522.2). PVC conduit also expands much more than metal conduit and so long runs require an expansion coupling to allow for conduit movement and to help prevent distortion during temperature changes.

All conduit installations must be erected first before any wiring is installed (IET Regulation 522.8.2). The radius of all bends in conduit must not cause the cables to suffer damage, and therefore the minimum radius of bends given in Table D5 of the *On-Site Guide* applies (IET Regulation 522.8.3). All conduits should terminate in a box or fitting and meet the boxes or fittings at right-angles, as shown in Fig. 3.62. Any unused conduit-box entries should be blanked off and all boxes covered with a box lid, fitting or accessory to provide complete enclosure of the conduit system. Conduit runs should be separate from other services, unless intentionally bonded, to prevent arcing from occurring from a faulty circuit within the conduit, which might cause the pipe of another service to become punctured.

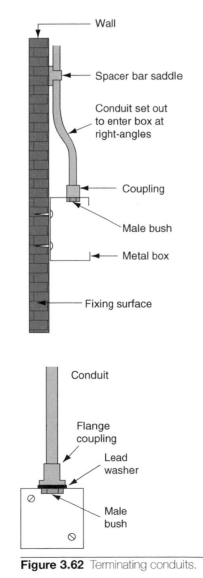

Figure 3.62 Terminating conduits.

Figure 3.63 Flexible conduit is made up of interlinked metal spirals.

Definition

Flexible conduit manufactured to BS 731-1: 1993 is made of interlinked metal spirals often covered with a PVC sleeving.

Definition

Single PVC insulated conductors are usually drawn into the installed conduit to complete the installation.

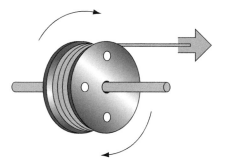

Cables *run off* will not twist; a short length of conduit can be used as an axle for the cable drum

Cables allowed to *spiral off* a drum will become twisted

Figure 3.64 Running off cable from a drum.

When drawing cables into conduit they must first be *run off* the cable drum. That is, the drum must be rotated as shown in Fig. 3.64 and not allowed to *spiral off*, which will cause the cable to twist.

Cables should be fed into the conduit in a manner that prevents any cable from crossing over and becoming twisted inside the conduit. The cable insulation must not be damaged on the metal edges of the draw-in box. Cables can be pulled in on a draw wire if the run is a long one. The draw wire itself may be drawn in on a fish tape, which is a thin spring steel or plastic tape.

A limit must be placed on the number of bends between boxes in a conduit run and the number of cables that may be drawn into a conduit to prevent the cables from being strained during wiring. Appendix E of the *On-Site Guide* gives a guide to the cable capacities of conduits and trunking.

Flexible conduit

Flexible conduit manufactured to BS 731-1: 1993 is made of interlinked metal spirals often covered with a PVC sleeving. The tubing must not be relied upon to provide a continuous earth path and, consequently, a separate CPC must be run either inside or outside the flexible tube (IET Regulation 543.2.7).

Flexible conduit is used for the final connection to motors so that the vibrations of the motor are not transmitted throughout the electrical installation and to allow for modifications to be made to the final motor position and drive belt adjustments.

Conduit capacities

Single PVC insulated conductors are usually drawn into the installed conduit to complete the installation. Having decided upon the type, size and number of cables required for a final circuit, it is then necessary to select the appropriate size of conduit to accommodate those cables. The objective is to fill any enclosure such as conduit to a maximum of 45% conductor space, which means that the remaining 55% is left to air space. This reduces the cumulative heating effect.

The tables in Appendix E of the *On-Site Guide* describe a 'factor system' for determining the size of conduit required to enclose a number of conductors. The tables are shown in Tables 3.5 and 3.6 and use of these tables ensures that the 45% rule mentioned above is adhered to. The method is as follows:

Table 3.5 Conduit cable factors

Cable factors for conduit in long straight runs over 3 m, or runs of any length incorporating bends	
Conductor CSA (mm²)	**Cable factor**
1	16
1.5	22
2.5	30
4	43
6	58
10	105
16	145

Table 3.6 Conduit factors

Length of run (m)	Conduit diameter (mm)											
	16	20	25	32	16	20	25	32	16	20	25	32
	Straight				One bend				Two bends			
3.5	179	290	521	911	162	263	475	837	136	222	404	720
4	177	286	514	900	158	256	463	818	130	213	388	692
4.5	174	282	507	889	154	250	452	800	125	204	373	667
5	171	278	500	878	150	244	442	783	120	196	358	643
6	167	270	487	857	143	233	422	750	111	182	333	600
7	162	263	475	837	136	222	404	720	103	169	311	563
8	158	256	463	818	130	213	388	692	97	159	292	529
9	154	250	452	800	125	204	373	667	91	149	275	500
10	150	244	442	783	120	196	358	643	86	141	260	474

- Identify the cable factor for the particular size of conductor; see Table 3.5.
- Multiply the cable factor by the number of conductors, to give the sum of the cable factors.
- Identify the appropriate part of the conduit factor table given by the length of run and number of bends (see Table 3.6).
- The correct size of conduit to accommodate the cables is that conduit which has a factor equal to or greater than the sum of the cable factors.

Example 1

Six 2.5mm² PVC insulated cables are to be run in a conduit containing two bends between boxes 10m apart. Determine the minimum size of conduit to contain these cables.
From Table 3.5:

$$\text{The factor for one 2.5mm}^2 \text{ cable} = 30$$
$$\text{The sum of the cable factors} = 6 \times 30$$
$$= 180$$

From Table 3.5, a 25mm conduit, 10m long and containing two bends, has a factor of 260. A 20mm conduit containing two bends only has a factor of 141 which is less than 180, the sum of the cable factors, and, therefore, 25mm conduit is the minimum size to contain these cables.

Example 2

Ten 1.0 mm² PVC insulated cables are to be drawn into a plastic conduit which is 6 m long between boxes and contains one bend. A 4.0 mm PVC insulated CPC is also included. Determine the minimum size of conduit to contain these conductors.

From Table 3.5:

The factor for one 1.0 mm cable = 16

The factor for one 4.0 mm cable = 43

The sum of the cable factors = $(10 \times 16) + (1 \times 43)$

$$= 203$$

From Table 3.5, a 20 mm conduit, 6 m long and containing one bend, has a factor of 233. A 16 mm conduit containing one bend only has a factor of 143 which is less than 203, the sum of the cable factors, and, therefore, 20 mm conduit is the minimum size to contain these cables.

Key fact

Trunking is generally easier to install additional cables to an existing circuit in comparison with conduit. Access only requires the removal of the protective lid in comparison to having to draw in further cables.

Definition

A *trunking* is an enclosure provided for the protection of cables that is normally square or rectangular in cross-section, having one removable side. Trunking may be thought of as a more accessible conduit system.

Definition

Metallic trunking is formed from mild steel sheet, coated with grey or silver enamel paint for internal use or a hot-dipped galvanized coating where damp conditions might be encountered.

Trunking installations

A trunking is an enclosure provided for the protection of cables that is normally square or rectangular in cross-section, having one removable side. Trunking may be thought of as a more accessible conduit system, and for industrial and commercial installations it is replacing the larger conduit sizes. A trunking system can have great flexibility when used in conjunction with conduit; the trunking forms the background or framework for the installation, with conduits running from the trunking to the point controlling the current-using apparatus. When an alteration or extension is required it is easy to drill a hole in the side of the trunking and run a conduit to the new point. The new wiring can then be drawn through the new conduit and the existing trunking to the supply point.

Trunking is supplied in 3 m lengths and various cross-sections measured in millimetres from 50 × 50 up to 300 × 150. Most trunking is available in either steel or plastic.

Metallic trunking

Metallic trunking is formed from mild steel sheet, coated with grey or silver enamel paint for internal use or a hot-dipped galvanized coating where damp conditions might be encountered and made to a specification defined by BS EN 500 85. A wide range of accessories is available, such as 45° bends, 90° bends, tee and four-way junctions, for speedy on-site assembly. Alternatively, bends may be fabricated in lengths of trunking, as shown in Fig. 3.66. This may be necessary or more convenient if a bend or set is non-standard, but it does take more time to fabricate bends than merely to bolt on standard accessories.

Insulated non-sheathed cables are permitted in a trunking system which provides at least the degree of protection IPXXD (which means total protection) or IP4X which means protection from a solid object greater than 1.0 mm such as a thin wire or strip. For site fabricated joints such as that shown in Fig 3.66, the installer must confirm that the completed item meets at least IPXXD (IET Regulation 521.10).

When fabricating bends the trunking should be supported with wooden blocks for sawing and filing, in order to prevent the sheet-steel from vibrating or becoming deformed. Fish-plates must be made and riveted or bolted to the trunking to form a solid and secure bend. When manufactured bends are used, the continuity of the earth path must be ensured across the joint by making all fixing screw connections very tight, or fitting a separate copper strap between the trunking and the standard bend. If an earth continuity test on the trunking is found to be unsatisfactory, an insulated CPC must be installed inside the trunking. The size of the protective conductor will be determined by the largest cable contained in the trunking, as described in Table 54.7 of the IET Regulations. If the circuit conductors are less than 16 mm², then a 16 mm² CPC will be required.

Non-metallic trunking

Trunking and trunking accessories are also available in high-impact PVC. The accessories are usually secured to the lengths of trunking with a PVC solvent adhesive. PVC trunking, like PVC conduit, is easy to install and is non-corrosive.

A separate CPC will need to be installed and non-metallic trunking may require more frequent fixings because it is less rigid than metallic trunking. All trunking fixings should use round-headed screws to prevent damage to cables since the thin sheet construction makes it impossible to countersink screw heads. The new fire proof support for non metallic cables systems must now be taken into consideration. See Choosing an Appropriate Wiring system earlier in this chapter.

Mini-trunking

Mini-trunking is very small PVC trunking, ideal for surface wiring in domestic and commercial installations such as offices. The trunking has a cross-section of 16 × 16 mm, 25 × 16 mm, 38 × 16 mm or 38 × 25 mm and is ideal for switch drops or for housing auxiliary circuits such as telephone or audio equipment wiring. The modern square look in switches and sockets is complemented by the mini-trunking, which is very easy to install (see Fig. 3.65).

Skirting trunking

Skirting trunking is a trunking manufactured from PVC or steel in the shape of a skirting board and is frequently used in commercial buildings such as hospitals, laboratories and offices. The trunking is fitted around the walls of a room at either the skirting board level or at the working surface level and contains the wiring for socket outlets and telephone points which are mounted on the lid, as shown in Fig. 3.65.

Top tip

When sawing metals, conduit trunking or tray, a safe and effective speed is 50 strokes per minute.

Definition

Multi-compartment metallic trunking is used to segregate Band I and II cables as well as cabling associated with emergency systems.

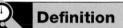

Definition

Mini-trunking is very small PVC trunking, ideal for surface wiring in domestic and commercial installations such as offices.

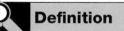

Definition

Skirting trunking is a trunking manufactured from PVC or steel in the shape of a skirting board and is frequently used in commercial buildings such as hospitals, laboratories and offices.

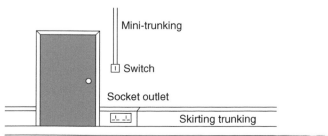

Mini-trunking

Switch

Socket outlet

Skirting trunking

Figure 3.65 Typical installation of skirting trunking and mini-trunking.

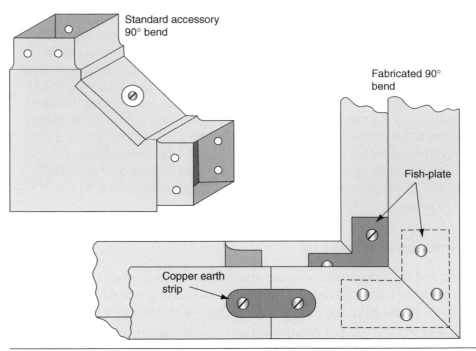

Figure 3.66 Alternative trunking bends.

Where any trunking passes through walls, partitions, ceilings or floors, short lengths of lid should be fitted so that the remainder of the lid may be removed later without difficulty. Any damage to the structure of the buildings must be made good with mortar, plaster or concrete in order to prevent the spread of fire.

Fire barriers must be fitted inside the trunking every 5 m, or at every floor level or room-dividing wall if this is a shorter distance, as shown in Fig. 3.67(a).

Where trunking is installed vertically, the installed conductors must be supported so that the maximum unsupported length of non-sheathed cable does not exceed 5 m. Figure 3.67(b) shows cables woven through insulated pin supports, which is one method of supporting vertical cables.

PVC insulated cables are usually drawn into an erected conduit installation or laid into an erected trunking installation. Table E4 of the *On-Site Guide* only gives factors for conduits up to 32 mm in diameter, which would indicate that conduits larger than this are not in frequent or common use. Where a cable enclosure greater than 32 mm is required because of the number or size of the conductors, it is generally more economical and convenient to use trunking.

Trunking capacities

The ratio of the space occupied by all the cables in a conduit or trunking to the whole space enclosed by the conduit or trunking is known as the space factor.

Where sizes and types of cable and trunking are not covered by the tables in the *On-Site Guide*, a space factor of 45% must not be exceeded. This means that the cables must not fill more than 45% of the space enclosed by the trunking.

The tables take this factor into account.

Definition

The *ratio* of the space occupied by all the cables in a conduit or trunking to the whole space enclosed by the conduit or trunking is known as the *space factor*.

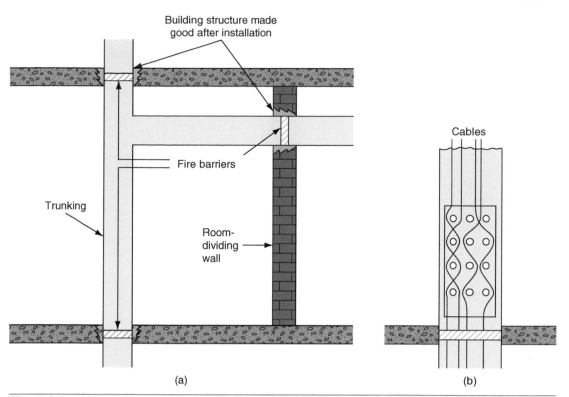

Figure 3.67 Installation of trunking (a) fire barriers in trunking, and (b) cable supports in vertical trunking.

To calculate the size of trunking required to enclose a number of cables:

- Identify the cable factor for the particular size of conductor (see Table 3.7).
- Multiply the cable factor by the number of conductors to give the sum of the cable factors.
- Consider the factors for trunking shown in Table 3.8. The correct size of trunking to accommodate the cables is trunking that has a factor equal to, or greater than, the sum of the cable factors.

Table 3.7 Trunking cable factors

Type of conductor	Conductor CSA (mm²)	PVC Cable factor	Thermosetting Cable factor
Solid	1.5	8	8.6
	2.5	11.9	11.9
Stranded	1.5	8.6	9.6
	2.5	12.6	13.9
	4	16.6	18.1
	6	21.2	22.9
	10	35.3	36.3

Table 3.8 Trunking factors

Dimensions of trunking (mm × mm)	Factor
50 × 38	767
50 × 50	1037
75 × 25	738
75 × 38	1146
75 × 50	1555
75 × 75	2371
100 × 25	993
100 × 38	1542
100 × 50	2091
100 × 75	3189
100 × 100	4252
150 × 38	2999
150 × 50	3091
150 × 75	4743
150 × 100	6394
150 × 150	9697

Example 3

Calculate the minimum size of trunking required to accommodate the following single-core PVC cables:

 20 × 1.5 mm solid conductors

 20 × 2.5 mm solid conductors

 21 × 4.0 mm stranded conductors

 16 × 6.0 mm stranded conductors

From Table 3.7, the cable factors are:

 for 1.5 mm solid cable – 8.0
 for 2.5 mm solid cable – 11.9
 for 4.0 mm stranded cable – 16.6
 for 6.0 mm stranded cable – 21.2

The sum of the cable terms is:

$(20 \times 8.0) + (20 \times 11.9) + (21 \times 16.6) + (16 \times 21.2) + 1085.8$. From Table 3.8, 75 × 38 mm trunking has a factor of 1146 and, therefore, the minimum size of trunking to accommodate these cables is 75 × 38 mm, although a larger size, say, 75 × 50 mm, would be equally acceptable if this was more readily available as a standard stock item.

Segregation of circuits

Where an installation comprises a mixture of low-voltage and very low-voltage circuits such as mains lighting and power, fire alarm and telecommunication circuits, they must be separated or *segregated* to prevent electrical contact (IET Regulation 528.1).

For the purpose of these regulations various circuits are identified by one of two bands as follows:

Band I: telephone, radio, bell, call and intruder alarm circuits, emergency circuits for fire alarm and emergency lighting.

Band II: mains voltage circuits.

When Band I circuits are insulated to the same voltage as Band II circuits, they may be drawn into the same compartment.

When trunking contains rigidly fixed metal barriers along its length, the same trunking may be used to enclose cables of the separate bands without further precautions, provided that each band is separated by a barrier, as shown in Fig. 3.68.

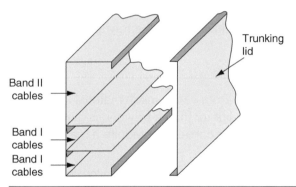

Trunking lid

Band II cables

Band I cables

Band I cables

Figure 3.68 Segregation of cables in trunking.

Multi-compartment PVC trunking cannot provide band segregation since there is no metal screen between the bands. This can only be provided in PVC trunking if screened cables are drawn into the trunking.

Cable tray installations

Cable tray is a sheet-steel channel with multiple holes. The most common finish is hot-dipped galvanized but PVC-coated tray is also available. It is used extensively on large industrial and commercial installations for supporting MI and SWA cables that are laid on the cable tray and secured with cable ties through the tray holes.

A cable tray should be adequately supported during installation by brackets that are appropriate for the particular installation. The tray should be bolted to the brackets with round-headed bolts and nuts, with the round head inside the tray so that cables drawn along the tray are not damaged.

The tray is supplied in standard widths from 50 to 900 mm, and a wide range of bends, tees and reducers is available. Fig. 3.69 shows a factory-made 90° bend at B. The tray can also be bent using a cable tray bending machine to create bends such as that shown at A in Fig. 3.69. The installed tray should be securely bolted with round-headed bolts where lengths or accessories are attached, so

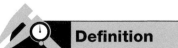

Definition

Cable tray is a sheet-steel channel with multiple holes. The most common finish is hot-dipped galvanized, but PVC-coated tray is also available. It is used extensively on large industrial and commercial installations for supporting MI and SWA cables that are laid on the cable tray and secured with cable ties through the tray holes.

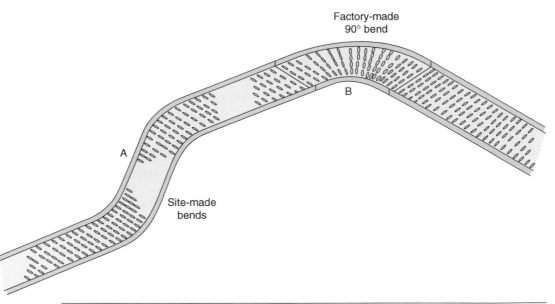

Figure 3.69 Cable tray with bends.

that there is a continuous earth path which may be bonded to an electrical earth. The whole tray should provide a firm support for the cables, and therefore the tray fixings must be capable of supporting the weight of both the tray and cables.

Cable basket installations

Definition

Cable basket is generally used to support data cables or emergency cables.

A cable basket is a form of cable tray made from a wire mesh material looking similar to a supermarket trolley. A basket is lighter than a tray and is generally used when there are numerous smaller, lighter cables that require a management system.

Cables are generally laid inside the basket rather than pulled in, allowing a quicker, more convenient installation.

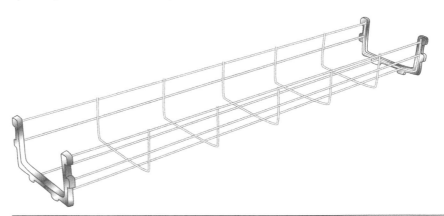

Figure 3.70 Cable basket.

A cable basket is much lighter than a tray and easier to install as well as being easier to carry out alterations to cabling after the installation due to the open nature of the basket.

Definition

Cable ladder tends to be used to support steel wire armour cables.

Cable ladder

Cable ladder is at the other end of the scale to cable basket and is a heavy duty version of support, constructed as the name suggests to look like a ladder.

Cable ladder is used to support heavier cables that do not need a continuous surface to be fixed to, but can span the spaces between the rungs. As these cables are usually high capacity cables, this type of support is usually used in industrial and large commercial installations.

As the cables are capable of carrying high currents, large forces can be exerted on the cables, especially under fault conditions. Due to this, cable clamps are used to attach the cable to the ladder to ensure a sufficiently robust fixing.

Cable ducting

Within the electrical industry, underground cable routes often need to be reusable to allow for alterations in the installation. Heavy duty corrugated plastic pipes are installed to facilitate this rather than bury a cable directly into the ground.

Within control panels a form of trunking with multiple outlets like a comb is also known as ducting. The ducting allows routeing of the cabling while also containing it in a neat manner.

Figure 3.71 Cable ladder.

Figure 3.72 Heavy duty corrugated plastic pipes.

Figure 3.73 Cable ducting used in panel wiring.

Modular wiring systems

Modular wiring systems are growing in popularity with designers and installers as a result of the reduction in installation times that can be up to 70% less than traditional methods. The idea of modular wiring is the provision of a factory manufactured and tested system that is fast to fit. The biggest benefit is the system can be manufactured off site in a quality controlled environment and supplied as a pluggable system, reducing onsite connections.

Many large projects such as the Olympics have used modular construction techniques.

Busbar and power track systems

In an electrical system, a busbar is usually a flat uninsulated copper strip contained inside switchgear or panels for carrying high currents. The flat shape of the conductor gives a large surface area that allows for heat dissipation and for circuits to branch off along the length.

Power track systems are similar to busbar systems in that they have conductors running through the containment and loads can be tapped off at multiple points. Whereas busbar systems are for high current distribution, power track systems are commonly used in offices under floors and similar installations feeding

Definition

Modular wiring systems are known as plug and play, and save on installation times since they have already been fully tested.

Definition

Busbar systems are used when high current usage is required. However, a disadvantage is that its many bolts and connections require maintenance tests to stay efficient.

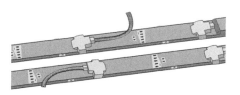

Figure 3.74 Power track system.

individual circuits such as socket outlets in floor boxes which provide flexibility for office reorganizations.

PVC/SWA cable installations

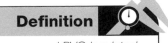

Steel wire armoured PVC insulated cables are now extensively used on industrial installations and often laid on cable tray. This type of installation has the advantage of flexibility, allowing modifications to be made speedily as the need arises. The cable has a steel wire armouring giving mechanical protection and permitting it to be laid directly in the ground or in ducts, or it may be fixed directly or laid on a cable tray. Figure 3.80 shows a PVC/SWA cable.

It should be remembered that when several cables are grouped together the current rating will be reduced according to the correction factors given in Appendix 4 (Table 4C1) of the IET Regulations.

The cable is easy to handle during installation, is pliable and may be bent to a radius of eight times the cable diameter. The PVC insulation would be damaged if installed in ambient temperatures over 70°C or below 0°C, but once installed the cable can operate at low temperatures.

The cable is terminated with a simple gland that compresses a compression ring onto the steel wire armouring to provide the earth continuity between the switchgear and the cable.

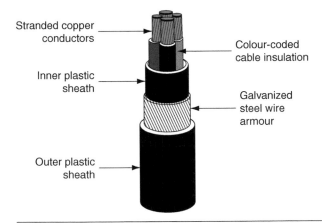

Figure 3.75 A four-core PVC/SWA cable.

MI cable installations

Mineral insulated cables are available for general wiring as:

- light-duty MI cables for voltages up to 600 V and sizes from 1.0 to 10 mm;
- heavy-duty MI cables for voltages up to 1000 V and sizes from 1.0 to 150 mm.

Figure 3.81 shows an MI cable and termination.

The cables are available with bare sheaths or with a PVC oversheath. The cable sheath provides sufficient mechanical protection for all but the most severe situations, where it may be necessary to fit a steel sheath or conduit over the cable to give extra protection, particularly near floor level in some industrial situations.

The cable may be laid directly in the ground, in ducts, on cable tray or clipped directly to a structure. It is not affected by water, oil or the cutting fluids used in engineering and can withstand very high temperatures or even fire. The cable diameter is small in relation to its current-carrying capacity and it should last indefinitely if correctly installed because it is made from inorganic materials.

These characteristics make the cable ideal for Band I emergency circuits, boiler houses, furnaces, petrol stations and chemical plant installations.

The cable is supplied in coils and should be run off during installation and not spiralled off, as described in Fig. 3.44 for conduit. The cable can be work hardened if over-handled or over-manipulated. This makes the copper outer sheath stiff and may result in fracture. The outer sheath of the cable must not be penetrated, otherwise moisture will enter the magnesium oxide insulation and lower its resistance. To reduce the risk of damage to the outer sheath during installation, cables should be straightened and formed by hammering with a hide hammer or a block of wood and a steel hammer. When bending MI cables the radius of the bend should not cause the cable to become damaged and clips should provide adequate support (IET Regulation 522.8.5); see Table D5 of the *On-Site Guide*.

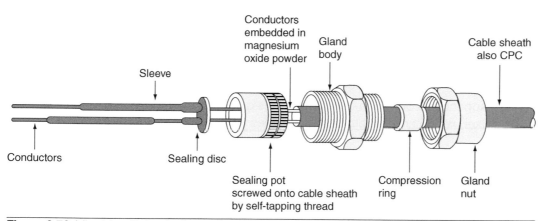

Figure 3.76 MI cable with terminating seal and gland.

The cable must be prepared for termination by removing the outer copper sheath to reveal the copper conductors. This can be achieved by using a rotary stripper tool or, if only a few cables are to be terminated, the outer sheath can be removed with side cutters, peeling off the cable in a similar way to peeling the skin from a piece of fruit with a knife. When enough conductor has been revealed, the outer sheath must be cut off square to facilitate the fitting of the sealing pot, and this can be done with a ringing tool. All excess magnesium oxide powder must be wiped from the conductors with a clean cloth. This is to prevent moisture from penetrating the seal by capillary action.

Cable ends must be terminated with a special seal to prevent the entry of moisture. Figure 3.76 shows a brass screw-on seal and gland assembly, which allows termination of the MI cables to standard switchgear and conduit fittings.

The sealing pot is filled with a sealing compound, which is pressed in from one side only to prevent air pockets from forming, and the pot closed by crimping home the sealing disc with an MI crimping tool.

Such an assembly is suitable for working temperatures up to 105°C. Other compounds or powdered glass can increase the working temperature up to 250°C.

The conductors are not identified during the manufacturing process and so it is necessary to identify them after the ends have been sealed. A simple continuity

Definition

Mineral-insulated seals are manufactured so that they can be matched to different explosive environments depending if the exposure is: continuous, likely or unlikely to occur.

or polarity test, as described later in this chapter, can identify the conductors, which are then sleeved or identified with coloured markers.

Connection of MI cables can be made directly to motors, but to absorb the vibrations a 360° loop should be made in the cable just before the termination. If excessive vibration is expected, the MI cable should be terminated in a conduit through box and the final connection made by flexible conduit.

Copper MI cables may develop a green incrustation or patina on the surface, even when exposed to normal atmospheres. This is not harmful and should not be removed. However, if the cable is exposed to an environment that might encourage corrosion, an MI cable with an overall PVC sheath should be used.

Selecting the appropriate type of wiring system for the environment

Chapter 52 of BS 7671 places requirements on the designer to select the appropriate wiring system for the type of installation giving consideration to:

- cables and conductors;
- their connections, terminations and/or joints;
- their associated supports or suspensions; and
- their enclosures or methods of protection against external influences.

Appendix C of the *On-Site Guide* gives guidance on the application of cables for fixed wiring and flexible cables in Tables C1 and C2.

Figure 3.77 Industrial plugs are often referred to as 'commando plugs'.

Industrial plugs, sockets and couplers

In 1968, the UK adopted the IEC 309 standard as BS 4343, and in 1999 replaced it with the European equivalent BS EN 60309.

Electricians often refer to these plugs as 'commando plugs' (refers to the MK Electric Company Commando range of connectors). The standard covers plugs, socket-outlets and couplers for industrial purposes.

IEC 60309-2 specifies a range of mains power connectors with circular housings, and different numbers and arrangements of pins for different applications. The 16 A single and three-phase variants are commonly used throughout Europe at campsites, marinas, workshops and farms. The colour of an IEC 60309 plug or socket indicates its voltage rating and the most common colours in use are yellow (110 V), blue (230 V) and red (400 V).

Cables to BS 7919: 2001 (2006)

Electric cables — Flexible cables for use with appliances and equipment intended for industrial and similar environments. Flexible cable, manufactured to BS 7919 Table 44 (not harmonized), commonly referred to as 3183A (Arctic Grade Flex), was specifically designed for use at 110 V a.c. from centre tapped transformers (55 V - 0 - 55 V).

The practice of using a 110 V centre tapped transformer is a UK practice, hence the lack of European harmonization of the standard.

The key feature of this cable is that it is designed to be suitable for installation and handling down to a temperature of −25°C, e.g. suitable for construction site installations.

As the standard applies to 110 V flexes only; strictly speaking, only a yellow flex should be referred to as an 'Artic' cable, however manufacturers supply other colours for other voltages such as blue for 230 V and often mark these as 'Artic' cables.

Terminating and connecting conductors

The entry of a cable end into an accessory, enclosure or piece of equipment is what we call a termination. Section 526 of the IET Regulations tells us that:

1 Every connection between conductors and equipment shall be durable, provide electrical continuity and mechanical strength and protection.
2 Every termination and joint in a live conductor shall be made within a suitable accessory, piece of equipment or enclosure that complies with the appropriate product standard.
3 Every connection shall be accessible for inspection, testing and maintenance.
4 The means of connection shall take account of the number and shape of the wires forming the conductor.
5 The connection shall take account of the cross-section of the conductor and the number of conductors to be connected.
6 The means of connection shall take account of the temperature attained in normal service.
7 There must be no mechanical strain on the conductor connections.

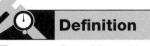

Definition

The entry of a cable end into an accessory, enclosure or piece of equipment is what we call a termination.

There is a wide range of suitable means of connecting conductors and we shall look at these in a moment. Whatever method is used to connect live conductors, the connection must be contained in an enclosed compartment such as an accessory; for example, a switch or socket box or a junction box. Alternatively, an equipment enclosure may be used; for example, a motor enclosure or an enclosure partly formed by non-combustible building material (IET Regulation 526.5). This is because faulty joints and terminations in live conductors can attain very high temperatures due to the effects of resistive heating. They might also emit arcs, sparks or hot particles with the consequent risk of fire or other harmful thermal effects to adjacent materials.

Types of terminal connection

Junction boxes

Junction boxes are probably the most popular method of making connections in domestic properties. Brass terminals are fixed inside a bakelite container. The two important factors to consider when choosing a junction box are the number of terminals required and the current rating. Socket outlet junction boxes have larger brass terminals than lighting junction boxes. See Fig. 3.78.

Key fact

Junction boxes are probably the most popular method of making connections in domestic properties.

Strip connectors

Strip connectors or a chocolate block is a very common method of connecting conductors. The connectors are mounted in a moulded plastic block in strips of 10 or 12. The conductors are inserted into the block and secured with the grub-screw. In order that the conductors do not become damaged, the screw connection must be firm but not overtightened. The size used should relate to the current rating of the circuit. Figure 3.84 shows a strip connector.

Key fact

Strip connectors or a chocolate block is a very common method of connecting conductors.

Pillar terminal

A pillar terminal is a brass pillar with a hole through the side into which the conductor is inserted and secured with a set-screw. If the conductor is small in relation to the hole it should be doubled back. In order that the conductor does not become damaged, the screw connection should be tight but not overtightened. Fig. 3.79 shows a pillar terminal.

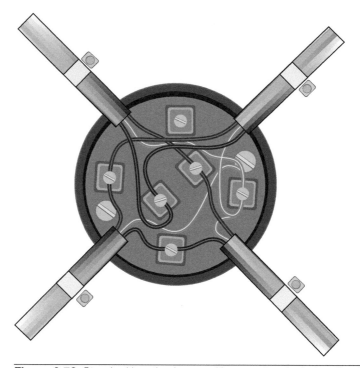

Figure 3.78 Standard junction boxes with screw terminals, junction box must be fixed and cables clamped.

Screwhead, nut and washer terminals

The conductor being terminated is formed into an eye, as shown in Fig. 3.79. The eye should be slightly larger than the screw shank but smaller than the outside diameter of the screwhead, nut or washer. The eye should be placed on the screw shank in such a way that the rotation of the screwhead or nut will tend to close the joint in the eye.

Claw washers

In order to avoid inappropriate separation or spreading of individual wires of multiwire, claw washers are used to obtain a good sound connection. The looped conductor is laid in the pressing as shown in Fig. 3.79, a plain washer is placed on top of the conductor and the metal points folded over the washer. When terminating very fine multiwire conductors, see also 526.9 of the IET Regulations.

Crimp terminals

Crimp terminals are made of tinned sheet copper. The chosen crimp terminal is slipped over the end of the conductor and crimped with the special crimping tool. This type of connection is very effective for connecting protective bonding conductors to approved earth clamps.

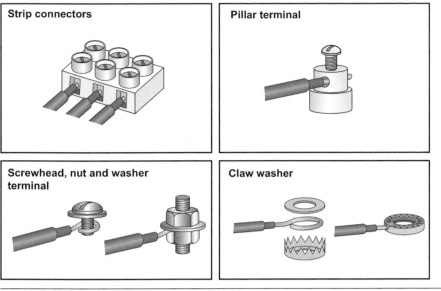

Strip connectors	Pillar terminal
Screwhead, nut and washer terminal	Claw washer

Figure 3.79 Types of terminal.

Soldered joints or compression joints

Although the soldering of large underground cables is still common today, joints up to about 100A are now usually joined with a compression joint. This uses the same principle as for the crimp termination above; it is just a little larger.

If a large SWA cable must be connected and the joint placed in a position which will be inaccessible for future inspection and testing, then a compression joint encased in a resin compound filled jacket will probably provide a solution.

Regulation 526.3 tells us that every connection must be accessible for inspection and testing except for the following:

1 A joint which is designed to be buried underground such as the one described above.

2 A compound filled or encapsulated joint.

3 A maintenance-free junction box marked with the symbol MF, as shown in Fig. 3.80.

The introduction of this maintenance-free junction box was a small but important change made by Amendment No 1: 2011 of the IET Regulations.

There has always been a debate as to when a junction box is accessible for inspection and testing. Is it accessible when installed under floorboards? Is a screwed down floorboard accessible? Does a fitted carpet make a difference, and who will know where it is?

A maintenance-free junction box does not have to be accessible for inspection and testing, and this new junction box will provide a solution to those difficult and unavoidable situations where there is doubt as to whether a junction box is inaccessible.

Whatever method is used to make the connection in conductors, the connection must be both electrically and mechanically sound if we are to avoid high-resistance joints, corrosion and erosion at the point of termination.

Key fact

Whatever method is used to make the connection in conductors, the connection must be both electrically and mechanically sound if we are to avoid high-resistance joints, corrosion and erosion at the point of termination.

All three images are reproduced here with permission from Wago.

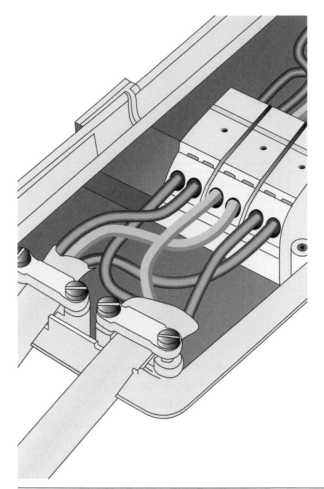

Figure 3.80 A maintenance-free junction box.

Safe terminations and connections

To ensure that all electrical terminations and connections are safe, the installing electrician should give consideration to the following good practice points:

- all connections must be both electrically and mechanically secure;
- all connections must be long lasting and not fail quickly;
- the method of connection must take account of:
  - the size of conductor and, therefore, the current carrying capacity of that conductor;
  - the material of the conductor; copper is a soft metal but aluminium is softer;
  - the number of conductors being connected;
  - the temperature to be attained at the point of connection in normal service;
  - the provision of adequate locking arrangements in situations subject to vibration.
- every connection must remain accessible for inspection and testing unless designed to be maintence free;
- every connection in a live conductor must be made within:
  - a suitable accessory such as a switch, socket ceiling rose or joint box; or
  - an equipment enclosure such as a luminaire; or
  - a non-combustible enclosure designed for this purpose;
- there must be no mechanical strain put on the conductors or connections.

Section 526 of the IET Regulations deals with electrical connections.

Figure 3.81 Both conductors and insulators are found in electrical cable.

132.15 Isolation and switching

Part 4 of the IET Regulations deals with the application of protective measures for safety and Chapter 53 with the regulations for switching devices or switchgear required for protection, isolation and switching of a consumer's installation.

The consumer's main switchgear must be readily accessible to the consumer and be able to:

- isolate the complete installation from the supply;
- protect against overcurrent;
- cut off the current in the event of a serious fault occurring.

The regulations identify four separate types of switching: switching for isolation, switching for mechanical maintenance, emergency switching and functional switching.

Isolation is defined as cutting off the electrical supply to a circuit or item of equipment in order to ensure the safety of those working on the equipment by making dead those parts that are live in normal service.

The purpose of isolation switching is to enable electrical work to be carried out safely on an isolated circuit or piece of equipment. Isolation is intended for use by electrically skilled or supervised persons.

An isolator is a mechanical device which is operated manually and used to open or close a circuit off load. An isolator switch must be provided close to the supply point so that all equipment can be made safe for maintenance. Isolators for motor circuits must isolate the motor and the control equipment, and isolators for discharge lighting luminaires must be an integral part of the luminaire so that it is isolated when the cover is removed or be provided with effective local isolation (IET Regulation 537.2). Devices that are suitable for isolation are isolation switches, fuse links, circuit-breakers, plugs and socket outlets. They must isolate all live supply conductors and provision must be made to secure the isolation (IET Regulation 537.2.5).

Definition

Subtle difference: A *switch* is an on-load device. An *isolator* is an off-load device.

Definition

Isolation is defined as cutting off the electrical supply to a circuit or item of equipment in order to ensure the safety of those working on the equipment by making dead those parts which are live in normal service.

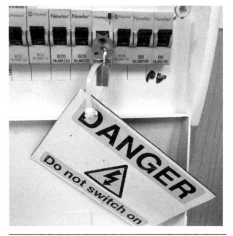

Figure 3.83 This circuit is now isolated and locked off.

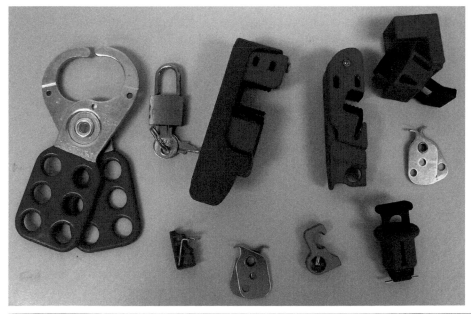

Figure 3.82 There are several different types of tool that can be used to 'lock-off' circuits.

Isolation at the consumer's service position can be achieved by a double pole switch that opens or closes all conductors simultaneously. On three-phase supplies the switch need only break the live conductors with a solid link in the neutral, provided that the neutral link cannot be removed before opening the switch.

The switching for mechanical maintenance requirements is similar to that for isolation except that the control switch must be capable of switching the full load current of the circuit or piece of equipment.

The purpose of switching for mechanical maintenance is to enable non-electrical work to be carried out safely on the switched circuit or equipment.

Mechanical maintenance switching is intended for use by skilled but nonelectrical persons. Switches for mechanical maintenance must be manually operated, not have exposed live parts when the appliance is opened, must be connected in the main electrical circuit and have a reliable on/off indication or visible contact gap (IET Regulation 537.3.3). Devices that are suitable for switching off for mechanical maintenance are switches, circuit-breakers, plugs and socket outlets.

Emergency switching involves the rapid disconnection of the electrical supply by a single action to remove or prevent danger. The purpose of emergency switching is to cut off the electrical energy *rapidly* to remove an unexpected hazard.

Emergency switching is for use by anyone. The device used for emergency switching must be immediately accessible and identifiable, and be capable of cutting off the full load current (see Fig. 3.84).

Electrical machines must be provided with a means of emergency switching, and a person operating an electrically driven machine must have access to an emergency switch so that the machine can be stopped in an emergency. A remote stop/start or emergency stop arrangement can be added to a direct on line starter to meet the requirements for an electrically driven machine as required by IET Regulation 537.3.3. Devices that are suitable for emergency switching are switches, circuit-breakers and contactors. Where contactors are operated by remote control they should *open* when the coil is de-energized, that is, fail safe. Pushbuttons used for emergency switching must be coloured red and latch in the stop or off position. They should be installed where danger may arise and be clearly identified as emergency switches. Plugs and socket outlets cannot be considered appropriate for emergency disconnection of supplies.

Functional switching involves the switching on or off, or varying the supply, of electrically operated equipment in normal service. The purpose of functional switching is to provide control of electrical circuits and equipment in normal service. Functional switching is for the user of the electrical installation or equipment. The Building Regulations require that switches and sockets be installed so that all persons, including those whose reach is limited, can reach them. The guidance is shown in Fig. 3.85 and applies to all new dwellings but not to rewires.

However, these recommendations will undoubtedly 'influence' the rewiring of existing dwellings. The device must be capable of interrupting the total steady current of the circuit or appliance. When the device controls a discharge lighting circuit it must have a current rating capable of switching an inductive load.

Definition

The *switching for mechanical maintenance* requirement is similar to that for isolation except that the control switch must be capable of switching the full load current of the circuit or piece of equipment.

Definition

Emergency switching involves the rapid disconnection of the electrical supply by a single action to remove or prevent danger.

Definition

Functional switching involves the switching on or off, or varying the supply, of electrically operated equipment in normal service.

Definition

The *nominated heights* of switches and sockets are determined through Part M (Access to buildings) of the Building Regulations not Part P. Part P is electrical safety.

The regulations acknowledge the growth in the number of electronic dimmer switches being used for the control and functional switching of lighting circuits. The functional switch must be capable of performing the most demanding duty it may be called upon to perform (IET Regulations 537.1 and 2).

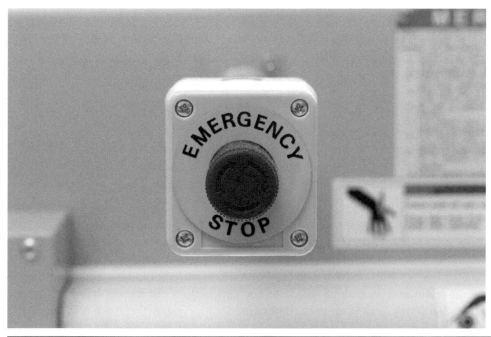

Figure 3.84 Emergency stop control.

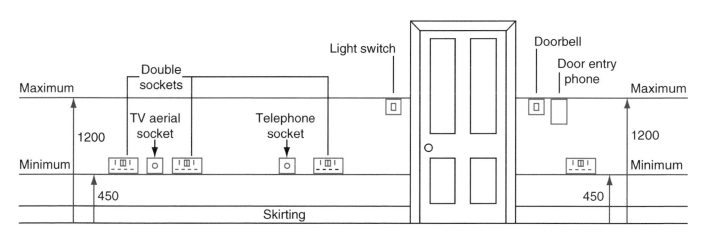

Figure 3.85 Fixing positions of switches and socket outlets.

Protection against fire

Chapter 42 in the wiring regulations contains information regarding protection against fire. The aim of this chapter therefore is to provide directives in order to prevent both fires and burns emanating from the installation of electrical equipment. It does so by examining actual ignition sources, how fire is spread and how burns are produced through processes such as arcing. Overall the heat generated through the equipment must not cause danger or harmful effects, including positioning fixed equipment that emanate heat within a sufficient distance away from any surface (421.1.4). The 18th Edition of the Regulations

Key fact

Amendment 3 of BS7671 states that consumer units and similar switchgear assemblies in domestic premises have to be of non-combustible materials such as metal.

now recommends the use of metal consumer units in household premises. Also, arc fault detection devises (AFDD) are 'recommended' as an additional means of protection against fire. (IET Regulations 421.1.201 and 421.1.7). Other considerations inherent on a designer of an installation concern the use of high-risk equipment such as welding equipment in that not only should consideration be given to equipment guarding but also the material used in the surrounding area should be scrutinized regarding if it is made from materials that might promote the fire. Equally, use of flammable liquid, which is defined by any material that can easily combust, must be positioned so that the burning liquid is contained within the area of use and is not allowed to escape and enlarge the fire.

When the possibility exists then the equipment should be: screened, mounted or supported within appropriate enclosures or from materials constructed with low thermal conductance (421.1.2). Where arcing is possible the equipment needs to be totally enclosed or protected by arc-resistant material or be positioned away from any material that the emission could have a detrimental effect on (421.1.3).

There are several interlinking elements within the wiring regulations concerning protection against fire. Regulation 532.2 states that where it is necessary to limit the effect of a fault condition from the perspective of a fire then an RCD must be fitted not only to comply with fault protection but should be installed at the circuit origin and be able to switch all live conductors but its rating shall not exceed 300 mA. This is an important consideration regarding fire alarm circuits, since there is potentially a conflict regarding meeting regulation 411.5.2 which specifies that RCD protection is required in TT earthing systems if high earth loop impedance values are recorded. On the other hand, installing an RCD within a fire alarm circuit is not recommended because of the possibility of nuisance tripping. Such a situation is resolved by installing an RCD dedicated to the alarm supply circuit but with a trip rating of 100 mA.

132.4 Electrical supplies for safety systems

Where a supply for safety services or standby electrical systems is required the designer must determine the characteristics of the supply and the services to be supplied by the safety source. Approved document B, fire safety, states that:

- there must be routes for people to escape to a place of safety;
- these routes must be protected from the effects of fire;
- the routes must be adequately illuminated;
- the exits must be suitably signed;
- there is sufficient means of giving early warning of fire to persons in the building.

All of the above must be applied to the extent that is dependent upon the use of the building, its size and its height.

Appendix C of the *Electrician's Guide to the Building Regulations* gives guidance on the provision required as follows:

- automatic fire detection and alarms complying with BS 5839 should be installed in institutional and other residential occupancies;
- it is essential that the fire detection systems are properly designed, installed and maintained;

Figure 3.86 Electrical equipment for fire safety.

- where services pass through walls, called fire separating elements, they must be sealed to prevent the passage of smoke and fire.

In dwellings not protected by automatic fire detection and alarm systems, including domestic homes, they are required to be fitted with a suitable number of smoke alarms. The building regulations require all new and refurbished dwellings to be fitted with mains operated smoke alarms. The requirements for a single family dwelling of not more than two storeys are:

- Smoke alarms should normally be positioned in the circulation spaces between sleeping spaces and where fires are most likely to start such as kitchens and living rooms.
- In a house or bungalow there should be at least one smoke alarm on every storey.
- Where more than one smoke alarm is installed they should be linked so that the detection of smoke by one unit operates the alarm in all units.
- Alarms should normally be ceiling-mounted and at least 300 mm from walls and light fittings.
- Where the kitchen area is not separated by a door there should be a heat detector in the kitchen.
- The power supply for a smoke alarm should be derived from the dwelling's mains electricity supply.
- The cable for the power supply to each self contained unit and the interconnecting cable need have no fire retardant properties and need no segregation from other circuits (BS 52661-1: 2011).
- Smoke alarms that include a standby power supply can operate during a mains failure and therefore may be connected to a regularly used local lighting circuit. This has the advantage that the circuit is unlikely to be disconnected for prolonged periods.

132.5 Environmental conditions

The electrical design must take into account the environmental conditions to which the installation will be subjected. Electrical equipment in surroundings susceptible to risk of fire or explosion shall be constructed or protected so as to prevent danger.

Appendix 5 of the IET Regulations gives us a concise list of external influences. Each condition is designated with a code; A is environment, B is utilization and C is construction of buildings. For example, a code AA4 signifies:

A = environment

AA = environment ambient temperature

AA4 = environment ambient temperature in the range –5°C to + 40°C

Section 522 of the IET Regulations details the installation requirements for:

- ambient temperature;
- external heat sources;
- the presence of water or/and humidity;
- the presence of corrosive or polluting substances;
- mechanical damage and stresses;
- vibration;
- presence of flora, fauna and mould growth;
- solar radiation.

Some installations require special consideration because of the inherent dangers listed above. Installations requiring special consideration are flameproof installations, construction sites, agricultural and horticultural buildings. All of these installations are described in detail in Part 7 of the IET Regulations and some are described in Chapter 3 of this book. Appendix C of the *On-Site Guide* gives guidance on the selection and types of cable for particular influences.

Figure 3.87 All electrical equipment must be suitable for the installed conditions.

This point in the design process might also be a good time to consider:

- the environmental impact of the design;
- the importance of sustainable design as described at the beginning of this chapter;
- the possible use of environmental technology systems and renewable energy systems as described in Chapter 1 of this book.

Electrical shock

Electric shock occurs when a person becomes part of the electrical circuit, as shown in Fig. 3.104. The lethal level is approximately 50 mA, above which muscles contract, the heart flutters and breathing stops. A shock above the 50 mA level is therefore fatal unless the person is quickly separated from the supply. Below 50 mA only an unpleasant tingling sensation may be experienced or you may be thrown across a room or fall from a roof or ladder, but the resulting fall may lead to serious injury.

To prevent people from receiving an electric shock accidentally, all circuits must contain protective devices to operate or isolate a circuit. But you must bear in mind that the primary job of the protective device is to protect the cable since left unchecked the cable insulation surrounding a conductor subjected to a very high fault current will melt and inevitably cause an electrical fire. Protecting the user or person, however, is through additional protection and specifically the use of RCDs. It is no coincidence that RCDs designed to protect power outlets are rated at 30 mA, in other words they will operate before the circuit current reaches the lethal level.

The *On-Site Guide* lists at section 3.6 when RCDs must be in place and includes:

- when earth loop impedance values are high such as TT earthing systems;
- socket outlets not exceeding 32A.
- low voltage circuits involving a bath or shower;
- mobile equipment not exceeding 32A for use outdoors.
- cables without earthed metallic sheathing/armour installed in walls to a depth less than 50 mm;
- cables not installed in steel conduit or similar.
- lighting circuits in domestic (household) premises.

Consumer units which comply with the 18th Edition requirements, utilise split boards that incorporate one main switch and two RCDs. This ensures that a single fault will only affect one half of the board and thus limit any inconvenience. This also complies with certain requirements that all power sockets and any instances of cables not being enclosed in metal enclosures or installed in walls to a depth of >50 mm are afforded 'additional protection'.

Construction workers and particularly electricians do receive electric shocks, usually as a result of carelessness or unforeseen circumstances. Temporary electrical supplies on construction sites can save many person-hours of labour by providing energy for fixed and portable tools and lighting. However, as stated previously in this chapter, construction sites are dangerous places and by definition temporary supplies are not as robust as permanent installations.

Certain safeguards must therefore be in place, such as ensuring that the input to temporary electrical supplies are protected by an RCD (704.410.3.10). Other

Definition

Electric shock occurs when a person becomes part of the electrical circuit.

measures include the use of a safety isolation transformer contained within SELV or PELV circuits, both of which limit the operating voltage to <50 V (typically 12 V). The assumption made is that being subjected to 50 V or touch voltage as it is known will not ordinarily be lethal. This is why when calculations are made regarding supplementary bonding or fusing current, the value used is touch voltage, so that it ensures that shock protection systems operate before dangerous voltages are allowed to build up. Such arrangements are known as a reduced voltage system and is mainly why the wiring regulations recommended that all portable hand equipment are supplied through a 110 V transformer. When working on a construction site, or in damp or wet conditions, a SELV system is strongly preferred.

Special locations

Special locations are listed in Part 7, Section 700 of the wiring regulations. They range from locations containing a bath or shower and temporary construction site installations to floor and ceiling heating systems. Each location is divided into zones, describes the level of index protection regarding external influences required, what kind of current-using equipment can be used and where they can be positioned. Additionally, use of supplementary bonding and additional protection through RCDs and protective measures such as SELV and PELV are also detailed. Some of these systems will now be discussed in detail.

Construction and Demolition site installations

Construction and demolition sites are potentially dangerous in many ways. The risk of electric shock is high because of the following factors:

- a construction site or demolition site is, by definition, a temporary state. Upon completion there will be either a building with all the necessary safety features, or a brown field site where the building previously stood.
- there is the possibility of damage to cables and equipment as a consequence of the temporary nature of the site and because the site is not always sealed in the early stages from the weather.
- mobile equipment such as electrical tools and hand lamps with trailing leads may be in use.
- there will be many extraneous conductive parts on site which cannot practically be bonded because of the changing nature of the construction process.

Temporary electrical supplies provided on construction sites can save many man hours of labour by providing the energy required for fixed and portable tools and lighting, which speeds up the completion of a project. However, construction sites are dangerous places and the temporary electrical supply that is installed to assist the construction process must comply with all of the relevant wiring regulations for permanent installations (IET Regulation 110.1). All equipment must be of a robust construction in order to fulfil the on-site electrical requirements while being exposed to rough handling, vehicular nudging, the wind, rain and sun. All equipment socket outlets, plugs and couplers must be of the industrial type to BS EN 60439 and BS EN 60309 and specified by IET Regulation 704.511.1 as shown in Fig. 3.88.

Where an electrician is not permanently on site, CBs are preferred so that overcurrent protection devices can be safely reset by an unskilled person.

The British Standards Code of Practice 1017, *The Distribution of Electricity on Construction and Building Sites*, advises that protection against earth faults may be obtained by first providing a low impedance path, so that overcurrent devices can operate quickly as described in Chapter 4, and secondly by fitting an RCD in addition to the overcurrent protection device (IET Regulation 704.410.3.10). The 18th Edition of the IET Regulations considers construction sites very special locations, devoting the whole of Section 704 to their requirements. A construction site installation should be tested and inspected in accordance with Part 6 of the IET Regulations every three months throughout the construction period. The source of supply for the temporary installation may be from a petrol or diesel generating set or from the local supply company. When the local electricity company provides the supply, the incoming cable must be terminated in a waterproof and locked enclosure to prevent unauthorized access and provide metering arrangements. IET Regulations in Section 704 and 411.8 tell us that reduced low voltage is strongly *preferred* for portable hand lamps and tools used on construction and demolition sites. The distribution of electrical supplies on a construction site would typically be as follows:

- 400 V three phase for supplies to major items of plant having a rating above 3.75 kW such as cranes and lifts. These supplies must be wired in armoured cables.
- 230 V single phase for supplies to items of equipment that are robustly installed such as flood lighting towers, small hoists and site offices. These supplies must be wired in armoured cable unless run inside the site offices.
- 110 V single phase for supplies to all mobile hand tools and all mobile lighting equipment. The supply is usually provided by a reduced voltage distribution unit, which incorporates splash proof sockets fed from a centre-tapped 110 V transformer. This arrangement limits the voltage to earth to 55 V, which is recognized as safe in most locations. A 110 V distribution unit is shown in Fig. 3.88. Edison screw lamps are used for 110 V lighting supplies so that they are not interchangeable with 230 V site office lamps.

There are occasions when even a 110 V supply from a centre-tapped transformer is too high; for example, supplies to inspection lamps for use inside damp or confined places. In these circumstances a safety extra-low voltage (SELV) supply would be required.

Industrial plugs have a keyway that prevents a tool from one voltage being connected to the socket outlet of a different voltage. They are also colour coded for easy identification as follows:

Safety first

Construction sites
- Low voltage or
- Battery tools must be used.

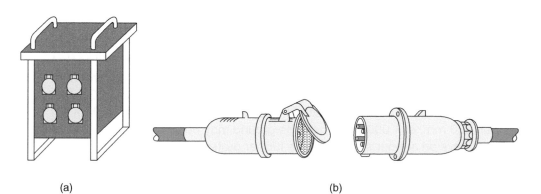

(a)   (b)

Figure 3.88 110 V distribution unit and cable connector, suitable for construction site electrical supplies (a) reduced-voltage distribution unit incorporating industrial sockets to BS EN 60309 and (b) industrial plug and connector.

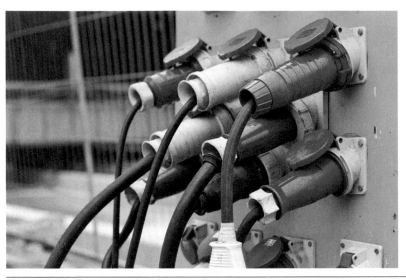

Figure 3.89 Temporary electrical supply.

- 400 V – red
- 230 V – blue
- 110 V – yellow
- 50 V – white
- 25 V – violet.

Agricultural and horticultural premises

Especially adverse installation conditions are to be encountered on agricultural and horticultural premises because of the presence of livestock, vermin, dampness, corrosive substances and mechanical damage. The 18th Edition of the IET Wiring Regulations considers these installations very special locations and has devoted the whole of Section 705 to their requirements. In situations accessible to livestock the electrical equipment should be of a type that is appropriate for the external influences likely to occur, and should have at least protection IP44; that is, protection against solid objects and water splashing from any direction (IET Regulation 705.512.2).

In buildings intended for livestock, all fixed wiring systems must be inaccessible to the livestock and cables liable to be attacked by vermin must be suitably protected (IET Regulation 705.513.2). PVC cables enclosed in heavy-duty PVC conduit are suitable for installations in most agricultural buildings. All exposed and extraneous metalwork must be provided with supplementary equipotential bonding in areas where livestock is kept (IET Regulation 705.415.2.1). In many situations, waterproof socket outlets to BS 196 must be installed. All socket outlet circuits must be protected by an RCD complying with the appropriate British Standard and the operating current must not exceed 30 mA. Cables buried on agricultural or horticultural land should be buried at a depth of not less than 600 mm, or 1,000 mm where the ground may be cultivated, and the cable must have an armour sheath and be further protected by cable tiles. Overhead cables must be insulated and installed so that they are clear of farm machinery or placed at a minimum height of 6 m to comply with IET Regulation 705.522. Horses and cattle are susceptible to an electric shock at voltages lower than 25 V r.m.s.

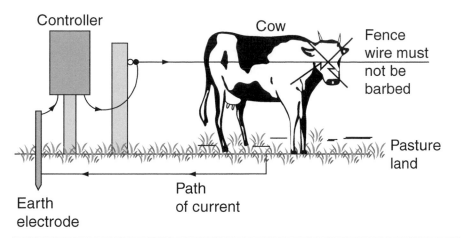

Controller   Cow   Fence wire must not be barbed

Pasture land

Earth electrode   Path of current

Figure 3.90 Farm animal control by electric fence.

The reasons for this are varied but include that:

- animals have a very low body resistance;
- a big potential difference exists between quadruped animals, in other words between the fore and hind legs;
- horses wear metal shoes;
- the ground can be wet;
- the electric path between fore and hind legs travels straight through the animals heart.

The sensitivity of farm animals to electric shock means that they can be contained by an electric fence. An animal touching the fence receives a short pulse of electricity, which passes through the animal to the general mass of earth and back to an earth electrode sunk near the controller, as shown in Fig. 3.90. The pulses are generated by a capacitor–resistor circuit inside the controller, which may be mains or battery operated (capacitor–resistor circuits are discussed in Chapter 6). There must be no risk to any human coming into contact with the controller, which should be manufactured to BS 2632. The output voltage of the controller must not exceed 10 kV and the energy must not be greater than 5 J. The duration of the pulse must not be greater than 1.5 ms and the pulse must never have a frequency greater than one pulse per second.

Figure 3.91 They say that the grass is greener on the other side of the fence, but one false move and this sheep will get an electric shock.

This shock level is very similar to that which can be experienced by touching a spark plug lead on a motor car. The energy levels are very low at 5 J. There are 3.6 million joules of energy in 1 kWh. Earth electrodes connected to the earth terminal of an electric fence controller must be separate from the earthing system of any other circuit and should be situated outside the resistance area of any electrode used for protective earthing. The electric fence controller and the fence wire must be installed so that they do not come into contact with any power, telephone or radio systems, including poles. Agricultural and horticultural installations should be tested and inspected in accordance with Part 6 of the Wiring Regulations every three years.

Caravans and caravan sites

The electrical installations on caravan sites, and within caravans, must comply in all respects with the wiring regulations for buildings. All the dangers that exist in buildings are present in and around caravans, including the added dangers associated with repeated connection and disconnection of the supply and the flexing of the caravan installation in a moving vehicle. The 18th Edition of the IET Regulations has devoted Section 721 to the electrical installation in caravans and motor caravans and Section 708 to caravan parks. Touring caravans must be supplied from a 16 A industrial type socket outlet adjacent to the caravan park pitch, having a degree of protection of at least IP44 similar to that shown in Fig. 3.92. Each socket outlet must be provided with individual overcurrent protection and an individual residual current circuit-breaker with a rated tripping current of 30 mA (IET Regulations 708.55.1 to 14). The distance between the caravan connector and the site socket outlet must not be more than 25 m (IET Regulation 708.530.3 Regulations Fig 708). These requirements are shown in Fig. 3.92. The supply cables must be installed outside the pitch area and be buried at a depth of at least 0.6 m (IET Regulation 708.521.7.2). The caravan or motor caravan must be provided with a mains isolating switch and an RCD to break all live conductors (IET Regulation 721.411). An adjacent notice detailing how to connect and disconnect the supply safely must also be provided, as shown in IET Regulation 721.514. Electrical equipment must not be installed in fuel storage compartments (IET Regulation 721.528.2.1). Caravans flex when being towed, and therefore the installation must be wired in flexible or stranded

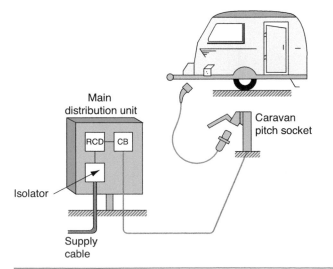

Figure 3.92 Electrical supplies to caravans.

conductors of at least 1.5 mm cross-section. The conductors must be supported on horizontal runs at least every 25 cm and the metalwork of the caravan and chassis must be bonded with 4.0 mm² cable.

The wiring of the extra low-voltage battery supply must be run in such a way that it does not come into contact with the 230 V wiring system (IET Regulation 721.528.1).

The caravan should be connected to the pitch socket outlet by means of a flexible cable, not longer than 25 m and having a minimum cross-sectional area of 2.5 mm² or as detailed in Section 708 Notes and Table 721 of the IET Regulations. Because of the mobile nature of caravans it is recommended that the electrical installation be tested and inspected at intervals considered appropriate, preferably not less than once every three years and annually if the caravan is used frequently (IET Regulation 721 Periodic Inspection Notes).

Bathroom installations

Rooms containing a fixed bath tub or shower basin are considered an area of increased shock risk and, therefore, additional regulations are specified in Section 701 of the IET Regulations. This is to reduce the risk of electric shock to people in circumstances where body resistance is lowered because of contact with water. The regulations can be summarized as follows:

- Socket outlets must not be installed and no provision is made for connection of portable appliances unless the socket outlet can be fixed 3 m horizontally beyond the zone 1 boundary within the bath or shower room (IET Regulation 701.512.3).
- Only shaver sockets that comply with BS EN 60742, that is those which contain an isolating transformer, may be installed in zone 2 or outside the zones in the bath or shower room (IET Regulation 701.512.3).
- All circuits in a bath or shower room, that is both power and lighting, circuits which are serving the location, or passing through zones 1 and 2 not serving the location, must be additionally protected by an RCD having a rated maximum operating current of 30 mA (IET Regulation 701.411.3.3).
- There are restrictions as to where appliances, switchgear and wiring accessories may be installed. See *Zones for bath and shower rooms* below.
- Local supplementary equipotential bonding (IET Regulation 701.415.2) must be provided to all gas, water and central heating pipes in addition to metallic baths, *unless the following two requirements are both met*:
  - all bathroom circuits, both lighting and power, are protected by a 30 mA RCD in addition to a circuit breaker or fuse; and
  - the bath or shower is located in a building with main protective equipotential bonding in place as described in Fig. 3.106 of this book and (IET Regulation 411.3.1.2 and 701.415.2).

Note: Local supplementary equipotential bonding as shown in Fig 3.93 may be an additional requirement of the Local Authority regulations in, for example, licensed premises, student accommodation and rented property.

Zones for bath and shower rooms

Locations that contain a bath or shower are divided into zones or separate areas as shown in Fig. 3.94.

Safety first

Electric shock
Animals and humans must be protected against electric shock.

Key fact

Caravans
- Every caravan pitch must have at least one 16 A industrial socket.
- Each socket must have RCD and overcurrent protection.

IET Regulation 708.533.

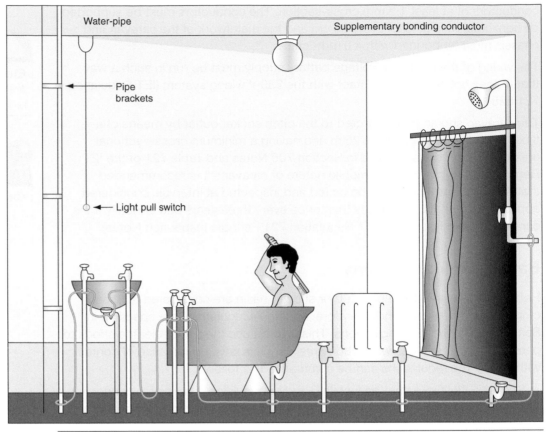

Figure 3.93 Supplementary protective bonding in bathrooms to metal pipework.

Zone 0 – the bath tub or shower basin itself, which can contain water and is, therefore, the most dangerous zone

Zone 1 – the next most dangerous zone in which people stand in water

Zone 2 – the next most dangerous zone in which people might be in contact with water

Outside
zones – people are least likely to be in contact with water but are still in a potentially dangerous environment and the general IET Regulations apply.

• Spaces under the bath that are accessible '*only with the use of a tool*' are outside zones;

• Spaces under the bath that are accessible '*without the use of a tool*' are zone 1.

Electrical equipment and accessories are restricted within the zones.

Zone 0 – being the most potentially dangerous zone, for all practical purposes no electrical equipment can be installed in this zone. However, the IET Regulations permit that whereas SELV fixed equipment with a rated voltage not exceeding 12 V a.c. cannot be located elsewhere, it may be installed in this zone (IET Regulation 701.55). The electrical equipment must have at least IPX7 protection against total immersion in water (IET Regulation 701.512.2).

Key fact

Bathrooms
All bathroom circuits, both power and lighting, must have additional RCD protection.

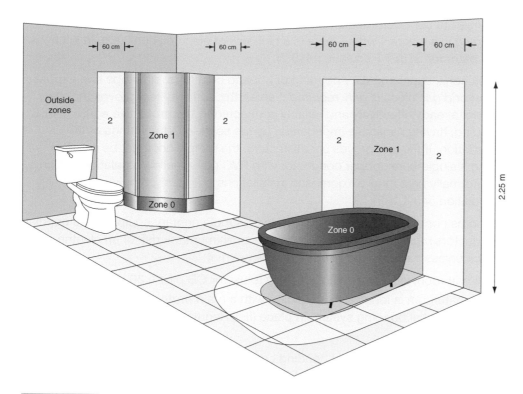

Figure 3.95 Zone 2 shaver socket.

Figure 3.94 Bathroom zone dimensions. (Any window recess is in zone 2.)

Zone 1 – water heaters, showers and shower pumps and SELV fixed equipment may be installed in zone 1. The electrical equipment must have at least IPX4 protection against water splashing from any direction. If the electrical equipment may be exposed to water jets from, for example, commercial cleaning equipment, then the electrical equipment must have IPX5 protection.

Zone 2 – luminaires and fans, and equipment from zone 1 plus shaver units to BS EN 60742 may be installed in zone 2. The electrical equipment must be suitable for installation in that zone according to the manufacturer's instructions and have at least IPX4 protection against splashing or IPX5 protection if commercial cleaning is anticipated.

Outside
zones – appliances are allowed plus accessories except socket outlets unless the location containing the bath or shower is very big and the socket outlet can be installed at least 3 m horizontally beyond the zone 1 boundary (IET Regulation 701.512.3) and has additional RCD protection (IET Regulation 701.411.3.3).

If underfloor heating is installed in these areas it must have an overall earthed metallic grid or the heating cable must have an earthed metallic sheath, which is connected to the protective conductor of the supply circuit (IET Regulation 701.753).

> **Key fact**
>
> Unlike Wales, Part P of the building regulations in England no longer considers kitchens as special locations.

Supplementary equipotential bonding

Modern plumbing methods make considerable use of non-metals (PTFE tape on joints, for example). Therefore, the metalwork of water and gas installations cannot be relied upon to be continuous throughout. The IET Regulations describe the need to consider additional protection by supplementary equipotential bonding in situations where there is a high risk of electric shock (e.g. in kitchens and bathrooms) (IET Regulation 415.2).

In kitchens, supplementary bonding of hot and cold taps, sink tops and exposed water and gas pipes *is only required* if an earth continuity test proves that they are not already effectively and reliably connected to the protective equipotential bonding, having negligible impedance, by the soldered pipe fittings of the installation. If the test proves unsatisfactory, the metalwork must be bonded using a single-core copper conductor with PVC green/yellow insulation, which will normally be 4 mm² for domestic installations but must comply with IET Regulation 543.1.1.

In rooms containing a fixed bath or shower, supplementary equipotential bonding conductors *must* be installed to reduce to a minimum the risk of an electric shock unless the following *two* conditions are met:

1   all bathroom circuits are protected by a fuse or CB plus a 30 mA RCD; and
2   the bathroom is located in a building with a main protective equipotential bonding system in place (IET Regulation 701.415.2) as shown in Fig 3.106.

Supplementary equipotential bonding conductors in domestic premises will normally be of 4 mm² copper with PVC insulation to comply with IET Regulation 543.1.1 and must be connected between all exposed metalwork (e.g. between metal baths, bath and sink taps, shower fittings, metal waste pipes and radiators).

The bonding connection must be made to a cleaned pipe, using a suitable bonding clip. Fixed at, or near, the connection must be a permanent label saying 'Safety electrical connection – do not remove' (IET Regulation 514.13.1).

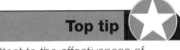

Top tip

To attest to the effectiveness of supplementary bonding, a low resistance measurement must be made from the earthing terminal to the extraneous pipes in the bathroom: reading to be <0.05 Ω.

Key fact

Band I and II circuits can be enclosed together, provided that the cables are insulated to the highest voltage present.

Segregation of circuits

Where an installation comprises a mixture of low-voltage and very low-voltage circuits such as mains lighting and power, fire alarm and telecommunication circuits, they must be separated or *segregated* to prevent electrical contact (IET Regulation 528.1).

For the purpose of these regulations various circuits are identified by one of two bands as follows:

* Band I: telephone, radio, bell, call and intruder alarm circuits, emergency circuits for fire alarm and emergency lighting.
* Band II: mains voltage circuits.

When Band I circuits are insulated to the same voltage as Band II circuits, they may be drawn into the same compartment. When trunking contains rigidly fixed metal barriers along its length, the same trunking may be used to enclose cables of the separate bands without further precautions, provided that each band is separated by a barrier, as shown in Fig. 3.96. Multi-compartment PVC trunking cannot provide band segregation since there is no metal screen between the bands. This can only be provided in PVC trunking if screened cables are drawn into the trunking.

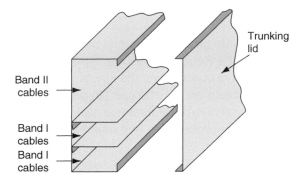

Figure 3.96 Segregation of cables in trunking.

Assessment criteria 2.1

Specify the operating principles of overcurrent protection devices

Assessment criteria 4.11

Verify the fault current capacities of protective devices

Protection against overcurrent

Excessive current may flow in a circuit as a result of an overload or a short circuit.

An overload or overcurrent is defined as a current that exceeds the rated value in an otherwise healthy circuit. A short-circuit is an overcurrent resulting from a fault of negligible impedance. For example, connecting a live part to earth.

Overload currents usually occur in a circuit because the circuit is being misused or overloaded, or because it was badly designed in the first place.

Short-circuits usually occur as a result of an accident that could not have been predicted before the event. An overload may result in currents of two or three times the rated current flowing in the circuit, while short-circuit currents may be hundreds of times greater than the rated current. In both cases, the basic requirement for protection is that the circuit should be interrupted quickly and the circuit isolated safely, before the fault causes a temperature rise or mechanical effect which might damage the insulation, connections, joints and terminations of the circuit conductors and their surroundings. (IET Regulations 131). If the device used for overload protection is also capable of breaking a prospective short-circuit current safely, then one device may be used to give protection from both faults (IET Regulation 432.1). Devices that offer protection from overcurrent are:

* semi-enclosed fuses manufactured to BS 3036;
* cartridge fuses manufactured to BS 88–3: 2010 (used to be 1361);
* high-breaking capacity fuses (HBC fuses) manufactured to BS 88–2: 2010;
* circuit breakers manufactured to BS EN 60898.

Key fact

An overload is not a fault but a situation where too much current is put on an otherwise healthy circuit. Please note that RCDs will not operate during an overload.

Figure 3.97 Certain cartridge fuses use a glass tube, which allows visibility of the fusing wire.

Overcurrent protection

The consumer's mains equipment must provide protection against overcurrent, that is, a current exceeding the rated value (IET Regulation 430.3). Fuses provide overcurrent protection when situated in the live conductors; they must not be connected in the neutral conductor. Circuit-breakers may be used in place of fuses, in which case the circuit-breaker may also provide the means of isolation, although a further means of isolation is usually provided so that maintenance can be carried out on the circuit-breakers themselves.

When selecting a protective device we must give consideration to the following factors:

- the prospective fault current;
- the circuit load characteristics;
- the current-carrying capacity of the cable;
- the disconnection time requirements for the circuit.

The essential requirements for a device designed to protect against overcurrent are:

- it must operate automatically under fault conditions;
- have a current rating matched to the circuit design current;
- have a disconnection time which is within the design parameters;
- have an adequate fault breaking capacity;
- be suitably located and identified.

We will look at these requirements below.

A *short circuit* is an overcurrent resulting from a fault of negligible impedance connected between conductors.

As already defined, an overcurrent may be an overload current, or a short-circuit current. An overload current is when we ask too much of a circuit and therefore we exceed the rated value in an otherwise healthy circuit. A short-circuit is an overcurrent resulting from a fault when, for example, the line conductors come in contact with the neutral, basically removing the load from the circuit. This is known as the prospective short circuit current.

Given that a short-circuit current may be hundreds of times greater than the rated current there is a need to ascertain if a protective device and switch gear can withstand this short circuit capacity without damage, and is known as ascertaining breaking capacity. In fact, there are two separate values defined: Icn which is marked within a rectangle on the device or through use of the letter M as in M6, which indicates the maximum value of current that the protective device can interrupt safely but may be un-useable thereafter. There is also Ics, which is known as the in-service short-capacity, which indicates the maximum fault current that can be interrupted without loss of performance.

In all cases, the basic requirement for protection is that any fault current is interrupted quickly and the circuit is isolated safely before the fault current causes a temperature rise or mechanical effects, which might damage the insulation, connections, joints and terminations of the circuit conductors or their surroundings (IET Regulations 131). It is well worth reminding the reader therefore that this process ties into the earthing system which forms one method of bringing about fault protection. In other words, any exposed conductive part must be connected to the earthing system via a CPC, whose resistance must be low in value so that the fault current is high. This means that the protective device will operate and disconnect the circuit quickly within the required time.

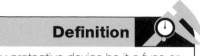

Definition

Any protective device be it a fuse or circuit-breaker is placed in the circuit to protect the circuit conductors.

Definition

An *overload current* can be defined as a current that exceeds the rated value in an otherwise healthy circuit.

The selected protective device should have a current rating that is not less than the full load current of the circuit but which does not exceed the cable current rating. The cable is then fully protected against both overload and short-circuit faults (IET Regulation 435.1). Devices that provide overcurrent protection are:

- High breaking capacity (HBC) fuses to BS 88-2: 2010. These are for industrial applications having a maximum fault capacity of 80 kA.
- Cartridge fuses to BS 88-3: 2010. These are used for a.c. circuits on industrial and domestic installations having a fault capacity of about 30kA.
- Cartridge fuses to BS 1362. These are used in 13 A plug tops and have a maximum fault capacity of about 6 kA.
- Semi-enclosed fuses to BS 3036. These were previously called rewirable fuses and are used mainly on domestic installations. Their fault capacity is small compared to other protective devices rated at about 4 kA.
- CBs to BS EN 60898. These are circuit-breakers (CBs) which may be used as an alternative to fuses for some installations. The British Standard includes ratings up to 100 A and maximum fault capacities of 9 kA. They are graded according to their instantaneous tripping currents – that is, the current at which they will trip within 100 ms. This is less than the time taken to blink an eye.

> **Definition**
>
> By definition a *fuse* is the weakest link in the circuit. Under fault conditions it will melt when an overcurrent flows, protecting the circuit conductors from damage.

Assessment criteria 2.3

Specify the advantages and limitations of different types of overcurrent protection devices

Semi-enclosed fuses (BS 3036)

The semi-enclosed fuse consists of a fuse wire, called the fuse element, secured between two screw terminals in a fuse carrier. The fuse element is connected in series with the load and the thickness of the element is sufficient to carry the normal rated circuit current. When a fault occurs an overcurrent flows and the fuse element becomes hot and melts or 'blows'.

This type of fuse is illustrated in Fig. 3.98. The fuse element should consist of a single strand of plain or tinned copper wire having a diameter appropriate to the current rating of the fuse. This type of fuse was very popular in domestic installations, but less so these days because of its disadvantages.

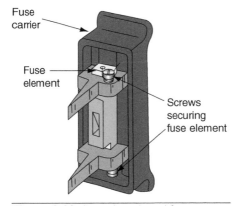

Figure 3.98 A semi-enclosed fuse.

Advantages of semi-enclosed fuses

- They are very cheap compared with other protective devices both to install and to replace.
- There are no mechanical moving parts.
- It is easy to identify a 'blown' fuse.

Disadvantages of semi-enclosed fuses

- The fuse element may be replaced with wire of the wrong size either deliberately or by accident.
- The fuse element weakens with age due to oxidization, which may result in a failure under normal operating conditions.

- The circuit cannot be restored quickly since the fuse element requires screw fixing.
- They have low breaking capacity since, in the event of a severe fault, the fault current may vaporize the fuse element and continue to flow in the form of an arc across the fuse terminals.
- They are not guaranteed to operate until up to twice the rated current is flowing.
- There is a danger from scattering hot metal if the fuse carrier is inserted into the base when the circuit is faulty.

Cartridge fuses (BS 88-3: 2012 (previously BS 1361))

The cartridge fuse breaks a faulty circuit in the same way as a semi-enclosed fuse, but its construction eliminates some of the disadvantages experienced with an open-fuse element. The fuse element is encased in a glass or ceramic tube and secured to end-caps that are firmly attached to the body of the fuse so that they do not blow off when the fuse operates. Cartridge fuse construction is illustrated in Fig. 3.99.

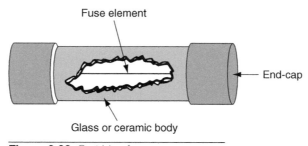

Fuse element

End-cap

Glass or ceramic body

Figure 3.99 Cartridge fuse.

With larger size cartridge fuses, lugs or tags are sometimes brazed on the end-caps to fix the fuse cartridge mechanically to the carrier. They may also be filled with quartz sand to absorb and extinguish the energy of the arc when the cartridge is brought into operation.

Advantages of cartridge fuses

- They have no mechanical moving parts.
- The declared rating is accurate.
- The element does not weaken with age.
- They have small physical size and no external arcing, which permits their use in plug tops and small fuse carriers.
- Their operation is more rapid than semi-enclosed fuses. Operating time is inversely proportional to the fault current, so the bigger the fault current the quicker the fuse operates.
- They are easy to replace.
- Larger valves have bolt-hole fixings.

Disadvantages of cartridge fuses

- They are more expensive to replace than fuse elements that can be rewired.
- They can be replaced with an incorrect cartridge.

- The cartridge may be shorted out by wire or silver foil in extreme cases of bad practice.
- It is not possible to see if the fuse element is broken.

Circuit-breakers (BS EN 60898)

The disadvantage of all fuses is that when they have operated they must be replaced. A circuit-breaker overcomes this problem since it is an automatic switch that opens in the event of an excessive current flowing in the circuit and can be closed when the circuit returns to normal.

Figure 3.100 CBs – B Breaker, fits Wylex standard consumer unit (courtesy of Wylex).

A circuit breaker of the type shown in Fig. 3.100 incorporates a thermal and magnetic tripping device. The load current flows through the thermal and the electromagnetic devices in normal operation but under overcurrent conditions they activate and trip the circuit-breaker.

The circuit can be restored when the fault is removed by pressing the ON toggle. This latches the various mechanisms within the CB and 'makes' the switch contact. The toggle switch can also be used to disconnect the circuit for maintenance or isolation, or to test the CB for satisfactory operation.

Advantages of circuit-breakers

- They have factory-set operating characteristics.
- Tripping characteristics and therefore circuit protection is set by the installer.
- They can distinguish between an overload and short circuit
- The circuit protection is difficult to interfere with.
- The circuit is provided with discrimination.
- A faulty circuit may be quickly identified by the position of the on/off toggle.
- A faulty circuit may be easily and quickly restored.
- The supply may be safely restored by an unskilled operator.

Disadvantages of CBs

- They are relatively expensive but look at the advantages to see why they are so popular these days.

- They contain mechanical moving parts and therefore require regular testing to ensure satisfactory operation under fault conditions.

Additional protection: RCDs

While not an overcurrent protective device, the residual current device is a device that provides earth leakage protection.

When it is required to provide the very best protection from electric shock and fire risk, earth fault protection devices are incorporated into the installation. The object of the regulations concerning these devices (411.3.2 to 411.3.3) is to remove an earth fault current very quickly, less than 0.4 s for all final circuits not exceeding 32 A, and limit the voltage that might appear on any exposed metal parts under fault conditions to not more than 50 V. They will continue to provide adequate protection throughout the life of the installation even if the earthing conditions deteriorate. This is in direct contrast to the protection provided by overcurrent devices, which require a low-resistance earth loop impedance path.

The regulations recognize RCDs as 'additional protection' in the event of failure of the provision for basic protection, fault protection or carelessness by the users of the installation (IET Regulation 415.1.1).

Key fact

- RCDs only operate with earth faults; they will not operate if a short circuit occurs.
- RCDs are known as additional protection because they cannot provide sole protection of a circuit and are therefore additional to a circuit protective device.

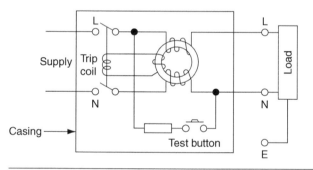

Figure 3.101 Construction of an RCD.

The basic circuit for a single-phase RCD is shown in Fig. 3.101. The load current is fed through two equal and opposing coils wound on to a common transformer core. The phase and neutral currents in a healthy circuit produce equal and opposing fluxes in the transformer core, which induces no voltage in the tripping coil. However, if more current flows in the line conductor than in the neutral conductor as a result of a fault between live and earth, an out-of-balance flux will result in an e.m.f. being induced in the trip coil, which will open the double-pole switch and isolate the load. Modern RCDs have tripping sensitivities between 10 and 30 mA, and therefore a faulty circuit can be isolated before the lower lethal limit to human beings (about 50 mA) is reached.

Consumer units are now supplied that incorporate one or more RCDs, as shown in the *On-Site Guide* Figs 3.6.3.

Wherever RCDs are installed a label shall be fixed near to each RCD stating 'The device must be tested 6 monthly'. (IET Regulation 514.12.2) Note In the 17th Edition the test period was three monthly.

RCBO

A residual current operated circuit-breaker with integral overcurrent protection (RCBO) provides protection against overload and/or short-circuits. RCBOs give the combined protection of a CB and an RCD in one device.

In a split-board consumer unit, about half of the total number of final circuits is protected by the RCD. A fault on any one final circuit will trip out all of the RCD protected circuits, which may cause inconvenience.

The RCBO gives the combined protection of a CB plus RCD for each final circuit so protected and, in the event of a fault occurring only the faulty circuit is interrupted. This arrangement is shown in the *On-Site Guide* at Fig 3.6.3 (i)

Finally, it should perhaps be said that a foolproof method of giving protection to people or animals who simultaneously touch both live and neutral has yet to be devised. The ultimate safety of an installation depends upon the skill and experience of the electrical contractor and the good sense of the user.

Assessment criteria 2.2

Distinguish the application of overcurrent protection devices

The selection of protective devices will depend on various factors, including:

1 Prospective fault current
2 Circuit load characteristics
3 The rated short circuit capacity of a protective device must not be less than the prospective fault current at the point it is installed. The *On-Site Guide* provides Table 7.2.7 (i) where it lists the rated short circuit capacities of overcurrent protective devices.

   * BS 3036 semi-enclosed (rewireable) fuses are still permitted, however cartridge fuses are preferred.
   * Cartridge fuses to BS1361 (now replaced by BS 88-3) can be found and used in domestic or similar premises.
   * Cartridge fuses to BS 88 are classified as:

     – gG general application;
     – gM motor circuits;
     – aM motor circuits.

Characteristics of Circuit Breakers (CBs)

CB Type B to BS EN 60898 will trip instantly at between three and five times its rated current and is also suitable for domestic and commercial installations.

CB Type C to BS EN 60898 will trip instantly at between five and ten times its rated current. It is more suitable for highly inductive commercial and industrial loads such as fluorescent lights.

CB Type D to BS EN 60898 will trip instantly at between 10 and 25 times its rated current. It is suitable for motors, welding and X-ray machines where large inrush currents may occur.

Assessment criteria 4.10

Verify discrimination between protective devices

Effective coordination of protective devices

IET Regulation 536 tells us that in the event of a fault occurring on an electrical installation only the protective device nearest to the fault should operate, leaving other healthy circuits unaffected. A circuit designed in this way would be considered to have effective co-ordination IET Regulation 536. The speed of operation of the protective device increases as the rating decreases.

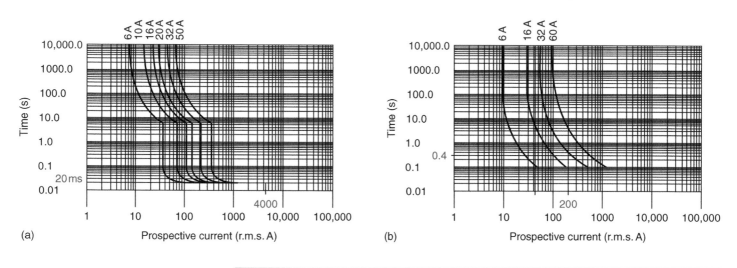

Figure 3.102 Time/current characteristics of (a) a Type B MCB to BS EN 60898 and (b) semi-enclosed fuse to BS 3036.

This can be seen in Fig. 3.102(b). A fault current of 200 A will cause a 16 A semi-enclosed fuse to operate in about 0.1 s, a 32 A semi-enclosed fuse in about 0.4 s and a 60 A semi-enclosed fuse in about 5.0 s. If a circuit is arranged as shown in Fig. 3.103 and a fault occurs on the appliance, effective co-ordination will be achieved because the 16 A fuse will operate more quickly than the other protective devices if they were, for example, all semi-enclosed type fuses with the characteristics shown in Fig. 3.102(b).

In general, when overcurrent protective devices are connected in series, only the device which the designer intended to operate, should operate. This is usually the device closest to the point at which the overcurrent occurs. In the case of Fig 3.103 a fault on the appliance will operate the 16 amp fuse protecting it, leaving the other circuits intact.

Security of supply, and therefore effective co-ordination, is an important consideration for an electrical designer and is also a requirement of the IET Regulations 536.

When existing installations are updated or modified it may be necessary to include additional protective devices, for example, in the following cases:

- when circuit conductors are reduced for instance 2.5mm² down to 1.5mm²;
- when wiring systems are changed for instance from mineral insulated to SWA;
- when a single circuit feed splits to supply multiple circuits.

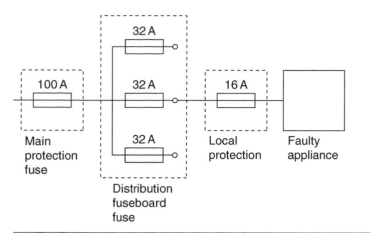

Figure 3.103 Effective co-ordination of the protective devices.

In these types of circumstances co-ordination does not necessary occur with smaller rated devices operating first. Protective devices allow a certain amount of energy through before they operate, therefore circuit designers would look at each device characteristics and specifically the energy limiting class number (1, 2 or 3) to ensure that a given amount of current will operate a certain device and not the others. Designers can also use time/current characteristics of protective devices included in BS 7671. For example, see Fig 3A4 on page 370 of the Regs book for a Type B circuit breaker.

These time current characteristics, which are designed specifically for a particular protective device, include inserted tables which denote the minimum amount of current required to isolate a circuit. For instance, a 20 A Type B circuit breaker requires 100 A to ensure safe disconnection. These values can therefore be used to ensure that co-ordination occurs.

Co-ordination is also a consideration regarding additional protection and the use of RCDs. Consideration must therefore be given to the position of RCDs that are upstream and downstream within the circuit, so that the tripping requirement of the upstream RCD is greater than the downstream, thus avoiding inadvertent isolation of the circuit.

Assessment criteria 3.1

Distinguish earthing systems

Assessment criteria 3.2

Interpret the requirements for protection against electric shock

Electrical supplies and earthing arrangements

We know from earlier chapters in this book that using electricity is one of the causes of accidents in the workplace. Using electricity is a hazard because it has the potential to cause harm. Therefore, the provision of protective devices in an electrical installation is fundamental to the whole concept of the safe use of electricity in buildings. The electrical installation as a whole must be protected against overload or short circuit, and the people using the building must be protected against the risk of shock, fire or other risks arising from their own misuse of the installation or from a fault. The installation and maintenance of adequate and appropriate protective measures is a vital part of the safe use of electrical energy. I want to look at protection against an electric shock by both basic and fault protection, at protection by equipotential bonding and automatic disconnection of the supply.

Let us first define some of the words we will be using. Chapter 54 of the IET Regulations describes the earthing arrangements for an electrical installation. It gives the following definitions:

Earth – the conductive mass of the earth whose electrical potential is taken as zero.

Earthing – the act of connecting the exposed conductive parts of an installation to the main protective earthing terminal of the installation.

Bonding conductor – a protective conductor providing equipotential bonding.

Bonding – the linking together of the exposed or extraneous metal parts of an electrical installation.

Circuit protective conductor (CPC) – a protective conductor connecting exposed conductive parts of equipment to the main earthing terminal. This is the green and yellow insulated conductor in twin and earth cable.

Exposed conductive parts – this is the metalwork of an electrical appliance or the trunking and conduit of an electrical system that can be touched because they are not normally live, but which may become live under fault conditions.

Extraneous conductive parts – this is the structural steelwork of a building and other service pipes such as gas, water, radiators and sinks. They do not form a part of the electrical installation but may introduce a potential, generally earth potential, to the electrical installation.

Shock protection – protection from electric shock is provided by basic protection and fault protection.

Basic protection – this is provided by the insulation of live parts in accordance with Section 416 of the IET Regulations.

Fault protection – this is provided by protective equipotential bonding and automatic disconnection of the supply (by a fuse or CB) in accordance with IET Regulations 411.3.

Protective equipotential bonding – this is equipotential bonding for the purpose of safety.

Basic protection and fault protection

The human body's movements are controlled by the nervous system. Very tiny electrical signals travel between the central nervous system and the muscles, stimulating operation of the muscles, which enable us to walk, talk and run; and remember that the heart is also a muscle.

If the body becomes part of a more powerful external circuit, such as the electrical mains, and current flows through it, the body's normal electrical operations are disrupted. The shock current causes unnatural operation of the muscles and the result may be that the person is unable to release the live conductor causing the shock, or the person may be thrown across the room. The current which flows through the body is determined by the resistance of the human body and the surface resistance of the skin on the hands and feet.

Figure 3.104 An electric shock can throw people across the room, often with devastating results.

This leads to the consideration of exceptional precautions where people with wet skin or wet surfaces are involved, and the need for special consideration in bathroom installations.

Two types of contact will result in a person receiving an electric shock. Direct contact with live parts involves touching a terminal or line conductor that is actually live. The regulations call this basic protection (131.2.1). Indirect contact results from contact with an exposed conductive part such as the metal structure of a piece of equipment that has become live as a result of a fault. The regulations call this fault protection (131.2.2).

The touch voltage curve in Fig. 3.105 shows that a person in contact with 230 V must be released from this danger in 40 ms if harmful effects are to be avoided. Similarly, a person in contact with 400 V must be released in 15 ms to avoid being harmed.

In installations operating at normal mains voltage, the primary method of protection against direct contact (basic protection) is by insulation. All live parts are enclosed in insulating material such as rubber or plastic, which prevent contact with those parts. The insulating material must, of course, be suitable for the circumstances in which they will be used and the stresses to which they will be subjected. Other methods of basic protection include the provision of barriers or enclosures which can only be opened by the use of a tool, or when the supply is first disconnected. Protection can also be provided by fixed obstacles such as a guardrail around an open switchboard or by placing live parts out of reach as with overhead lines.

Fault protection

Protection against indirect contact, called fault protection (IET Regulation 131.2.2), is achieved by connecting exposed conductive parts of equipment to the main earthing terminal of the installation as shown in Figs. 3.106 to 3.108. In Chapter 13 of the IET Regulations we are told that where the metalwork of

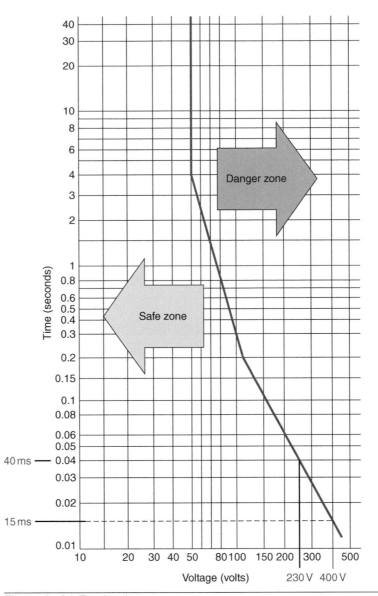

Figure 3.105 Touch voltage curve.

electrical equipment may become charged with electricity in such a manner as to cause danger, that metalwork will be connected with earth so as to discharge the electrical energy without danger.

The application of equipotential bonding is one of the important principles for safety.

There are five methods of protection against contact with metalwork that has become unintentionally live, that is, indirect contact with exposed conductive parts recognized by the IET Regulations. These are:

1 Protective equipotential bonding coupled with automatic disconnection of the supply.
2 The use of Class II (double insulated) equipment.
3 The provision of a non-conducting location.
4 The use of earth free equipotential bonding.
5 Electrical separation.

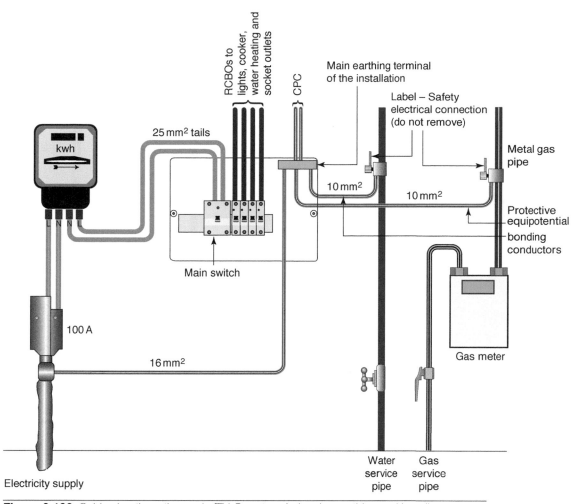

RCBOs to lights, cooker, water heating and socket outlets

CPC

Main earthing terminal of the installation

Label – Safety electrical connection (do not remove)

Metal gas pipe

25 mm² tails

kwh

10 mm²

10 mm²

L N N L

Protective equipotential bonding conductors

Main switch

Gas meter

100 A

16 mm²

Water service pipe

Gas service pipe

Electricity supply

Figure 3.106 Cable sheath earth supply (TN-S systems) showing earthing and bonding arrangements.

Methods 3 and 4 are limited to special situations under the effective supervision of trained personnel.

Method 5, electrical separation, is little used but does find an application in the domestic electric shaver supply unit, which incorporates an isolating transformer.

Method 2, the use of Class II insulated equipment, is limited to single pieces of equipment such as tools used on construction sites, because it relies upon effective supervision to ensure that no metallic equipment or extraneous earthed metalwork enters the area of the installation.

The method that is most universally used in the United Kingdom is, therefore, Method 1 – protective equipotential bonding coupled with automatic disconnection of the supply.

This method relies upon all exposed metalwork being electrically connected together to an effective earth connection. Not only must all the metalwork associated with the electrical installation be so connected, that is, conduits, trunking, metal switches and the metalwork of electrical appliances, but also IET Regulation 411.3.1.2 tells us to connect the extraneous metalwork of water service pipes, gas and other service pipes and ducting, central heating and air conditioning systems, exposed metallic structural parts of the building and

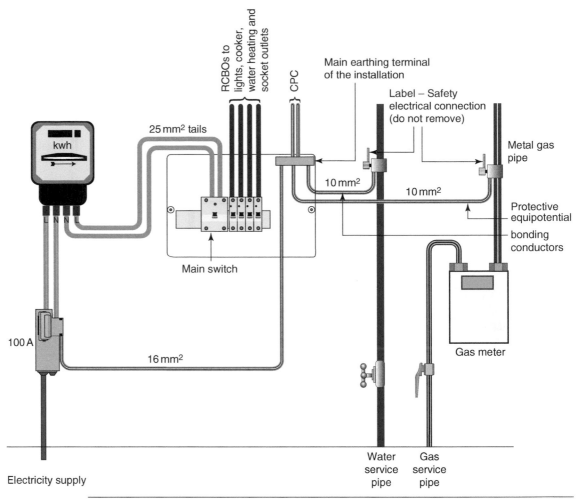

Figure 3.107 Protective multiple earth supply (TN-C-S system) showing earthing and bonding arrangements.

Key fact

When carrying out earthing and bonding activities:

- use a suitable bonding clamp;
- connect to a cleaned pipe;
- make sure all connections are tight;
- fix a label 'Safety Electrical Connection' close to the connection.

IET regulation 514.13.

lightning protective systems to the main protective earthing terminal. In this way the possibility of a voltage appearing between two exposed metal parts is removed. Protective equipotential bonding is shown in Figs 3.106 to 3.108.

Figures 3.106 to 3.108 show protective equipotential bonding conductors connected to gas and water supplies. This is necessary if the incoming supplies are metallic. However, the 18th Edition of the IET Regulations at 411.3.1.2 tells us that metal pipes entering a building having an insulated section at their entry to the building need not be connected to the equipotential bonding of the installation.

However, the premises gas, water and electricity supplies must be bonded on the consumers hard metal pipe work, at the point of entry to the building as shown in Figures 3.106 to 3.108 of this book, and figures 2.1 (i) to 2.1 (iii) of the *On Site Guide* (IET Regulation 544.1.2).

You will see in Figs 3.106 to 3.108 that a consumer unit is being used for the control and distribution of electrical energy, however recent fire statistics have shown that a large number of domestic fires involved plastic consumer units as the source of the fire. Consumer units are often located at the entrance or exit door of the home or under the stairs, raising the possibility that a fire starting as a result of faulty wiring could block the emergency exit route. Regulation 421.1.201 of the 18th Edition of the Regulations now requires that consumer units be

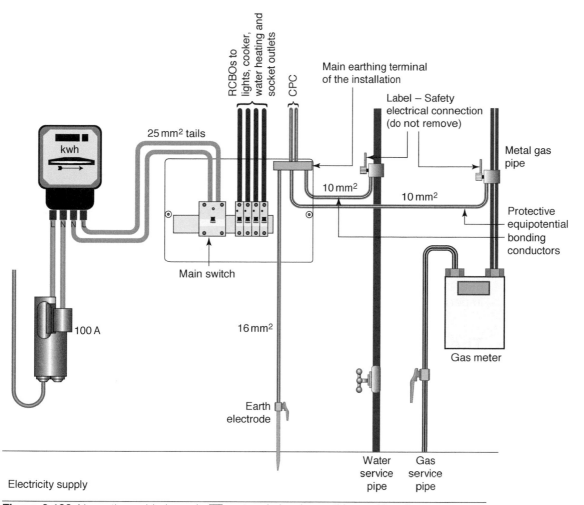

Figure 3.108 No earth provided supply (TT systems) showing earthing and bonding arrangements.

manufactured from non-combustible material, for example, metal, or be enclosed in a non-combustible enclosure.

The 18th Edition of the Regulations at 421.1.7 now also recommends the use of arc fault detection devices (AFDD) as a means of providing additional protection against fire caused by arc faults in AC final circuits.

Supply system earthing arrangements

The British government agreed on 1 January 1995 that the electricity supplies in the United Kingdom would be harmonized with those of the rest of Europe. Thus the voltages used previously in low-voltage supply systems of 415 and 240 V have become 400 V for three-phase supplies and 230 V for single-phase supplies. The Electricity Supply Regulations 1988 have also been amended to permit a range of variation from the newly declared nominal voltage. From January 1995 the permitted tolerance is the nominal voltage +10% and –6%. Previously it was +/– 6%. This gives a voltage range of 216–253 V for a nominal voltage of 230 V and 376–440 V for a nominal voltage of 400 V (IET Regulations Appendix 2).

It is further proposed that the tolerance levels will be adjusted to +10% of the declared nominal voltage. All EU countries will have to adjust their voltages to comply with a nominal voltage of 230 V single-phase and 400 V three-phase. The supply to a domestic, commercial or small industrial consumer's installation is usually protected at the incoming service cable position with a 100 A BS 88-3 high breaking capacity (HBC) fuse. Other items of equipment at this position are the energy meter and the consumer's distribution unit, providing the protection for the final circuits and the earthing arrangements for the installation.

The supply earthing

Five earthing systems are described in the definitions but only the TN-S, TN-C-S and TT systems are suitable for public supplies. A system consists of an electrical installation connected to a supply. Systems are classified by a capital letter designation.

The supply earthing arrangements are indicated by the first letter, where T means one or more points of the supply are directly connected to earth and I means the supply is not earthed or one point is earthed through a fault-limiting impedance.

The installation earthing

The installation earthing arrangements are indicated by the second letter, where T means the exposed conductive parts are connected directly to earth and N means the exposed conductive parts are connected directly to the earthed point of the source of the electrical supply.

The earthed supply conductor

The earthed supply conductor arrangements are indicated by the third letter, where S means a separate neutral and protective conductor and C means that the neutral and protective conductors are combined in a single conductor.

Cable sheath earth supply (TN-S system)

This is one of the most common types of supply system to be found in the United Kingdom where the electricity companies' supply is provided by underground cables. The neutral and protective conductors are separate throughout the system. The protective earth conductor (PE) is the metal sheath and armour of the underground cable, and this is connected to the consumer's main earthing terminal. All exposed conductive parts of the installation, gas pipes, water pipes and any lightning protective system are connected to the protective conductor via the main earthing terminal of the installation. These arrangements are shown in Figs 3.109 to 3.113

(PME) Protective multiple earthing supply (TN-C-S system)

This type of underground supply is becoming increasingly popular to supply new installations in the United Kingdom. It is more commonly referred to as protective multiple earthing (PME). The supply cable uses a combined protective earth and neutral (PEN) conductor. At the supply intake point a consumer's main earthing terminal is formed by connecting the earthing terminal to the neutral conductor. All exposed conductive parts of the installation, gas pipes, water pipes and any lightning protective system are then connected to the main earthing terminals. Thus phase to earth faults are effectively converted into phase to neutral faults. These arrangements are shown in Figs. 3.109 to 3.111.

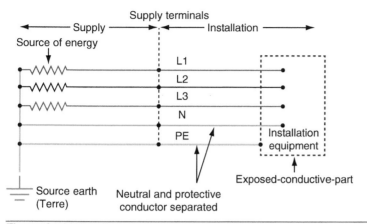

Figure 3.109 TN-S system.

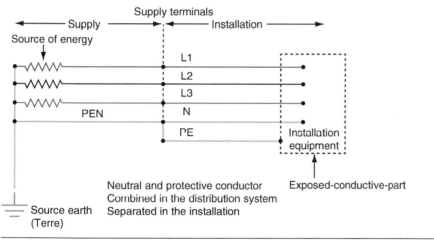

Figure 3.110 TN-C-S system (PME).

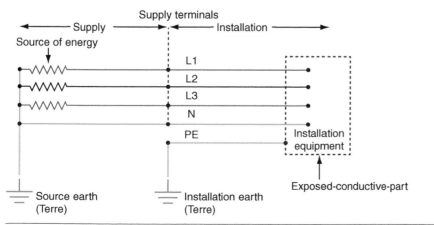

Figure 3.111 TT system.

No earth provided supply (TT system)

This is the type of supply more often found when the installation is fed from overhead cables. The supply authorities do not provide an earth terminal and the installation's CPCs must be connected to earth via an earth electrode provided by the consumer. Regulation 542.2.3 lists the type of earth rod, earth plates or earth tapes recognized by BS 7671. An effective earth connection is sometimes difficult to obtain and in most cases a residual current device (RCD) is provided when this type of supply is used. The arrangement is shown in Fig. 3.108 and 3.111 which show the layout of a typical domestic service position for these supply systems. They show circuits protected by RCBOs. The use of RCBOs will minimize inconvenience because, in the event of a fault occurring, only the faulty circuit will disconnect. However, alternative consumer unit arrangements using CBs and RCDs are shown in the *On-Site Guide* at Section 3.6.3. The TN-C and IT systems of supply do not comply with the supply regulations and therefore cannot be used for public supplies. Their use is restricted to private generating plants and for this reason I shall not include them here, but they can be seen in Part 2 of the IET Regulations.

Assessment criteria 3.3

Select suitably sized protective conductors in accordance with BS 7671

The general layout of a domestic supply is shown clearly within section 2 of the *On-Site Guide* including the size of the main earthing conductor (16 mm²) and main equipotential bonding conductors (10 mm²). However, these values are in respect to both the earthing system in use, whether they are buried and protected and the size of the meter tails. All such details are specified in Chapter 54 of BS 7671 or Section 4 of the *On-Site Guide*. Another consideration regarding protective conductors comes with thermal constraint since Table 54.7 of Chapter 54 specifies that protective conductors and line conductors up to ≤ 16 mm must be the same. However, when PVC flat profile cabling is used aside from 1 mm², all other size of conductors include a smaller sized CPC. Thermal constraint therefore is a calculation to ensure that the size of CPC used can safely carry any prospective fault current. This is covered in greater detail in learning outcome 4.

Part 1 of Chapter 13, section 132 of the IET Regulations details the fundamental principles for the design of an electrical installation. Let us now look at electrical design in some detail as required by the EAL syllabus and Section 132 of the IET Regulations.

Electrical systems design – the process

The designer of an electrical installation must ensure that the design meets the requirements of the IET Wiring Regulations for electrical installations and any other regulations that may be relevant to a particular installation. The designer may be a professional technician or engineer whose job is to design electrical installations for a large contracting firm. In a smaller firm, the designer may also be the electrician who will carry out the installation to the customer's

Definition

The *designer* of an electrical installation is the person who interprets the electrical requirements of the customer within the regulations.

requirements. The designer of any electrical installation is the person who interprets the electrical requirements of the customer within the regulations, identifies the appropriate types of installation, the most suitable methods of protection and control and the size of cables to be used.

Chapter 13, Section 132 of the regulations gives us 16 regulations that make up the design process. The headings are as follows:

Table 3.9

132.1	General
132.2	Characteristics of the supply
132.3	The nature of the demand
132.4	Supplies for safety systems
132.5	Environmental conditions
132.6	Conductor size calculations
132.7	Type of wiring and method of installation
132.8	Protective equipment
132.9	Emergency control
132.10	Disconnecting devices
132.11	Prevention of mutual detrimental influences
132.12	Accessibility of electrical equipment
132.13	Documentation for the electrical installation
132.14	Protective devices and switches
132.15	Isolation and switching
132.16	Additions and alterations to an installation

Energy Efficiency – a new Part 8 of the Regulations

The 18th Edition of the Regulations gives us a new Appendix 17, Energy Efficiency. It is intended that this new Appendix will become a new Part 8 in future Amendments to the regulations. The new Appendix provides recommendations for the design and erection of electrical installations. All new electrical installation design, and modifications to existing installations must continue to meet the required levels of safety and capacity which we will calculate in this section, but must NOW ALSO optimise electrical efficiency, providing the lowest levels of electricity consumption.

The new Appendix considers design and maintenance from the context of improving the efficiency of the electrical installation. There is a change of emphasis in this new section, to incorporate energy efficiency into the electrical installation design as a prerequisite, and not just as an aspiration.

The standard makes it clear that any measure taken to make the electrical installation more efficient, must not compromise the safety of any occupants, property or livestock, and that much of this appendix will not apply to domestic and similar installations.

Let us now consider some of the design process regulations.

132.1 General

In the introduction, the IET Regulation tells us that the electrical installation must be designed to provide for:

* the protection of people, livestock and property; and
* to meet the intended use required of the installation.

132.2 Characteristics of the supply

One of the first considerations for the designer is to determine the nature or characteristics of the supply and the earthing arrangements. Inspection of the existing supply or by requesting details from the supply distributor will be necessary to determine:

* the supply voltage;
* the type of supply and earthing arrangements, that is TN-S, TN-C-S or TT system;
* the value of the earth fault loop impedance, Ze for the supply;
* the location of the supply.

For new electrical installations or for significant alterations to existing ones, the Electricity Safety Quality and Continuity Regulations require the electricity distributor to install the cut out and meter in a 'safe location' where they can be mechanically protected and safely maintained. In compliance with this 'safe location' requirement the electricity distributor and installer may in some areas be required to take account of the risk of flooding. To comply, the distributors' equipment and the installation's consumer unit or fuse boards must be installed above the flood level. Upstairs power and lighting circuits and downstairs lighting should be installed above the flood level. Upstairs and downstairs circuits should have separate protective devices. Consumer units or fuse boards must be generally accessible for use by responsible persons in the household. They should not be installed where young people might interfere with them (IET Guide to the Building Regs 2.1 and 3.1).

Figure 3.112 The electricity grid power supply mostly travels along conductors supported by pylons.

Upon request, the supply distributor will give the following information for a typical domestic supply:

- single phase a.c. supply;
- maximum value of the distributors protective device is 100 A;
- the normal voltage will be 230 V to earth;
- frequency 50 cycles per second;
- the maximum or worst case external earth fault loop impedance (Ze) will be

  - 0.8 ohms for cable sheath (TN-S) supplies;
  - 0.35 ohms for PME (TN-C-S) supplies;
  - for no earth provided (TT) supplies. The value given by the distributor for the earth electrode resistance at the supply transformer is usually 21 ohms. The resistance of the consumer's installation earth electrode should be as low as possible but a value exceeding 200 ohm might not be stable (IET *On-Site Guide* 1.1);

- the maximum or worst case prospective fault current at the origin of the installation will normally be 16 kA. If a protective device is to operate safely, its rated short-circuit capacity must be not be less than the prospective fault current at the point where it is installed. It is recommended that the electrical installation design should be based upon the maximum fault current value provided by the distributor and not on measured values. Except for London and some other major city centres, the maximum fault current for 230 V single phase supplies up to 100 A will not exceed 16 kA (IET *On-Site Guide* Table 7.2.1(i)).

These characteristics are very important, but equally is the process behind the selection and erection of the wiring and associated equipment including enclosures.

We will now look at the electrical design process by first examining how much current the installation requires.

Assessment criteria 4.1

Determine the design current

Assessment criteria 4.2

Select a suitably rated protective device

132.3 The nature of demand

The designer must identify the number of circuits required and the expected load on these circuits. The total current demand of any final circuit is estimated by adding together the current demands of all points of utilization such as socket and lighting points and equipment outlets. This involves using the main formula for power for an electrician $P = V \times I$.

Final circuit current demand

In this chapter we will only look at straightforward household installations but the same principles apply to shops, hotels and guest houses. All of these premises are dealt with in Appendix A of the *On-Site Guide*. Let us begin by looking at the current demand to be assumed for points of utilization given in Table A1 of the *On-Site Guide* and shown here at Table 3.10.

Table 3.10 Assumed current demand for electrical equipment and circuits

Current using equipment	Assumed current demand
Lighting points	Minimum of 100 watts/lampholder
Discharge lighting such as fluorescent tubes	Lamp watts × 1.8 to take account of the control gear
Electric clock, shaver unit, bell transformer	May be neglected in this assessment
2A socket outlet	0.5A
Standard household −13A ring circuit	30A or 32A – max floor area 100 m² wired in 2.5 mm cable
Standard household – 13A radial circuit	30A or 32A – max floor area 75 m² wired in 4.0 mm cable
Standard household – 13A radial circuit	20A – max floor area 50 m² wired in 2.5 mm cable
Cooking appliances	The first 10A of the rated current plus 30% of the remainder of the rated current plus 5A if control unit incorporates a socket outlet
All other stationary equipment such as shower or immersion heater	British Standard rated current

Example 1

Calculate the current demand of a 7.36kW electric shower connected to the 230 V mains supply.

The power $\quad$ P = 7.36kW $\qquad$ or $\qquad$ 7360 watts
The voltage $\quad$ V = 230 volts
The power factor = cos ϕ = 1 for a resistive load

Now power = V I cos ϕ $\quad$ so transposing for current $I = \dfrac{\text{Power}}{V \cos \phi}$ amps

So the current demand $I = \dfrac{7360}{230 \times 1} = 32$ amps

Example 2

Calculate the current demand of an electric cooker comprising

A hob with 4 2.5kW rings	= 10kW
A main oven rated at 2kW and	= 02kW
A grill/top oven rated at 2kW	= 02kW total 14kW

The cooker control unit incorporates a 13A socket outlet

So the total capacity of the cooker is 14kW at 230 volts

As in example 1 the current $I = \dfrac{\text{Power}}{V \cos \phi}$

(Continued)

Example 2 (continued)

And cos $\phi = 1$ for a resistive load

$$\text{Therefore } I = \frac{14000}{230 \times 1} = 60.87 \text{ amps}$$

From Table 3.10 above we take the first 10A of the rated current, plus 30% of the remainder plus 5A if the control unit incorporates a socket outlet.

$$\text{Therefore } I = 10A + \left[\frac{30}{100} \times (60.87 - 10)\right] + 5A$$

$$I = 10A + \left[\frac{30}{100} \times 50.87\right] + 5A$$

$$I = 10 + 15.26 + 5$$

$$I = 30.26 \text{ amps}$$

Therefore a 32 amp CB would adequately protect this circuit.

Example 3

A lighting circuit consists of 10 points. Calculate the current demand.

From Table 3.10 we must assume a minimum of 100 watts per point.

As in the previous examples $I = \dfrac{\text{Power}}{V \cos \phi}$ amps

$$\text{Therefore } I = \frac{10 \times 100}{230} = 4.35 \text{ amps}$$

Therefore a 5 amp fuse or a 6 amp CB would protect this circuit if ordinary GLS lamps were being used. However, if extra-low voltage or discharge lighting is connected to this circuit and protected by a type 'B' CB we must take account of the inrush current which occurs at switch-on and sometimes causes unwanted or nuisance tripping of the CB. To avoid this, the circuit will be adequately protected by a 10 amp CB.

Example 4

The kitchen work surface of a domestic kitchen is to be illuminated by 10 30 watt fluorescent tubes fixed to the underside of the cupboards above the work surface. Calculate the current demand of this circuit.

From Table 3.10 the demand is taken as the lamp watts multiplied by 1.8

As in the previous examples $\quad I = \dfrac{\text{power}}{V \cos \phi}$ amps

$$\text{Therefore } I = \frac{10 \times 30W \times 1.8}{230 \times 1} = 2.34 \text{ amps}$$

Therefore a 5 amp fuse or a 6 amp CB would protect this circuit.

Example 5

Calculate the current demand of a 3kW immersion heater installed in a 30-litre water storage vessel. We know from Table 3.10 that we should take the British Standard rated current for a water heater such as this, or we can calculate the current rating as before.

$$\text{Current } I = \frac{\text{power}}{V \cos \phi} \qquad \text{therefore } I = \frac{3000W}{230\,V \times 1} = 13.04 \text{ amps}$$

A 15 amp fuse or 16 amp CB will protect this circuit. We also know from Appendix H5 of the *On-Site Guide* that water heaters fitted to vessels in excess of 15 litres must be supplied on their own separate circuit.

Diversity between final circuits

The current demand of a circuit is the current taken by that circuit over a period of time, say 30 minutes. Some loads make a constant demand all the time they are switched on. A 100 watt light bulb will make a constant current demand of (100 watts ÷ 230 volts) 0.43 amperes whenever it is switched on. However, an automatic washing machine is made up of a variable speed motor, a pump and a water heater all controlled by a programmer. The washer load will not be constant for the whole time it is switched on, but will vary depending upon the wash cycle. Diversity makes an allowance on the basis that not all of the load or connected items will be in use at the same time. The design would be wasteful if it did not take advantage of the diversity between different loads. Let us now look at diversity as it applies to a straightforward household installation but the same principles will apply to shops, offices, business premises and hotels. Appendix A of the *On-Site Guide* gives the diversity for all these different types of premises.

The allowances for diversity shown in Table A2 of the *On-Site Guide* are for very specific situations and can only provide guidance. The diversity allowances for lighting circuits should be applied to 'items of equipment' connected to the consumer unit at a rate of 66%. However, standard power circuit arrangements for households, as described in Appendix H of the *On-Site Guide*, can be applied to the rated current of the overcurrent protective device for the circuit. The diversity allowance for standard socket outlet circuits is 100% of the largest circuit, plus 40% of all other power circuits. It is important to ensure that the consumers unit is of sufficient rating to take the total load connected without the application of diversity.

Example 6

Calculate the current demand, including diversity, for a six-way consumer unit comprising the following circuits in a domestic dwelling.

Circuit 1. A lighting circuit comprising 10 points as described in Example 3, previously having a maximum demand of 4.35A and protected by a 6 amp CB. (66% diversity allowed for lighting circuits)

Circuit 2. A lighting circuit comprising 10, 30 watt fluorescents as described in Example 4, previously having a maximum demand of 2.3A and protected by a 6 amp CB. (Again 66% diversity allowed.)

(Continued)

Example 6 (continued)

Circuit 3. A thermostatically controlled 3kW immersion heater as described in Example 5, previously having a maximum demand of 13.04A and protected by a 16 amp CB. (No diversity allowed)

Circuit 4. A ring circuit of 13A socket outlets installed in accordance with Section H of the *On-Site Guide* and protected by a 32 amp CB. (This is one of the largest socket circuits, so, no diversity allowed.)

Circuit 5. A radial circuit of 13A socket outlets installed in accordance with section H of the *On-Site Guide* and protected by a 32 amp CB. (40% diversity allowed.)

Circuit 6. A radial circuit of 13A socket outlets installed in accordance with section H of the *On-Site Guide* and protected by a 20 amp CB. (40% diversity allowed.)

Table A2 line 9 of the *On-Site Guide* tells us that for standard household circuits as described in Appendix H of the OSG, (for example ring and radial circuits) **the allowable diversity is to be calculated as follows: 100% of the current demand of the largest circuit plus 40% of the current demand of every other standard socket outlet circuit.** The diversity for all lighting circuits is 66%. The IET Design Guide follows this format, so let us now consider our own Example 6.

From the question

Circuit 1 has a maximum demand of	4.35 amps
Circuit 2 has a maximum demand of	2.34 amps
Circuit 3 has a maximum demand of	13.04 amps
Circuit 4 has a maximum demand of	32.00 amps
Circuit 5 has a maximum demand of	32.00 amps
Circuit 6 has a maximum demand of	20.00 amps

We can see from the above that the largest circuit in the consumers unit is 32 amps, so let us nominate circuit 4 as the largest circuit and apply diversity to all other standard final circuits.

Circuit 1 $= 4.35A \times \dfrac{66}{100} =$ 2.87A

Circuit 2 $= 2.34A \times \dfrac{66}{100} =$ 1.54A

Circuit 3 $= 13.04A$ (no diversity) 13.04A

Circuit 4 $= 32.00A$ (no diversity) 32A

Circuit 5 $= 32.00A \times \dfrac{40}{100} =$ 12.80A

Circuit 6 $= 20.00A \times \dfrac{40}{100} =$ 8.00A

Adding the values 25.21A + 45.04A

Total installed demand with diversity = 70.25 amps

To summarize:

- Dividing the power rating of the load by the working voltage tells us the amount of basic current the circuit requires. This basic current is known as the design current Ib.
- Certain circuits are not on all the time and are given an allowance which when applied reduces the current demand and will therefore allow us to use a smaller circuit conductor. This is known as applying diversity (Appendix A of the *On-Site Guide*).

Once Ib has been established BS 7671 (Chapter 43) states that there must exist a certain coordination between the conductor and the nominal value of the protective device In. The first of these being that:

$$In \geq Ib$$

In other words the protective device needs to be the same value as the design current or bigger. The designer also needs to look at the size of conductor required and the manner in which the circuit cable is to be installed and this is known as establishing the reference method.

Assessment criteria 4.3

Establish the installation method reference

Assessment criteria 4.4

Apply appropriate rating factors

Assessment criteria 4.5

Determine the minimum current carrying capacity of live conductors

Assessment criteria 4.6

Establish if the voltage drop is acceptable

132.6 Conductor size calculations

The size of a cable to be used for an installation depends upon:

- the current rating of the cable under defined installation conditions; and
- the maximum permitted drop in voltage as defined by IET Regulation 525. The current carrying capacity of a conductor is known as Iz and the designer is also influenced by a second circuit coordination if there is a requirement for the circuit to withstand overloads as well as a short circuit. The requirement is this:

$$In \leq Iz$$

In other words, the current carrying capacity of the cable Iz must be equal to or greater than the nominal value In of the protective device.

Once this has been established there are other factors that influence the current rating:

1 *Design current*: cable must carry the full load current.

2 *Type of cable*: PVC, MICC, copper conductors or aluminium conductors.

3 *Installed conditions*: clipped to a surface or installed with other cables in a trunking.

4 *Surrounding temperature*: cable resistance increases as temperature increases and insulation may melt if the temperature is too high.

5 *Type of protection*: for how long will the cable have to carry a fault current?

IET Regulation 525 states that the drop in voltage from the supply terminals to the fixed current-using equipment must not exceed 3% for lighting circuits and 5% for other uses of the mains voltage. That is, a maximum of 6.9 V for lighting and 11.5 V for other uses on a 230 V installation. The volt drop for a particular cable may be found from:

$$VD = \text{Factor} \times \text{Design current} \times \text{Length of run}$$

The factor is given in the tables of Appendix 4 of the IET Regulations and Appendix F of the *On-Site Guide.* They are also given in Table 3.13 in this book. The cable rating, denoted I_t, may be determined as follows:

$$I_t = \frac{\text{Current rating of protective device}}{\text{Any applicable correction factors}}$$

The cable rating must be chosen to comply with IET Regulation 433.1. The correction factors that may need applying are given below as:

Ca – the ambient or surrounding temperature correction factor, which is given in Tables 4B1 and 4B2 of Appendix 4 of the IET Regulations or Table F1 of the *On-Site Guide*. They are also shown in Table 3.11 of this book.

Cg – the grouping correction factor given in Tables 4C1 to 4C5 of the IET Regulations and Table F3 of the *On-Site Guide*.

Cf – the 0.725 correction factor to be applied when semi-enclosed fuses protect the circuit as described in item 5.1.1 of the preface to Appendix 4 of the IET Regulations and Appendix F of the *On-Site Guide*.

Ci – the correction factor to be used when cables are enclosed in thermal insulation. For example by examining Table F2 of the *On-Site Guide* we can determine that where the cable is *totally* surrounded by thermal insulation over a length greater than 0.5m we must apply a factor of 0.5.

Where the cable is *totally* surrounded over a short length, the appropriate factor given in Table 52.2 of the IET Regulations or Table F2 of the *On-Site Guide* should be applied.

Note: A cable should preferably not be installed in thermal insulations.

Having calculated the cable rating, I_t the smallest cable should be chosen from the appropriate table which will carry that current. This cable must also meet the voltage drop (IET Regulation 525) and this should be calculated as described earlier. When the calculated value is less than 3% for lighting and 5% for other uses of the mains voltage the cable may be considered suitable. If the calculated value is greater than this value, the next larger cable size must be tested until a cable is found that meets both the current rating and voltage drop criteria.

Table 3.11 Ambient temperature correction factors

Type of insulation	Conductor Operating temperature	Ambient temperature (°C)			
		25	30	35	40
Thermoplastic (general purpose PVC)	70°C	1.03	1.0	0.94	0.87

Example

A house extension has a total load of 6kW installed some 18m away from the mains consumer unit for lighting. A PVC insulated and sheathed twin and earth cable will provide a sub-main to this load and be clipped to the side of the ceiling joists over much of its length in a roof space which is anticipated to reach 35°C in the summer and where insulation is installed up to the top of the joists. Calculate the minimum cable size if the circuit is to be protected by a type B CB to BS EN 60898. Assume a TN-S supply; that is, a supply having a separate neutral and protective conductor throughout.

Let us solve this question using only the tables given in the *On-Site Guide*. The tables in the regulations will give the same values, but this will simplify the problem because we can refer to Tables 3.11, 3.12 and 3.13 in this book which give the relevant *On-Site Guide* tables.

$$\text{Design current}, I_t = \frac{\text{Power}}{\text{Volts}} = \frac{6000W}{230V} = 26.09A$$

Nominal current setting of the protection for this load I_n = 32 A.

The cable rating I_t is given by:

$$I_t = \frac{\text{Current rating of protective device } (I_n)}{\text{The product of the correction factors}}$$

The correction factors to be included in this calculation are:

Ca ambient temperature; as shown in Table 3.11 the correction factor for 35°C is 0.94.

Cg grouping factors need not be applied.

Cf, since protection is by CB no factor need be applied.

Ci thermal insulation demands that we assume installed Method A in Table 3.12

The design current is 26.09A and we will therefore choose a 32A CB for the nominal current setting of the protective device, I_n.

$$\text{Current rating}, I_t = \frac{32}{0.94} = 34.047A$$

(Continued)

Example (continued)

From column 2 in Table 3.12, a 10mm cable, having a rating of 43A, is required to carry this current.

Now test for volt drop: from Table 3.13 the volt drop per ampere per metre for a 10mm cable is 4.4mV from column 3. So the volt drop for this cable length and load is equal to:

$$4.4 \times 10^{-3} \text{ V} \times 26.09\text{A} \times 18\text{m} = 2.06\text{V}$$

Since this is less than the maximum permissible value for a lighting circuit of 6.9V, a 10mm cable satisfies the current and drop in voltage requirements when the circuit is protected by a CB. This cable is run in a loft that gets hot in summer and has thermal insulation touching one side of the cable. We must, therefore, use installed reference Method A of Table 3.12. If we were able to route the cable under the floor, clipped direct or in conduit or trunking on a wall, we may be able to use a 6mm cable for this load. You can see how the current-carrying capacity of a cable varies with the installed method by looking at Table 3.12. Compare the values in column 2 with those in column 6. When the cable is clipped direct on to a wall or surface the current rating is higher because the cable is cooler. If the alternative route was longer, you would need to test for volt drop before choosing the cable. These are some of the decisions the electrical contractor must make when designing an installation which meets the requirements of the customer and the IET Regulations.

If you are unsure of the standard fuse and CB rating of protective devices, you can refer to Fig. 3A4 of Appendix 3 of the IET Regulations.

Key fact

Volt drop
Maximum permissible volt drop on 230V supplies:

- 3% for lighting – 6.9V
- 5% for other uses –11.5V

IET Regulation 525 and Table 4Ab Appendix 4.

Cable size for standard domestic circuits

Section 3.2 of Chapter 3 of the IET *Electrical Installation Design Guide* tells us that the basic design intent is to use standard final circuits wherever possible to avoid repeated design. Provided earth fault loop impedances are below 0.35 ohms for TN-C-S supplies and 0.8 ohms for TN-S supplies the standard circuits can be used as the basis of all final circuits. Appendix 4 of the IET Regulations (BS 7671) and Appendix F of the IET *On-Site Guide* contain tables for determining the current carrying capacities of conductors, which we looked at in the last section. However, for standard domestic circuits, Table 3.14 gives a guide to cable size. In this table, I am assuming a standard 230 V domestic installation, having a sheathed earth or PME supply terminated in a 100 A HBC fuse at the mains position. Final circuits are fed from a consumer unit, having Type B CB protection and wired in PVC insulated and sheathed cables with copper conductors having a grey thermoplastic PVC outer sheath or a white thermosetting cable with LSF (low smoke and fume properties). I am also assuming that the surrounding temperature throughout the length of the circuit does not exceed 30°C and the cables are run singly and clipped to a surface.

Table 3.12 Current carrying capacity of cables

Conductor cross-sectional area	Reference Method A (enclosed in conduit in an insulated wall, etc.)		Reference Method B (enclosed in conduit on a wall or ceiling, or in trunking)		Reference Method C (clipped direct)		Reference Method E (on a perforated cable tray) or in free air	
	Two cables single-phase a.c. or d.c.	Three or four cables three-phase a.c.	One two-core cable, single-phase a.c. or d.c.	One three-core cable or one four-core cable, three-phase a.c.	One two-core cable, single-phase a.c. or d.c.	One three-core cable or one four-core cable, three-phase a.c.	One two-core cable, single-phase a.c. or d.c.	One three-core cable or one four-core cable, three-phase a.c.
1 mm²	2 A	3 A	4 A	5 A	6 A	7 A	8 A	9 A
1	11	10	13	11.5	15	13.5	17	14.5
1.5	14	13	16.5	15	19.5	17.5	22	18.5
2.5	18.5	17.5	23	20	27	24	30	25
4	25	23	30	27	37	32	40	34
6	32	29	38	34	46	41	51	43
10	43	39	52	46	63	57	70	60
16	57	52	69	62	85	76	94	80
25	75	68	90	80	112	96	119	101
35	92	83	111	99	138	119	148	126

Table 3.13 Voltage drop in cables factor

Voltage drop (per ampere per metre)			Conductor operating temperature: 70°C
Conductor cross-sectional area (mm²) 1	Two-core cable, d.c. (mV/A/m) 2	Two-core cable, single-phase a.c. (mV/A/m) 3	Three- or four-core cable, three-phase (mV/A/m) 4
1	44	44	38
1.5	29	29	25
2.5	18	18	15
4	11	11	9.5
6	7.3	7.3	6.4
10	4.4	4.4	3.8
16	2.8	2.8	2.4
25	1.75	1.75	1.50
35	1.25	1.25	1.10

Table 3.14 Cable size for standard domestic circuits

Type of final circuit	Cable size (twin and earth)	MCB rating, Type B (A)	Maximum floor area covered by circuit (m²)	Maximum length of cable run (m)
Fixed lighting	1.0	6	–	40
Fixed lighting	1.5	6	–	60
Immersion heater	2.5	16	–	30
Storage radiator	2.5	16	–	30
Cooker (oven only)	2.5	16	–	30
13A socket outlets (radial circuit)	2.5	20	50	30
13A socket outlets (ring circuit)	2.5	32	100	90
13A socket outlets (radial circuit)	4.0	32	75	35
Cooker (oven and hob)	6.0	32	–	40
Shower (up to 7.5 kw)	6.0	32	–	40
Shower (up to 9.6 kw)	10	40	–	40

Assessment criteria 4.7

Verify if disconnection times have been achieved

Assessment criteria 4.9

Establish the maximum disconnection times for different circuits

The general earthing arrangements has already been covered in learning outcome 3.1 but the principle behind earth fault loop impedance will now be covered.

Earth fault loop impedance Zs

In order that an overcurrent protective device can operate successfully it must meet the required disconnection times of IET Regulation 411.3.2.2, that is, final circuits not exceeding 32 A shall have a disconnection time not exceeding 0.4 s. To achieve this, the earth fault loop impedance value measured in ohms must be less than those values given in Appendix B of the *On-Site Guide* and Tables 41.2 and 41.3 of the IET Regulations.

The actual fault path to earth is known as the earth loop impedance, impedance being used when we combine an inductor and resistance. Very simply it can be thought of as alternating resistance and opposes any rate of change in an alternating signal such as the mains which operates at a frequency of 50 Hz. The mains, of course, comes from a sub-station and is stepped down through the use of a transformer. The transformer therefore uses mutual induction and R1 and R2 are the line and CPC, respectively. Therefore by combining both the inductor within the transformer and the conductors we bring about the term impedance.

The value of the earth fault loop impedance may be verified by means of an earth fault loop impedance test as described in Chapter 4 of this book. The formula is:

$$Z_S = Z_E + (R_1 + R_2)(\Omega)$$

Here Z_E is the impedance of the supply side of the earth fault loop. The actual value will depend upon many factors: the type of supply, the ground conditions, the distance from the transformer, etc. The value can be obtained from the area electricity companies, but typical values are 0.35 Ω for TN-C-S (protective multiple earthing, PME) supplies and 0.8 Ω for TN-S (cable sheath earth) supplies as described a little earlier in this chapter. Also in the above formula, R_1 is the resistance of the line conductor and R_2 is the resistance of the earth conductor. The complete earth fault loop path is shown in Fig. 3.113.

Values of $R_1 + R_2$ have been calculated for copper and aluminium conductors and are given in Table I1 of the *On-Site Guide* as shown in Table 3.15.

The maximum permitted value given in Table 41.3 of the IET Regulations for a 20 A CB protecting a socket outlet is 2.19 Ω as shown by Table 3.16. The circuit earth fault loop impedance is less than this value and therefore the protective device will operate within the required disconnection time of 0.4 s. The value of the earth fault loop impedance can be measured using the test described in IET Regulation 619.2 and compared with the tables of 41.42 and 41.3 of the IET Regulations and Appendix B of the *On-Site Guide*. The test is also described in Chapter 4 of this book.

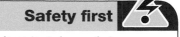

Safety first

A very important change that came about through Amendment 3 of BS 7671 (2015) is that earth loop impedance values were lowered.

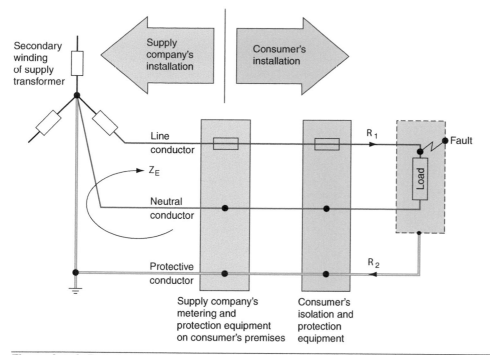

Figure 3.113 Earth fault loop path for a TN-S system.

Example

A 20 A radial socket outlet circuit is wired in 2.5 mm² PVC cable incorporating a 1.5 mm² CPC. The cable length is 30 m installed in an ambient temperature of 20°C and the consumer's protection is by 20 A CB Type B to BS EN 60898. The earth fault loop impedance of the supply is 0.5 Ω. Calculate the total earth fault loop impedance Z_S, and establish that the value is less than the maximum value permissible for this type of circuit.
We have:

$$Z_S = Z_E + (R_1 + R_2)\ (\Omega)$$
$$Z_E = 0.5\,\Omega\ \text{(value given in the question)}$$

From the value given in Table I1 of the *On Site Guide* and also shown in Table 3.15, a 2.5 mm phase conductor with a 1.5 mm protective conductor has an $(R_1 + R_2)$ value of $19.51 \times 10^{-3}\,\Omega/\text{m}$.

$$\text{For 30m cable } (R_1 + R_2) = 19.51 \times 10^{-3}\ \Omega/\text{m} \times 30\text{m} = 0.585\,\Omega$$

However, under fault conditions, the temperature and therefore the cable resistance will increase. To take account of this, we must multiply the value of cable resistance by the factor given in Table I3 of the *On-Site Guide*. In this case the factor is 1.20 and therefore the cable resistance under fault conditions will be:

$$0.585\,\Omega \times 1.20 = 0.702\,\Omega$$

The total earth fault loop impedance is therefore:

$$Z_S = 0.5\,\Omega + 0.702\,\Omega = 1.202\,\Omega$$

Table 3.15 value of resistance/metre for copper conductors and of $R_1 + R_2$ per metre at 20°C in milliohms/metres

Cross-sectional area (mm²)		Resistance/metre or $(R_1 + R_2)$/metre (mΩ/m)
Phase conductor	**Protective conductor**	**Copper**
1	–	18.10
1	1	36.20
1.5	–	12.10
1.5	1	30.20
1.5	1.5	24.20
2.5	–	7.41
2.5	1	25.51
2.5	1.5	19.51
2.5	2.5	14.82
4	–	4.61
4	1.5	16.71

Table 3.16 Maximum earth fault loop impedance Z_s (Ω) when the overcurrent protective device is an MCB Type B to BS EN 60898

	MCB rating (A)					
	6	10	16	20	25	32
For 0.4 s disconnection Z_s(Ω)	7.28	4.37	2.73	2.19	1.75	1.37

The maximum permitted value given in the IET Regulations for a 20 amp circuit breaker protecting a socket outlet is 2.19 ohms as shown in Table 3.16. The circuit earth fault loop impedance in the example is 1.202 ohms, which is less than 2.19 ohm, and therefor the protective device will operate within the required disconnection time of 0.4 seconds. The value of earth fault loop impedance can be measured using the test described in IET Regulation 643.7.3 and in Chapter 4 of this book.

Assessment criteria 4.8

Evaluate thermal constraints

Protective conductor size

The CPC forms an integral part of the total earth fault loop impedance, so it is necessary to check that the cross-section of this conductor is adequate. If the cross-section of the CPC complies with Table 54.7 of the IET Regulations, there is no need to carry out further checks. Where line and protective conductors are made from the same material, Table 54.7 tells us that:

- for line conductors equal to or less than 16 mm², the protective conductor should equal the line conductor;
- for line conductors greater than 16 mm² but less than 35 mm², the protective conductor should have a cross-sectional area of 16 mm²;
- for line conductors greater than 35 mm², the protective conductor should be half the size of the line conductor.

However, where the conductor cross-section does not comply with this table, then the formula given in IET Regulation 543.1.3 must be used:

$$S = \frac{\sqrt{I^2 t}}{k} \, (\text{mm}^2)$$

where

S = cross-sectional area in mm²

I = value of maximum fault current in amperes

t = operating time of the protective device

k = a factor for the particular protective conductor

Finally, as has been stated previously, a foolproof method of giving protection to people or animals who simultaneously touch both live and neutral has yet to be devised. The ultimate safety of an installation depends upon the skill and experience of the electrical contractor and the good sense of the user.

Given the complexity of the process involved it will now be covered again step by step with reference to both BS 7671 and the *On-Site Guide*. To assist with referencing, each specific element is signposted below to indicate where certain information can be found:

Example 1

A 230 V ring main circuit of socket outlets is wired in 2.5 mm single PVC copper cables in a plastic conduit with a separate 1.5 mm CPC. An earth fault loop impedance test identifies Z_s as 1.15 Ω. Verify that the 1.5 mm CPC meets the requirements of IET Regulation 543.1.3 when the protective device is a 30 A semi-enclosed fuse.

$$I = \text{Maximum fault current} = \frac{V}{Z_s} \, (\text{A})$$

$$\therefore \frac{230}{1.15} = 200 \, \text{A}$$

t = maximum operating time of the protective device for a circuit not exceeding 32 A is 0.4 s from IET Regulation 411.3.2.2. From Fig. 3A2(a) in Appendix 3 of the IET Regulations you can see that the time taken to clear a fault of 200 A is about 0.4 s

k = 115 (from Table 54.3)

$$S = \frac{\sqrt{I^2 t}}{k} \, (\text{mm}^2)$$

$$S = \frac{\sqrt{(200 \, \text{A})^2 \times 0.4 \, \text{s}}}{115} = 1.10 \, \text{mm}^2$$

A 1.5 mm² CPC is acceptable since this is the nearest standard-size conductor above the minimum cross-sectional area of 1.10 mm² found by calculation.

Example 2

A TN supply feeds a domestic immersion heater wired in 2.5 mm^2 PVC insulated copper cable and incorporates a 1.5 mm^2 CPC. The circuit is correctly protected with a 16A semi-enclosed fuse to BS 3036. Establish by calculation that the CPC is of an adequate size to meet the requirements of IET Regulation 543.1.3. The characteristics of the protective device are given in IET Regulation Fig. 3A2(a) of Appendix 3.

For final circuits less than 32A the maximum operating time of the protective device is 0.4 s. From Fig. 3A2(a) of Appendix 3 it can be seen that a current of about 90A will trip the 15A fuse in 0.4 s. The small insert table on the top right of Fig. 3A2(a) of Appendix 3 of the IET Regulations gives the value of the prospective fault current required to operate the device within the various disconnection times given.

So, in this case the table states that 90A will trip a 16A semi-enclosed fuse in 0.4 s

$$\therefore I = 90 \, A$$
$$t = 0.4 \, s$$
$$k = 115 \, (\text{from Table 54.3})$$
$$S = \frac{\sqrt{I^2 t}}{k} \, (\text{mm}^2) \, (\text{from IET Regulation 543.1.3})$$
$$S = \frac{\sqrt{(90 \, A)^2 \times 0.4s}}{115} = 0.49 \, \text{mm}^2$$

The CPC of the cable is greater than 0.49 mm^2 and is therefore suitable. If the protective conductor is a separate conductor, that is, it does not form part of a cable as in this example and is not enclosed in a wiring system as in Example 1, the cross-section of the protective conductor must be not less than 2.5 mm^2 where mechanical protection is provided or 4.0 mm^2 where mechanical protection is **not** provided in order to comply with IET Regulation 544.2.3.

Design from first principles

- Design current: Power/Volts
- Protective device (rating available): *On-Site Guide* Appendix B
- Installation method: *On-Site Guide* Table 7.1 (ii)
- Circuit correction factors and formula: *On-Site Guide* Appendix F
- Select cable size (current carrying capacity): *On-Site Guide*, Table F6
- Volt drop formula: *On-Site Guide* Appendix F
- Confirm volt drop: *On-Site Guide*, Table F4 (ii) or F6
- Confirm shock protection is met through loop impedance values: Appendix B Table B6
- Thermal constraint formula from BS 7671: Part 5, Chapter 54.

Extracting design data

Answers are in the back of the book.

1 State how we calculate design current.

2 State the rule regarding In (size of protective device) in relation to design current.

3 Where in the *On-Site Guide* can you find a list of protective device ratings normally available?

4 Write out the formula used when calculating the tabulated current-carrying capacity of the selected cable.

5 When using a BS 3036 fuse (Cf Correction) what correction factor should we use during the cable calculation stage?

6 What correction factor should we use for 3 PVC/PVC twin core cables that are grouped together and installed using Reference Method C? (Look in Appendix F.)

7 What correction factor should we use for any type of cable that is surrounded by thermal insulation for 0.5 m?

8 What correction factor should we use for a PVC/PVC twin core cable that is exposed to an ambient temperature of 35°C?

9 If the protective device of a circuit is rated at 20 A and Ca (0.8) and Ci (0.65) corrections have to be applied, calculate It: the tabulated current carrying capacity.

10 With reference to Appendix F of the *On-Site Guide*, what ambient temperature does not need correcting?

11 What is the correct description of Reference (Installation) Method C?

12 Given the reference method above and using Table F6 of the *On-Site Guide*, determine what size cable is required to carry 47 A.

13 State the tolerances for voltage drop regarding lighting and power. (Look in Appendix F of the *On-Site Guide* or volt drop in the index.)

14 What is the formula used for calculating voltage drop? (Look in Appendix F of the *On-Site Guide* or volt drop in the index.)

15 Using Table F6 of the *On-Site Guide*, the size of cable at step 12, a design current of 27 A, a circuit that is 20 m long, calculate if this circuit complies with the volt drop requirements for a lighting circuit.

16 What do we mean by shock protection?

17 Look in the *On-Site Guide* and determine the maximum *Zs* value for a 40 A Type B circuit breaker.

18 Given that this value has been extracted from the *On-Site Guide* do we have to correct the value by multiplying by 0.8?

19 Given the value above in step 17, if the measured loop impedance value Zs of a circuit is 0.91Ω, what does this mean regarding shock protection?

20 Given the Z_s value above calculate the PFC.

21 What do we mean by thermal constraint?

22 A 6 mm² PVC flat profile cable has a 2.5 mm² CPC and has a K constant of 115. If the PFC of the circuit is 1.7 KA and the circuit needs to be disconnected in 0.4 s, calculate if the thermal constraint requirement is met.

Design exercise

A 10 kW kiln is to be fed from a domestic consumer unit that is positioned 15m away. The wiring type is PVC/PVC flat profile cable that is clipped direct to the surface and the circuit should be protected by a Type B circuitbreaker. From enquiry the DNO state that the PFC is 329 A and Zs has been measured at 0.7 Ω. The cable is grouped with one other circuit in an area with an ambient temperature of 35°C.

Work out the following:

Answers are in the back of the book.

1 Design current (Ib).

2 Size of protective device (In).

3 Having established any correction factors, calculate the tabulated current carrying capacity (It).

4 Size of cable required.

5 Is the volt drop OK?

6 Is the shock protection OK?

7 Is the thermal constraint OK?

Assessment criteria 5.1

Specify the health and safety requirements in establishing a safe working environment

The Health and Safety at Work Act 1974 was passed as a result of recommendations made by the Royal Commission and, therefore, the Act puts legal responsibility for safety at work on *both* the employer and employee.

Figure 3.114 It is the duty of the employer to ensure adequate health and safety systems have been put in place.

In general terms, the employer must put adequate health and safety systems in place at work and the employee must use all safety systems and procedures responsibly.

In specific terms the employer must:

- provide a Health and Safety Policy Statement if there are five or more employees;
- display a current employer's liability insurance certificate as required by the Employers' Liability (Compulsory Insurance) Act 1969;
- report certain injuries, diseases and dangerous occurrences to the enforcing authority such as the Health & Saftey Executive;
- provide adequate first aid facilities;
- provide PPE;
- provide information, training and supervision to ensure staffs' health and safety;
- provide adequate welfare facilities;
- put in place adequate precautions against fire, provide a means of escape and means of fighting fire;
- ensure plant and machinery are safe and that safe systems of operation are in place;
- ensure articles and substances are moved, stored and used safely;
- make the workplace safe and without risk to health by keeping dust, fumes and noise under control.

In specific terms the employee must:

- take reasonable care of his/her own health and safety and that of others who may be affected by what they do;
- cooperate with his/her employer on health and safety issues by not interfering with or misusing anything provided for health, safety and welfare in the working environment;
- report any health and safety problem in the workplace to, in the first place, a supervisor, manager or employer.

Assessment criteria 5.2

Specify how to ensure that worksites are correctly prepared

Safe system of work

Employers' responsibilities are specifically defined through the Management of Health and Safety at Work Regulations 1999, which detail how employers must systematically examine the workplace, the work activity and the management of safety, ensuring that a safe system of work is in place.

Therefore a safe system of work (SSOW) is simply a means of ensuring that employees operate in a safe and controlled environment. In the electrical installation industry, a safe system of work is underpinned by three primary elements: risk assessments, method statements and permit to work systems. They are designed to:

- Identify hazards and exposure to the hazards (risk): **formal risk assessment**
- Detail how a procedure will be carried out: **method statement**
- Work in hazardous areas has to be authorized: **permit to work**

Let us look at each in detail.

The process of risk assessment for the electrical installation industry would include:

- working at heights;
- using electrical power tools;
- falling objects;
- working in confined places;
- electrocution and personal injury;
- working with 'live' equipment;
- using hire equipment;
- manual handling – pushing – pulling – lifting;
- site conditions – falling objects – dust – weather – water – accidents and injuries.

Aside from considering personnel, the same consideration should be given to the property and any animals and livestock including reflecting on the safest route to and from a site or specific location. Any particular hazardous activity would need to be controlled through a permit to work, which instils structure through a formal authorization pro forma, but also requires the person undertaking the job declaring that they are fully conversant with the task in hand. A properly defined method statement will detail how a procedure or job will be carried out referencing any risk assessments, special tools, access equipment as well as how any waste material will be disposed of.

All these procedures make up a safe system of work but knowing where they are stored and how to gain access to them is also important. This information should be shared with all employees including apprentices.

Definition

- A hazard is something that might cause harm.
- A risk is the chance that harm will be done.

Assessment criteria 5.3

Clarify the preparations that should be completed before electrical installation work starts

Assessment criteria 5.4

Outline how to check for any pre-existing damage to customer/client property

Assessment criteria 5.5

State the actions that should be taken if pre-existing damage to customer/client property is identified

Assessment criteria 5.6

Specify methods for protecting the fabric and structure of the property before and during installation work

To make the work area safe before starting work, and during work activities, it may be necessary to:

- use barriers or tapes to screen off potential hazards;
- place warning signs as appropriate;
- inform those who may be affected by any potential hazard;
- use a safe isolation procedure before working on live equipment or circuits;
- obtain any necessary permissions or 'Permits to Work' before work begins.

Get into the habit of always working safely and being aware of the potential hazards around you when working. If pre-existing damage to the structure or fabric of the building is observed by the electrical team upon arrival at the site, then a decision must be taken, before the electrical work commences, as to who will be responsible for the repairs. If the building structure is damaged, it may be unsafe to commence the electrical work before remedial work has been undertaken. If the building fabric or structure is damaged as a result of the electrical work activities, for example where conduits or trunking pass through the floor or walls, then it must be made good by the electrical team or their representative.

The electrical team must always be respectful of the customers property and contents when carrying out their electrical activities. This might involve laying dust sheets over some items, or moving others to a safe area while work is carried out, or wearing plastic protectors over your shoes so that carpets are not damaged.

Always be polite to the customer. The customer will probably see more of the electrican and electrical trainee than the managing director of the company, and therefore the image presented by them to the customer will be assumed to reflect the company policy. You are therefore your companies most important representative at that time, and indirectly, it is the customer who pays your wages.

Finally, when the job is completed, clean up and dispose of any waste material responsibly. This is an important part of your companies 'good customer relationship' with the client. We also know that we have a 'duty of care' for the waste that we produce as an electrical company.

Test your knowledge

When you have completed the questions check out the answers at the back of the book.

Note: more than one multiple-choice answer may be correct.

Learning outcome 1

1   Which of the following statutory requirements is most important when planning an installation?
    a.   Health and Safety Act 1974
    b.   BS 7671 Wiring Regulations
    c.   Building Regulation 2000
    d.   Electricity at Work Regulation 1989.

2   Which of the following is an important consideration regarding the movement of vehicles and adequate flood lighting?
    a.   safe access and egress
    b.   training facilities
    c.   adequate welfare facilities
    d.   a safety policy.

3   According to the Electricity at Work Act 1989 when can you carry out a 'live' repair?
    a.   only if you are supervised
    b.   only if you feel that you are 'competent' to do so
    c.   never, you are not allowed to
    d.   only if you are experienced.

4   As fitted, drawings form part of the:
    a.   fitting element
    b.   timetabling element
    c.   scheduling element
    d.   completion element.

5   Band 1 and Band 2 cabling can be enclosed in the same enclosure if:
    a.   they are insulated to the highest voltage present
    b.   at 90° to each other
    c.   separated by metal divisions
    d.   wrapped around each other.

6   What kind of drawing is this?
    a.   block
    b.   schematic
    c.   detail
    d.   layout.

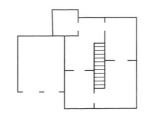

7   Which document details the duties of employers to employees in providing a safe place of work?

    a.   IET Wiring Regulations

    b.   Building Regulations

    c.   Electricity at Work Regulations

    d.   Health and Safety at Work Act.

8   Ignoring BS 7671 would most likely cause:

    a.   an electrical fire

    b.   a real fire

    c.   a solid fuel fire

    d.   a gas fire.

9   Removing all elements of electrical supply to a piece of equipment in order to ensure the safety of those working on it is one definition of:

    a.   emergency switching

    b.   functional switching

    c.   switching for isolation

    d.   switching for mechanical maintenance.

10   Cutting off the full load electrical supply to a piece of equipment in order to ensure the safety of those working on it is one definition of:

    a.   emergency switching

    b.   functional switching

    c.   switching for isolation

    d.   switching for mechanical maintenance.

11   The rapid disconnection of the electrical supply to remove or prevent danger is one definition of:

    a.   emergency switching

    b.   functional switching

    c.   switching for isolation

    d.   switching for mechanical maintenance.

12   Switching electrical equipment in normal service is:

    a.   emergency switching

    b.   functional switching

    c.   switching for isolation

    d.   switching for mechanical maintenance.

13 Looking at the diagram, the numbers 2, 3 and 4 represent:

  a. service meter, meter tails, service fuse

  b. service meter, service tails, service fuse

  c. service head, meter tails, service fuse

  d. service head, service meter, service fuse.

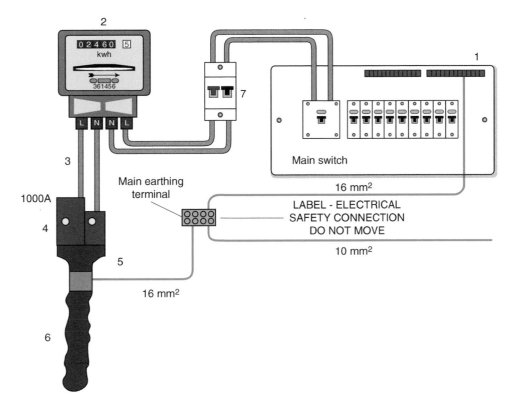

14 The maximum working temperature of PVC/PVC thermoplastic is:

  a. 40°C

  b. 50°C

  c. 60°C

  d. 70°C.

15 A light fitting or small spot light must be fixed at an adequate distance from:

  a. combustible material

  b. combustible walls

  c. combustible ceilings

  d. combustible joists.

16 Which of the following **CANNOT** be used as an isolator?

  a. switch

  b. circuit-breaker

  c. semi-conductor (electronic device)

  d. switch disconnector.

17 What equipment type does this symbol relate to?

  a. Class I

  b. Class II

  c. Class III

  d. Class IV.

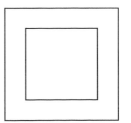

18 Three factors that would affect the choice of wiring system would include:
   a. temperature
   b. animals and plants
   c. architect
   d. impact damage.

19 Additional protection applies to:
   a. cables located within metal enclosures
   b. SWA
   c. cables embedded >50 mm
   d. cables embedded <50 mm
   e. special locations.

20 What is the normal size and rating of a lighting cable and protective device?
   a. 6 mm² and 6 A
   b. 2.5 mm² and 6 A
   c. 4 mm² and 6 A
   d. 1.5 mm² and 6 A.

21 Equipment installed in a flour mill needs to be protected against entry of particles such as:
   a. dust stopped through IP protection
   b. dust stopped through fault protection
   c. dust stopped through basic protection
   d. dust stopped through additional protection.

22 The protection provided by insulating live parts is called:
   a. extraneous conductive parts
   b. basic protection
   c. exposed conductive parts
   d. fault protection.

23 The typical cabling for use in electrical installations in domestic premises would be:
   a. SWA
   b. PVC/PVC
   c. MICC
   d. PVC single.

24 Ladder racking is a system for:
   a. allowing access to storage racks
   b. allowing multiple section extension ladders
   c. supporting armoured cables
   d. supporting vertical cable basket systems.

25 The type of wiring for use in a factory with high ambient temperatures is:
   a. PVC insulated and sheathed
   b. PVC insulated cables in galvanized conduit
   c. mineral insulated metal sheathed
   d. rubber insulated in PVC conduit.

26 Mineral insulated cables are sometimes provided with an overall covering of PVC to:

a. make the cable waterproof

b. protect the cable from sunlight and harmful rays

c. prevent corrosion of the sheath

d. make the cable suitable for use in low ambient temperatures.

27 Where would it be unsuitable to use a rate of rise temperature type heat detector?

a. kitchen

b. dining room

c. hallway

d. garage.

28 An appropriate wiring method for an underground feed to a remote building would be a:

a. metal conduit installation

b. trunking and tray installation

c. PVC cables

d. PVC/SWA cables.

29 An appropriate wiring system for a three-phase industrial installation would be:

a. PVC cables

b. PVC conduit

c. one that meets the requirements of Part 2 of the IET Regulations

d. one that meets the requirements of Part 5 of the IET Regulations.

30 Which wiring system is non-ageing and will ensure that the space between its conductors is kept the same even if flattened?

a. mineral insulated

b. SWA

c. PVC/PVC

d. PVC singles.

31 Which conductor uses the principle of total internal refraction?

a. FP cable

b. fibre-optic

c. mineral insulated

d. PVC/PVC flat profile cable.

32 An alternative to mineral insulated cable that does not require specialized tools would be:

a. FP

b. fibre-optic

c. SWA

d. PVC flex.

33   Which of the following is the most appropriate containment system for industrial high current applications?
a.   busbar systems
b.   flexible conduit
c.   cable tray
d.   power track.

34   Which of the following would be most appropriate regarding offsetting the adverse effects of vibration?
a.   PVC conduit
b.   plastic conduit
c.   metal conduit
d.   flexible conduit.

35   Which of the following apply to electrical equipment in bathrooms?
a.   supplementary bonding
b.   index protection
c.   earthing
d.   additional protection.

36   The maximum floor area that a ring circuit can feed is:
a.   50 m²
b.   75 m²
c.   100 m²
d.   25 m².

37   Combining a copper cable with an aluminium crimp will result in:
a.   increased fault current
b.   lower termination resistance
c.   dissimilar metal corrosion
d.   an explosion.

38   Single core non-armoured cables (apart from bonding conductors) must always be installed:
a.   in free air partially covered
b.   in conduit or trunking
c.   secured with plastic clips
d.   in cavity walls.

39   A non-fused spur from a ring may feed:
a.   any number of socket outlets
b.   any number of fixed appliances
c.   one single or twin socket outlet
d.   two fixed appliances.

40   An A1 type circuit is associated with:
a.   2.5 mm with 30/32 A fuse
b.   4 mm radial with protective device
c.   1.5 mm radial with 20 A fuse
d.   4 mm ring with 20 A fuse.

41 Ten 1.0 mm single conductors are to be drawn into a conduit. Use Table E3 of the *On-Site Guide* or Table 3.5 of this book to find the sum of the cable factors for these cables.

    a.   16

    b.   30

    c.   160

    d.   300.

42 A three-metre length of conduit with two bends is to contain the ten 1.0 mm cables. Determine the size of conduit required using Table E4 of the *On-Site Guide* or Table 3.6 of this book.

    a.   16 mm

    b.   20 mm

    c.   25 mm

    d.   32 mm.

43 Ten 2.5 mm single conductors are to be drawn into a conduit. Use Table E3 of the *On-Site Guide* or Table 3.5 of this book to find the sum of the cable factors for these cables.

    a.   16

    b.   30

    c.   160

    d.   300.

44 A three-metre length of conduit with two bends is to contain the ten 2.5 mm cables. Determine the size of conduit required using Table E4 of the *On-Site Guide* or Table 3.6 of this book.

    a.   16 mm

    b.   20 mm

    c.   25 mm

    d.   32 mm.

45 Ten 1.5 mm solid PVC cables are to be drawn into trunking. Use Table E5 or Table 3.7 of this book to find the sum of the cable factors for these cables.

    a.   80

    b.   60

    c.   40

    d.   20.

46 Ten 2.5 mm solid PVC cables are to be drawn into trunking. Use Table E5 or Table 3.8 of this book to find the sum of the cable factors for these cables.

    a.   116

    b.   117

    c.   118

    d.   119.

47 Refer to Tables E1 and E2 of the *On-Site Guide* and by applying cable and conduit factors, work out how many 4 mm² cables can be inserted into a 25 mm piece of conduit used in a short straight run.

    a.   14

    b.   13

    c.   12

    d.   11.

Learning outcome 2

48 Additional protection is applicable to which of the following?
   a. HRC fuse
   b. re-wireable fuse
   c. RCBO
   d. CB.

49 Which of the following will not operate if the line conductor bypasses the load and shorts out against the neutral conductor?
   a. HRC fuse
   b. RCD
   c. RCBO
   d. CB.

50 An Industrial welder would need to be protected by:
   a. an RCD
   b. a Type C circuit-breaker
   c. a Type D circuit-breaker
   d. a HBC fuse.

51 Coordination of fuses during fault or overload conditions is known as:
   a. diversity
   b. diverging
   c. discrimination
   d. deviation.

52 A circuit-breaker has two modes of operation:
   a. overload due to thermal effect or short circuit due to a magnetic trip
   b. overload due to a magnetic trip or short circuit due to thermal effect
   c. overload due to a chemical reaction or short circuit due to a magnetic trip
   d. overload due to a thermal trip or chemical reaction due to a magnetic trip.

53 An CB rated at 40 A will trip:
   a. only when a short circuit occurs
   b. only when an overload occurs
   c. when an overload, short circuit or earth fault occurs
   d. when 41 A flows through it.

54 An RCD rated at 30 mA will trip:
   a. when enough current flows to earth
   b. through an earth fault or an overload
   c. through an overload, short circuit or earth fault
   d. when 30 mA flows.

55 What type of CB is associated with commercial applications especially fluorescent lighting?
   a. Type 3
   b. Type B
   c. Type C
   d. Type D.

56   Which device will detect an imbalance between line and neutral, overloads and all types of overcurrent faults?

     a.   RCD

     b.   RCBO

     c.   CB

     d.   BBC.

57   A fuse protects:

     a.   the person

     b.   the switch

     c.   the cable

     d.   the load.

58   An RCD protects:

     a.   the person

     b.   the switch

     c.   the cable

     d.   the load.

59   A value that ensures that a circuit-breaker can withstand a fault current is known as:

     a.   breaking capacity

     b.   rating

     c.   thermal constraint

     d.   shock protection.

Learning outcome 3

60   Which earthing system is not recommended for petrol stations, caravans and marina pontoons?

     a.   TN-S

     b.   TT

     c.   TNC

     d.   PME.

61   What is the maximum tabulated value of Z_e for a TN-C-S system?

     a. $0.8 \, \Omega$

     b. $0.35 \, \Omega$

     c. $20 \, \Omega$

     d. $1.75 \, \Omega$.

62   Which earthing system due to the possibility of high Z_e values must have additional protection?

     a.   TN-S

     b.   TT

     c.   TN-C-S

     d.   PME.

63   Which uncommon earthing system is used in hospitals?

     a.   TN-S

     b.   IT

     c.   TNC

     d.   PME.

64   What supply system uses a PEN conductor?
    a.   TT
    b.   TN-S
    c.   IT
    d.   TN-C-S.

65   Which earthing system uses an earth electrode and is common in rural areas?
    a.   TN-S
    b.   TT
    c.   TN-C-S
    d.   PME.

66   The metalwork that does *not* form part of an electrical installation is called:
    a.   extraneous conductive parts
    b.   basic protection
    c.   exposed conductive parts
    d.   fault protection.

67   Which of the following is an extraneous conductive part?
    a.   metal casing of a washing machine
    b.   metal switch back plate
    c.   metal casing of an electric motor
    d.   metal casing of a gas pipe.

68   A current that goes beyond the rated value in an otherwise healthy circuit is one definition of:
    a.   earthing
    b.   bonding
    c.   overload
    d.   short circuit.

69   What is the maximum disconnection time for TT earthing systems > 32 A?
    a.   5 s
    b.   1 s
    c.   0.2 s
    d.   0.4 s.

70   A TN-S earthing system operates through the:
    a.   earthing electrode
    b.   PEN conductor
    c.   supply cable armouring
    d.   earth itself.

71   Earthing ensures that:
    a.   a low resistance path, will safeguard that the PFC is high and ensure disconnection is quick
    b.   a high resistance path, will safeguard that the PFC is high and ensure disconnection is quick
    c.   a low resistance path, will safeguard that the PFC is low and ensure disconnection is slow
    d.   a high resistance path, will safeguard that the PFC is low and ensure disconnection is quick.

72 Which of the following symbols represents the external earth loop impedance value?

   a. Zx

   b. Zs

   c. Ze

   d. ZT.

73 Which of the following symbols represents the total earth loop impedance value?

   a. Zx

   b. Zs

   c. Ze

   d. ZT.

74 The protection provided by equipotential bonding and automatic disconnection of the supply is called:

   a. extraneous conductive parts

   b. basic protection

   c. exposed conductive parts

   d. fault protection.

75 Equipotential bonding is carried out to:

   a. earth all equipment

   b. earth all exposed conductive parts

   c. link all extraneous conductive parts

   d. stop fuses operating.

76 Which earthing system changes line to earth faults to line to neutral faults?

   a. TN-S

   b. TT

   c. TN-C-S

   d. IT.

77 A TN-S earthing system will follow which path?

   a. CPC, main earthing terminal, supply armouring back to the transformer

   b. earth electrode, CPC, main earthing terminal back to the transformer

   c. main earthing terminal, CPC, PEN conductor, back to the transformer

   d. CPC, high impedance load, main earthing terminal back to the transformer.

78 Which earthing system has to be earthed itself along its various supply routes?

   a. TN-S

   b. TT

   c. PME

   d. IT.

Learning outcome 4

79   What device is described here: The 'It' rating must be equal to or above the normal operating current of the circuit?

a.   a fuse

b.   fuse-board

c.   distribution board

d.   junction box.

80   What device is described here: The 'It' rating must be equal to or above the normal current carrying capacity of the protective device?

a.   cable

b.   switch

c.   load

d.   consumer unit.

81   Calculate the total current demand for a domestic lighting circuit if it contains 20 × 100 W GLS filaments and the factor for diversity is 66%.

a.   7.8 A

b.   8.69 A

c.   6 A

d.   5.7 A.

82   When cabling is routed through insulation, the rated:

a.   current carrying ability will be decreased

b.   current carrying ability will be increased

c.   current carrying ability will stay exactly the same

d.   current carrying ability will be doubled per metre.

83   If voltage drop in a final circuit is found to be excessive the corrective action is to:

a.   limit the time the load is connected

b.   boost the voltage through a transformer

c.   increase the size of the supply cable

d.   decrease the size of the supply cable.

84   When overload and overcurrent protection is necessary which of the following is correct?

a. $Ib \geq In \geq Iz$

b. $Ib \leq In \geq Iz$

c. $Ib \leq In \leq Iz$

d. $Ib \geq In \leq Iz$.

85   Which of the following is correct regarding earth fault loop impedance?

a.   Zs can be obtained by enquiry

b.   Ze can be obtained by enquiry

c.   Zs can be obtained by calculation

d.   Zs can be obtained by measurement.

86   One disadvantage of using the tabulated values of Ze in the *On-Site Guide* is:

     a.   they are hard to read

     b.   the values could be high

     c.   the values are unreliable

     d.   the values could be low.

87   When checking earth loop impedance values which of the following can be used directly without correction?

     a.   BS 7671

     b.   wiring regulations

     c.   GN3

     d.   *On-Site Guide*.

88   A domestic cooker rated at a maximum load of 60 A is controlled by a cooker control unit, which does not have a 13 A socket outlet. Using Table A1 of the *On-Site Guide*, and after allowing for diversity, calculate the design current.

     a.   20 A

     b.   25 A

     c.   18 A

     d.   30 A.

89   Correction factors as applied to de-rate system cables are:

     a.   thermal insulation, grouping, re-wireable fuses and ambient temperature

     b.   outside insulation, grouping, re-wireable fuses and ambient temperature

     c.   using SWA cable, grouping, re-wireable fuses and ambient temperature

     d.   thermal insulation, bare wires, re-wireable fuses and ambient temperature.

90   Which of the following should be recorded regarding the PFC?

     a.   the prospective short circuit current

     b.   the prospective earth fault current

     c.   the highest value PFC at the point of supply

     d.   the higher value PFC at the furthermost point of the circuit.

91   Why is thermal constraint particularly important when installing with PVC flat profile cable?

     a.   it is cheap in comparison to other types

     b.   its CPC is smaller than the live conductors

     c.   its CPC is bigger than the live conductors

     d.   its CPC does not come with earthing sleeving.

92   To ensure the effective operation of the overcurrent protective devices, the earth fault loop path must have:

     a.   a 230 V supply

     b.   very low resistance

     c.   fuses or CBs in the live conductor

     d.   very high resistance.

93 Shock protection ensures that:
   a. the protective device operates within a specified time
   b. the actual loop impedance values are less than specified
   c. the PFC is low enough to operate the protective devices in time
   d. the circuit is switched off in time.

94 According to IET Regulation 411.3.2.2 all final circuits not exceeding 32 A in a building supplied with a 230 V TN supply shall have a maximum disconnection time not exceeding:
   a. 0.2 s
   b. 0.4 s
   c. 5.0 s
   d. unlimited.

95 The current demand of 10–65 W fluorescent tubes connected to a 230 V supply is:
   a. 5.08 mA
   b. 5.09 A
   c. 150 W
   d. 650 W.

96 Use the *On-Site Guide* (Appendix A) to calculate the assumed current demand of a cooker having 4–3 kW boiling rings and a 2 kW oven connected to a 230 V supply. The cooker control unit includes a socket outlet.
   a. 60.86 A
   b. 30.26 A
   c. 25.25 A
   d. 10 A.

97 The total current demand of a lighting circuit in an individual household is calculated to be 5 A. The current demand applying the diversity factor given in Table A2 of the *On-Site Guide* will be:
   a. 3.3 A
   b. 3.75 A
   c. 4.5 A
   d. 5.0 A.

98 The volt drop in mV per ampere per metre of a two-core single-phase 6.0 mm cable as given in Table F5 (ii) of the *On-Site Guide* or Table 3.13 of this book is:
   a. 18
   b. 11
   c. 7.3
   d. 4.4.

99 The volt drop in mV per ampere per metre of a two-core single-phase 10.0 mm cable as given in Table F5 (ii) of the *On-Site Guide* or Table 3.13 of this book is:
   a. 18
   b. 11
   c. 7.3
   d. 4.4.

100 With respect to the volt drop values in the last two questions (98–99), which of the following applies.

    a.   volt drop value is less if cable size is increased

    b.   volt drop value is bigger if cable size is increased

    c.   volt drop value is less if cable size is decreased

    d.   volt drop value is less if cable size is bigger

101 The current carrying capacity of a two-core single-phase 2.5 mm non-armoured PVC cable enclosed in trunking as given in Table F5 (i) of the *On-Site Guide* or Table 3.12 of this book is:

    a.   18 A

    b.   23 A

    c.   27 A

    d.   30 A.

102 The current carrying capacity of a two-core single-phase 2.5 mm non-armoured PVC cable clipped direct to a wall as given in Table F5 (i) of the *On-Site Guide* or Table 3.12 of this book is:

    a.   18 A

    b.   23 A

    c.   27 A

    d.   30 A.

103 The current carrying capacity of a two-core single-phase 2.5 mm non-armoured PVC cable laid on a perforated cable tray or cable basket as given in Table F5 (i) of the *On-Site Guide* or Table 3.12 of this book is:

    a.   18 A

    b.   23 A

    c.   27 A

    d.   30 A.

104 What % volt drop is associated with a power circuit?

    a.   3%

    b.   4%

    c.   5%

    d.   6%.

105 A cooker can be fitted with a smaller cable after applying?

    a.   discrimination

    b.   diversity

    c.   circuit correction

    d.   correction due to a semi-enclosed fuse.

106 Which of the following is the correction factor regarding semi-enclosed fuses?

    a. C_f

    b. C_i

    c. C_a

    d. C_g.

107 Which of the following represents design current?
   a. Ib
   b. I²
   c. Iz
   d. Ie.

108 A domestic instantaneous heater is rated at 3 kW. Given that the voltage is 230 V, work out its design current.
   a. 76.6 A
   b. 0.07 A
   c. 13.04 A
   d. 690 A.

109 From the example above using Table B6 of the *On-Site Guide* select a realistic and appropriate value for a circuit-breaker to act as a protective device (In).
   a. 13.04 A
   b. 15 A
   c. 16 A
   d. 12 A.

110 A PVC/PVC cable is to be clipped direct to the wall and routed alongside two other circuits but also comes into contact with 50 mm of thermal insulation. Using your value of In from question 109 and Tables F2 and F3 of the *On-Site Guide*, calculate the total tabulated current It.
   a. 15 A
   b. 13.04 A
   c. 17.045 A
   d. 21.57A.

111 Given that as already stated above the cable has been clipped direct to the surface (Reference method C), using Appendix F (Table F5 (i)) of the *On-Site Guide*, determine the size of the conductor required.
   a. 4 mm²
   b. 2.5 mm²
   c. 10 mm²
   d. 6 mm².

112 The cable length of the circuit is 25 m long. Using Appendix F to obtain the formula and Table F5 (ii) of the *On-Site Guide*, calculate the circuit volt drop.
   a. 5.86 V
   b. 6.75 V
   c. 4.89 V
   d. 9.7 V.

113 Given your volt drop calculations above, which of the following is correct?
   a. circuit volt drop complies with % volt drop tolerance required for power circuits
   b. circuit volt drop would also comply with % volt drop tolerance required for lighting circuits
   c. circuit volt drop does not comply with % volt drop tolerance required for lighting circuits
   d. circuit volt drop does not comply.

114 The prospective fault current has been measured at 1.7 kA, calculate Zs.
   a. Zs = 135 Ω
   b. Zs = 0.135 A
   c. Zs = 0.135 Ω
   d. Zs = 0.0135 Ω.

115 Use the rating of your protective device (In) from question 109 in combination with your calculated values of Zs above in order to compare them to a Type B circuit breaker Table B6 of the *On-Site Guide* to determine if shock protection is met.
   a. calculated is more than *On-Site Guide* values therefore shock protection is not met
   b. calculated is more than *On-Site Guide* values therefore shock protection is met
   c. calculated is equal to or more than *On-Site Guide* values therefore shock protection is met
   d. calculated is equal to or less than *On-Site Guide* values therefore shock protection is met.

116 If the same process is applied to BS 7671, which of the following is correct?
   a. calculated is more than 80% of BS 7671 values therefore shock protection is met
   b. calculated is less than 80% of BS 7671 values therefore shock protection is met
   c. calculated is equal to or more than 80% of BS 7671 values therefore shock protection is met
   d. calculated is equal to or less than BS 7671 values therefore shock protection is not met.

117 Given that PVC/PVC has a K constant of 115, alongside the requirement that the circuit needs to be disconnected in 0.1s. Using a PFC value given in question 114, work out the size of CPC required.
   a. 100 mm²
   b. 6 mm²
   c. 3.736 mm²
   d. 4.67 mm².

118 Using your answer above, the *On-Site Guide* Table 7.1 (i) (under cable size (mm²)) determine what size CPC is associated with a 2.5 mm² cable in order to work out if thermal constraint is met.
   a. cable conductor is bigger than calculated therefore thermal constraint is met
   b. cable conductor is smaller than calculated therefore thermal constraint is met
   c. cable conductor is bigger than calculated therefore thermal constraint is not met
   d. cable conductor is smaller than calculated therefore thermal constraint is not met.

Learning outcome 5

119   Prior to starting work there is an expectation that all individuals:

    a.   carry out a risk assessment of their immediate area

    b.   check their tools before use

    c.   hydrate themselves

    d.   complete their time sheet.

120   The most important way an employer can ensure that workplaces are safe is through:

    a.   first aid provisions

    b.   first aid qualified personnel

    c.   safe systems of work

    d.   annual appraisals.

121   The correct wiring system required will be shown on:

    a.   a method statement

    b.   a permit to work

    c.   as fitted drawings

    d.   the specification.

122   All equipment and tooling must be:

    a.   new

    b.   checked daily

    c.   secured

    d.   accounted for.

123   Risk assessment will highlight control measures such as:

    a.   access equipment

    b.   lifting equipment

    c.   wiring type

    d.   appropriate PPE.

124   A survey of the building must be carried out:

    a.   after completion of the installation

    b.   prior to commencement of the installation

    c.   mid way through installation

    d.   at any time.

125   Briefly explain why an electrical installation needs protective devices.

126   List the four factors on which the selection of a protective device depends.

127   List the five essential requirements for a device designed to protect against overcurrent.

128   Briefly describe the action of a fuse under fault conditions.

129   State the meaning of 'discrimination' or 'coordination' as applied to circuit protective devices.

130   Use a sketch to show how 'discrimination' or 'coordination' can be applied to a piece of equipment connected to a final circuit.

131   List typical 'exposed parts' of an installation.

132 List typical 'extraneous parts' of a building.

133 Use a sketch to show the path taken by an earth fault current.

134 Use bullet points and a simple sketch to briefly describe the operation of an RCD.

135 State the need for RCDs in an electrical installation:
   a.   supplying socket outlets with a rated current not exceeding 20 A and
   b.   for use by mobile equipment out of doors as required by IET Regulation 411.3.3.

136 Briefly describe an application for RCBOs.

137 In your own words state the meaning of circuit overload and short circuit protection. What will provide this type of protection?

138 State the purpose of earthing and earth protection. What do we do to achieve it and why do we do it?

139 In your own words state the meaning of earthing and bonding. What type of cables and equipment would an electrician use to achieve earthing and bonding on an electrical installation?

140 In your own words state what we mean by 'basic protection' and how it is achieved.

141 In your own words state what we mean by 'fault protection' and how it is achieved.

Classroom activities

Statutory and non-statutory regulations

Match one of the numbered boxes to each lettered box on the left.

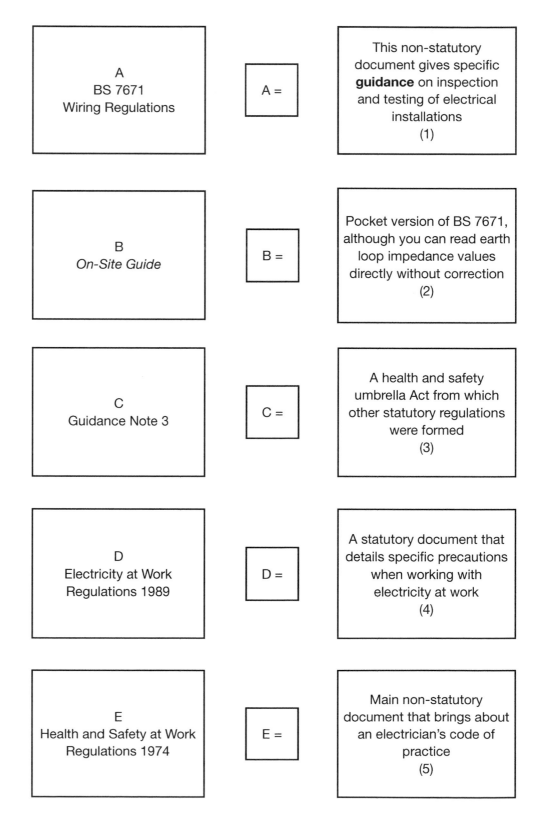

A BS 7671 Wiring Regulations	A =	This non-statutory document gives specific **guidance** on inspection and testing of electrical installations (1)
B *On-Site Guide*	B =	Pocket version of BS 7671, although you can read earth loop impedance values directly without correction (2)
C Guidance Note 3	C =	A health and safety umbrella Act from which other statutory regulations were formed (3)
D Electricity at Work Regulations 1989	D =	A statutory document that details specific precautions when working with electricity at work (4)
E Health and Safety at Work Regulations 1974	E =	Main non-statutory document that brings about an electrician's code of practice (5)

ELEC3/04A Electrical Installation Planning, Preparing and Designing

Chapter checklist

Learning outcome	Assessment criteria	Page number
1. Understand how to plan for the installation of wiring systems and equipment.	1.1 Interpret statutory and non-statutory requirements for the planned installation.	112
	1.2 Specify the criteria that affect the selection of wiring systems and equipment.	122
	1.3 Select wiring systems and equipment appropriate to the situation and use.	126
	1.4 Interpret sources of information relevant to planning of the installation of wiring systems and equipment.	148
	1.5 Plan wiring system routes.	148
	1.6 Specify the positional requirements of wiring systems and equipment.	148
	1.7 Interpret the requirements of BS 7671 for the planned installation.	153
2. Understand protection against overcurrent.	2.1 Specify the operating principles of overcurrent protection devices.	199
	2.2 Distinguish the application of overcurrent protection devices.	205
	2.3 Specify the advantages and limitations of different types of overcurrent protection devices.	201
3. Understand earthing and protection.	3.1 Distinguish earthing systems.	207
	3.2 Interpret the requirements for protection against electric shock.	207
	3.3 Select suitably sized protective conductors in accordance with BS 7671.	216
4. Understand the electrical design procedure.	4.1 Determine the design current.	219
	4.2 Select a suitably rated protective device.	219
	4.3 Establish the installation method reference.	224
	4.4 Apply appropriate rating factors.	224
	4.5 Determine the minimum current carrying capacity of live conductors.	224
	4.6 Establish if the voltage drop is acceptable.	224
	4.7 Verify if the disconnection times have been achieved.	230
	4.8 Evaluate thermal constraints.	232
	4.9 Establish the maximum disconnection times for different circuits.	230
	4.10 Verify discrimination between protective devices.	206
	4.11 Verify the fault current capacities of protective devices.	199

Learning outcome	Assessment criteria	Page number
5. Understand how to prepare the worksite.	5.1 Specify the health and safety requirements in establishing a safe working environment.	239
	5.2 Specify how to ensure that worksites are correctly prepared.	241
	5.3 Clarify the preparations that should be completed before electrical installation work starts.	242
	5.4 Outline how to check for any pre-existing damage to customer/client property.	242
	5.5 State the actions that should be taken if pre-existing damage to customer/client property is identified.	242
	5.6 Specify methods for protecting the fabric and structure of the property before and during installation work.	242

CHAPTER 4

EAL Unit QELTK3/006

Understanding the principles, practices and legislation for the inspection, testing, commissioning and certification of electrotechnical systems

EAL Electrical Installation Work – Level 3, 2nd Edition 978 0 367 19564 9
© 2019 T. Linsley. Published by Taylor & Francis. All rights reserved.
https://www.routledge.com/9780367195649

Learning outcomes

When you have completed this chapter you should:

1. Understand the principles, regulatory requirements and procedures for completing the safe isolation of an electrical circuit and complete electrical installations in preparation for inspection, testing and commissioning.
2. Understand the principles and regulatory requirements for inspecting, testing and commissioning electrical systems, equipment and components.
3. Understand the regulatory requirements and procedures for completing the inspection of electrical installations.
4. Understand the regulatory requirements and procedures for the safe testing and commissioning of electrical installations.
5. Understand the procedures and requirements for the completion of electrical installation certificates and related documentation.

Assessment criteria 1.1

State the requirements of the Electricity at Work Regulations 1989 for the safe inspection of electrical systems and equipment, in terms of those carrying out the work and those using the building during the inspection

Safety first

The EWR expects all electricians to work on dead supplies unless it is unreasonable for them not to be live. In other words all supplies must be dead unless it can be justified why they remain live.

The Electricity at Work Regulations 1989 (EWR)

This legislation came into force in 1990 and is made under the Health and Safety at Work Act 1974, and enforced by the Health and Safety Executive. The purpose of the regulations is to 'require precautions to be taken against the risk of death or personal injury from electricity in work activities'. This regulation gives a clear statement of who the EWR applies to and makes a statement:

> It shall be the duty of every employee while at work to comply with the provisions of these regulations in so far as they relate to matters which are within his/her control.

This means that a trainee electrician, although not fully qualified, must adhere to the EWR as well as always cooperating with their employer. The EWR also applies to non-technical personnel such as an office worker for instance, who must realize what they can or cannot do regarding electrical equipment and the importance of not interfering with safety equipment.

Regulation 13 Precautions for work on equipment made dead

This regulation ensures that adequate precautions shall be taken to prevent electrical equipment which has been made dead from becoming live when work is carried out on or near that equipment. In essence, what this regulation is proposing is that electricians always carry out a safe isolation procedure. The regulation also states that where reasonable a written procedure known as a permit to work is used to authorize and control electrical maintenance.

Regulation 14 Work on or near live conductors

This regulation ensures that no person shall be engaged in any work activity on or so near any live conductor (other than one suitably covered with insulating material so as to prevent danger) that danger may arise unless:

(a) it is unreasonable in all the circumstances for it to be dead; and

(b) it is reasonable in all the circumstances for him to be at work on or near it while it is live; and

(c) suitable precautions (including where necessary the provision of suitable protective equipment) are taken to prevent injury.

In other words, there is an expectation that electricians only work on dead supplies unless there is a reason against it and you can justify that reason.

The regulation also specifies that the test instrument used to establish that a circuit is dead has to be fit for purpose or 'approved' as laid out through GS38 specifically written by the Health and Safety Executive.

Assessment criteria 1.2

Specify and undertake the correct procedure for completing safe isolation with regard to carrying out safe working practices etc.

Secure electrical isolation

Electric shock occurs when a person becomes part of the electrical circuit. The level or intensity of the shock will depend upon many factors, such as age, fitness and the circumstances in which the shock is received. The lethal level is approximately 50 mA, above which muscles contract, the heart flutters and breathing stops. A shock above the 50 mA level is therefore fatal unless the person is quickly separated from the supply. Below 50 mA only an unpleasant tingling sensation may be experienced or you may be thrown across a room or shocked enough to fall from a roof or ladder, but the resulting fall may lead to serious injury.

To prevent people from receiving an electric shock accidentally, all circuits contain protective devices. All exposed metal is earthed; fuses and circuit-breakers (CBs) are designed to trip under fault conditions, and residual current devices (RCDs) are designed to trip below the fatal level as described in Chapter 3, page 189.

Construction workers and particularly electricians do receive electric shocks, usually as a result of carelessness or unforeseen circumstances. As an electrician working on electrical equipment you must always make sure that the equipment is switched off or electrically isolated before commencing work. Every circuit must be provided with a means of isolation (IET Regulation 132.15). When working on portable equipment or desk top units it is often simply a matter of unplugging the equipment from the adjacent supply. Larger pieces of equipment and electrical machines may require isolating at the local isolator switch before work commences. To deter anyone from re-connecting the supply while work is being carried out on equipment, a sign 'Danger – Electrician at Work' should be displayed on the isolator and the isolation 'secured' with a small padlock or the fuses removed so that no one can re-connect while work is being carried out on that piece of equipment. The Electricity at Work Regulations 1989 are very specific at Regulation 12(1) that we must ensure the disconnection and separation of electrical equipment from every source of supply and that this disconnection and separation is secure. Where a test instrument or voltage indicator is used to prove the supply is dead, Regulation 4(3) of the Electricity at Work Regulations 1989 recommends that the following procedure is adopted.

1 First, connect the test device such as that shown in Fig. 4.2 to the supply, which is to be isolated. The test device should indicate mains voltage.

2 Next, isolate the supply and observe that the test device now reads zero volts.

Safety first

Isolation

- Never work 'live'.
- Isolate.
- Secure the isolation.
- Prove the work supply 'dead' before starting work.

Figure 4.1 Electric shock occurs when a person becomes part of a live circuit.

Definition

Electrical isolation – we must ensure the disconnection and separation of electrical equipment from every source of supply and that this disconnection and separation is secure.

Definition

GS38 is a method statement specifically written by the HSE detailing approved equipment and exacting processes required when undertaking safe isolation procedures.

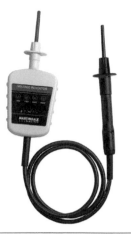

Figure 4.2 Typical voltage indicator.

Figure 4.4 Modern proving unit.

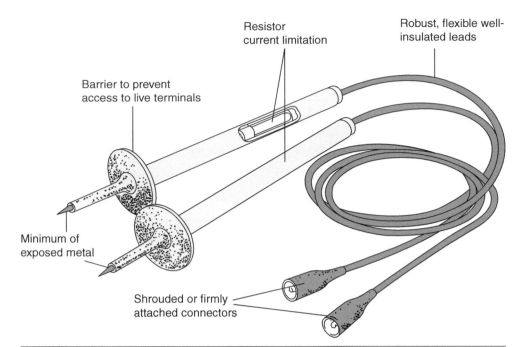

Resistor current limitation

Robust, flexible well-insulated leads

Barrier to prevent access to live terminals

Minimum of exposed metal

Shrouded or firmly attached connectors

Figure 4.3 Recommended type of test probe and leads.

3 Then connect the same test device to a known live supply or proving unit such as that shown in Fig. 4.4 to 'prove' that the tester is still working correctly.

4 Finally, secure the isolation and place warning signs; only then should work commence.

The test device being used by the electrician must incorporate safe test leads that comply with the Health and Safety Executive Guidance Note 38 (fourth edition, 2015) on electrical test equipment. These leads should incorporate barriers to prevent the user from touching live terminals when testing, but most testers do not include a protective fuse because they can add significant resistance to the test leads. While protection is still afforded through electronic circuitry, the meter leads should be well insulated and robust, such as those shown in Figure 4.3 with exposed metal tips optimally extending by 2mm but never more than 4mm in length. To isolate a piece of equipment or individual circuit successfully, competently, safely and in accordance with all the relevant regulations, we must follow a procedure such as that given by the flow diagram of Fig. 4.6. Start at the top and work down the flow diagram.

When the heavy outlined amber boxes are reached, pause and ask yourself whether everything is satisfactory up until this point. If the answer is 'yes', move on. If the answer is 'no', go back as indicated by the diagram.

Live testing

The Electricity at Work Regulations 1989 at Regulation 4(3) tell us that it is preferable that supplies be made dead before work commences. However, it does acknowledge that some work, such as fault finding and testing, may require the electrical equipment to remain energized. Therefore, if the fault finding and testing can only be successfully carried out live, then the person carrying out the fault diagnosis must:

Safety first

Your first objective when carrying out safe isolation procedures is to carry out a risk assessment of your working area.

Safety first

An inherent danger of applying testing procedures is that electricians have to periodically work on live supplies.

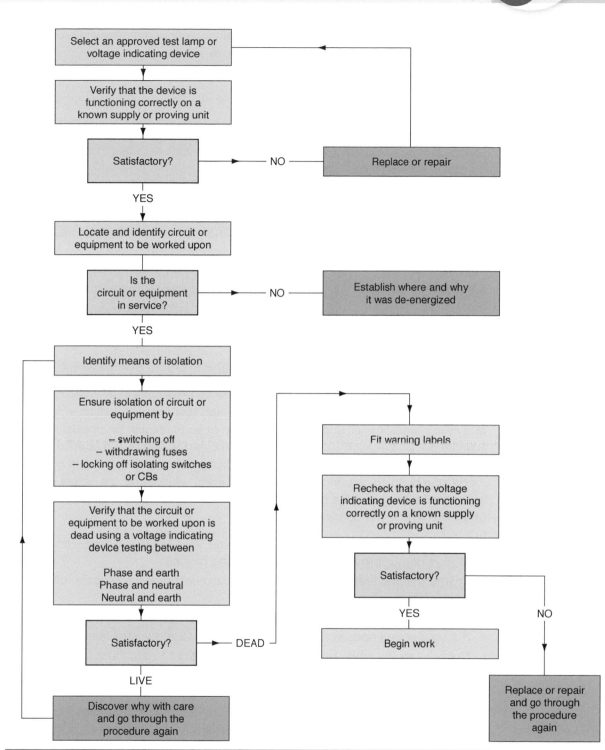

Figure 4.5 Flowchart for secure isolation procedure.

- be trained so that they understand the equipment and the potential hazards of working live and can, therefore, be deemed 'competent' to carry out that activity;
- only use approved test equipment;
- set up appropriate warning notices and barriers so that the work activity does not create a situation dangerous to others.

While live testing may be required by workers in the electrical industries in order to find a fault, live repair work must not be carried out. The individual circuit or piece of equipment must first be isolated before work commences in order to comply with the Electricity at Work Regulations 1989.

Assessment criteria 1.3

State the implications of carrying out safe isolations

Assessment criteria 1.4

State the implications of not carrying out safe isolations

Assessment criteria 1.5

Identify all health and safety requirements that apply when inspecting, testing and commissioning electrical installations and circuits

Safe working procedures – inspection and testing

Whether you are carrying out the inspection and test procedure (i) as a part of a new installation, (ii) upon the completion of an extension to an existing installation, (iii) because you are trying to discover the cause of a fault on an installation or (iv) because you are carrying out an electrical installation condition report of a building, you must always be aware of your safety, the safety of others using the building and the possible damage that your testing might cause to other systems in the building.

For your own safety:

- Always use 'approved' test instruments and probes.
- Ensure that the test instrument carries a valid calibration certificate, otherwise the results may be invalid.
- Secure all isolation devices in the 'off' position.
- Put up warning notices so that other workers will know what is happening.
- Notify everyone in the building that testing is about to start and for approximately how long it will continue.
- Obtain a 'permit to work' if this is relevant.
- Obtain approval to have systems shut down that might be damaged by your testing activities. For example, computer systems may 'crash' when supplies are switched off. Ventilation and fume extraction systems will stop working when you disconnect the supplies.

For the safety of other people:

- Fix warning notices around your work area.
- Use cones and highly visible warning tape to screen off your work area.
- Make an effort to let everyone in the building know that testing is about to begin. You might be able to do this while you carry out the initial inspection of the installation.
- Obtain verbal or written authorization to shut down information technology, emergency operation or standby circuits.

Safety first

A breakdown in communication either through a failure of procedures such as permits, or messages not being recorded or passed on could not only affect production but endanger life.

To safeguard other systems:

- Computer systems can be severely damaged by a loss of supply or the injection of a high test voltage from, for example, an insulation resistance test. Computer systems would normally be disconnected during the test period but this will generally require some organization before the testing begins. Commercial organizations may be unable to continue to work without their computer systems and, in these circumstances, it may be necessary to test outside the normal working day.
- Any resistance measurements made on electronic equipment or electronic circuits must be achieved with a battery-operated ohmmeter in order to avoid damaging the electronic circuits.
- Farm animals are creatures of habit and may become very grumpy to find you testing their milking parlour equipment at milking time.

Hospitals and factories may have emergency standby generators that reenergize essential circuits in the event of a mains failure. Your isolation of the circuit for testing may cause the emergency systems to operate. Discuss any special systems with the person authorizing the work before testing begins.

Assessment criteria 2.1

State the purpose of and requirements for initial verification and periodic inspection of electrical installations

Assessment criteria 2.2

Identify and interpret the requirements of the relevant documents associated with the inspection, testing and commissioning of an electrical installation

---

Why inspect and test an electrical installation?

The purpose of carrying out an initial verification of a new building is to ensure that the installation complies with the design, and construction aspects of BS 7671 that are reasonably practicable (GN3, Chapter 2). In doing so the inspection and testing of an electrical installation will:

1 ensure the safety of people and livestock;
2 ensure the protection of property from fire and heat;
3 confirm that the property has not been damaged or deteriorated so as to reduce its safety;
4 identify any defects in the installation and the subsequent need for improvement.

All electrical equipment deteriorates with age as well as with wear and tear from being used. It also deteriorates as a result of excessive loading and environmental influences leading to, for example, corrosion. The electrical installation must therefore be inspected and tested periodically during its lifetime to confirm that it remains safe to use at least until the next inspection and test is carried out. Organizations insuring buildings often require evidence that the electrical installation is safe and that the insurer's risk is therefore low. For example, insurance companies and mortgage lenders ask for electrical inspection and test reports when:

• there has been a change of building occupancy;
• there is a change of use of the building;
• alterations have been carried out;
• there is reason to believe that damage may have been caused to the building as a result of, for example, flooding.

Figure 4.6 Like most things electrical equipment deteriorates with age, so it is important to have it periodically tested during its lifetime.

In the event of an injury or fire alleged to have been caused by the electrical installation itself, the production of past certificates and reports will provide documentary evidence that the installation has been installed and subsequently maintained to a satisfactory standard of safety by skilled or instructed persons. Statutory and non-statutory regulations also clearly state the requirements for inspection and testing as a part of a maintenance programme to reduce danger. Regulation 4 (2) of the Electricity at Work Regulations advises that 'as may be necessary to prevent danger, all systems shall be maintained so as to prevent such danger'.

The Guidance Notes on the Electricity at Work Regulations advise that this regulation is concerned with the need for maintenance to ensure the safety of 'the system' and not the actual 'doing of maintenance' in a safe manner. The regular inspection of equipment including the electrical installation is an essential part of any preventative maintenance programme.

There is no specific requirement to test the installation on every inspection. Where testing requires dismantling, the tester should consider whether the risks associated with dismantling and reassembly are justified. Dismantling, and particularly disconnection of cables and components, introduces the possible risk of unsatisfactory reassembly.

IET Regulations 641 advise us that every installation shall, during erection and on completion before being put into service, be inspected and tested to verify that the regulations have been met. Regulation 651 also advises that where required, periodic inspection and testing shall be carried out in order to determine if the installation is in a satisfactory condition for continued service. Regulation 642.3 gives us an inspection checklist and tells us that the inspection must include at least these 20 or more items plus, if appropriate, the particular requirements for special installations such as bathrooms described in Part 7 of the regulations. The *On-Site Guide* at Section 9.2.2 gives the same inspection checklist. Guidance Note 3 gives us guidance in using the IET checklist given in Regulation 642.3. We will look at this checklist later in this chapter under the sub-heading 'Visual inspection'.

The *Electrician's Guide to the Building Regulations* Part P at Section 6.2 follows the requirements given in Section 641 of the IET Regulations. Appendix 6 of the IET Regulations gives us a Condition Report Inspection Schedule (schedule means a planned programme of work) for a domestic or similar premises with an up to 100 A supply. This inspection schedule checklist is to be used when carrying out a periodic inspection of an existing installation for the purpose of completing an Electrical Installation Condition Report (see page 476 of Blue 18th Edition).

Assessment criteria 2.3

Specify the information that is required to correctly conduct the initial verification of an electrical installation in accordance with the IET Wiring Regulations and IET Guidance Note 3

Both the IET Wiring Regulations and IET Guidance Note 3 specify that certain information must be made available to the inspector when carrying out initial verification. First, any relevant diagrams, charts or tables that contain information such as utilization points, type and composition of circuits and how such elements as fault and basic protection are provided. Furthermore, details should be provided of the general characteristics such as:

- maximum demand;
- number and type of conductors;
- earthing system used;
- nominal voltage;
- supply frequency and current;
- prospective short circuit current at the origin;
- Z_e (suppliers earth loop impedance value);
- type and rating of overcurrent protection devices.

Assessment criteria 3.1

Identify the items to be checked during the inspection process for given electrotechnical systems and equipment, and their locations as detailed in the IET Wiring Regulations

Assessment criteria 3.3

State the items of an electrical installation that should be inspected in accordance with IET Guidance Note 3

Sampling when carrying out inspections

The inspection required of an electrical installation must be a 'detailed examination' of the installation without dismantling or with only partial dismantling as is required (IET Regulation 651.2). A thorough visual inspection should be made of all electrical equipment that is not concealed and should include the accessible internal condition of a sample of the electrical equipment. It is not practicable to inspect every joint and termination in an electrical installation. Nevertheless, a detailed examination of the installation, without dismantling, should be made of all switches, switchgear, distribution boards, luminaire points and socket outlets to ensure that all terminal connections of the conductors are properly installed and secure. Any signs of overheating of conductors, terminations or equipment should be thoroughly investigated and included in the report (Guidance Note 3 section 3.8).

Considerable care and engineering judgement must be exercised when deciding upon the extent of the sample to receive the 'covers removed' visual inspection. The factors to be considered when determining the size of the sample are:

- the availability of previous certificates;
- the age and condition of the installation;
- any evidence of ongoing maintenance;
- the time elapsed since the previous inspection;
- the size of the installation.

If no previous certificates are available and the installation is looking tired or abused it would be necessary to inspect and test a larger percentage of the installation and in some cases 100% of the installation. The degree or extent of the sampling must be agreed with the person ordering the work before work commences. It may be impractical to inspect and test the whole of a large installation at one time. If this is the case and sampling is agreed with the person ordering the work, it is important to examine different parts of the installation in subsequent inspections. It would not be acceptable for the same parts of the installation to be repeatedly inspected to the exclusion of other parts.

Figure 4.7 Past certificates may not always be available, but you should always complete the appropriate paperwork when testing installations.

Assessment criteria 3.2

State how human senses (sight, touch etc.) can be used during the inspection process

The senses of sight, hearing, smell and touch are powerful human forces that we can use when carrying out the 'covers removed' electrical inspection or when we are attempting to trace a fault on a circuit. Connections can become hot under fault conditions, which may be as a result of loose connections or a circuit overload. The heat generated by the electrical fault might heat up the insulation of the conductors and the terminal block holding the connections. In these circumstances, the Bakelite insulation used in the construction of terminal blocks and switchgear becomes hot and gives off a pungent smell, like rotting fish. So, we can use our powerful human sense of smell to guide us to the source of possible faults. Conductor insulation blackens when it becomes very hot and becomes brittle and stiff. Once more, we can use the human senses of sight and touch to detect these faults when carrying out an internal inspection of socket outlets, switches and luminaries.

Assessment criteria 3.4

Specify the requirements for inspection

Visual inspection

The installation must be visually inspected before testing begins. The aim of the **visual inspection** is to confirm that all equipment and accessories are undamaged and comply with the relevant British and European Standards, and also that the installation has been securely and correctly erected. IET Regulation 642.3 gives a checklist or schedule for the initial visual inspection of an installation, including:

- connection of conductors;
- identification of conductors;
- routeing of cables in safe zones;
- selection of conductors for current-carrying capacity and volt drop;
- connection of single-pole devices for protection or switching in phase conductors only;
- correct connection of socket outlets, lampholders, accessories and equipment;
- presence of fire barriers, suitable seals and protection against thermal effects;
- methods of 'basic protection' against electric shock, including the insulation of live parts and placement of live parts out of reach by fitting appropriate barriers and enclosures;
- methods of 'fault protection' against electric shock, including the presence of earthing conductors for both protective bonding and supplementary bonding;
- prevention of detrimental influences (e.g. corrosion);
- presence of appropriate devices for isolation and switching;
- presence of undervoltage protection devices;
- choice and setting of protective devices;
- labelling of circuits, fuses, switches and terminals;

- selection of equipment and protective measures appropriate to external influences;
- adequate access to switchgear and equipment;
- presence of danger notices and other warning notices;
- presence of diagrams, instructions and similar information;
- appropriate erection method.

The checklist is a guide. It is not exhaustive or detailed, and should be used to identify relevant items for inspection, which can then be expanded upon. For example, the first item on the checklist, **connection of conductors**, might be further expanded to include the following:

- Are connections secure?
- Are connections correct? (conductor identification)
- Is the cable adequately supported so that no strain is placed on the connections?
- Does the outer sheath enter the accessory?
- Is the insulation undamaged?
- Does the insulation proceed up to, but not *into*, the connection?

This is repeated for each appropriate item on the checklist.

Assessment criteria 4.1

State the tests to be carried out on an electrical installation in accordance with the IET Wiring Regulations and IET Guidance Note 3

Assessment criteria 4.4

Explain why it is necessary for test results to comply with standard values and state the actions to take in the event of unsatisfactory results being obtained

Assessment criteria 4.5

Explain why testing is carried out in the exact order as specified in the IET Wiring Regulations and IET Guidance Note 3

Electrical testing

The electrical contractor is charged with a responsibility to carry out a number of tests on an electrical installation and electrical equipment. The individual tests are dealt with in Part 6 of the IET Regulations and Chapter 2 of GN3, described later in this chapter.

The reasons for testing the installation are:

- to ensure that the installation complies with the IET Regulations;
- to ensure that the installation meets the specification;
- to ensure that the installation is safe to use.

Those who are to carry out the electrical tests must first consider the following safety factors:

- an assessment of safe working practice must be made before testing begins;
- all safety precautions must be put in place before testing begins;
- everyone must be notified that the test process is about to take place, for example the client and other workers who may be affected by the tests;
- 'permits to work' must be obtained where relevant;
- all sources of information relevant to the tests have been obtained;
- the relevant circuits and equipment have been identified;
- safe isolation procedures have been carried out – care must be exercised here, in occupied premises, not to switch off computer systems without first obtaining permission;
- those who are to carry out the tests are competent to do so.

Chapter 2 of GN3 states that initial tests should be carried out in the following sequence where relevant and practical. The tests are devised so that safety critical elements such as continuity of protective conductors and insulation resistance are carried out whilst the circuit is made dead, before live testing of the installation takes place. The test is listed as follows:

(a) Test for continuity of protective conductors, including protective equipotential and supplementary bonding

(b) Test the continuity of all ring final circuit conductors

(c) Test for insulation resistance

(d) Protection by SELV, PELV, or by electrical separation

(e) Protection by barriers or enclosures

(f) Insulation resistance of non-conducting locations

(g) Polarity

(h) Earth electrode resistance

(I) Protection by automatic disconnection of the supply

(j) Test the earth fault loop impedance

(k) Additional protection

(l) Prospective fault current

(m) Check of phase sequence

(n) Functional testing

(o) Verification of volt drop

If any test fails to comply with the IET Regulations, then *all* the preceding tests must be repeated after the fault has been rectified. This is because the earlier test results may have been influenced by the fault (IET Regulation 643.1). There is an increased use of electronic devices in electrical installation work; for example, in dimmer switches and ignitor circuits of discharge lamps. These devices should temporarily be disconnected so that they are not damaged by the test voltage of, for example, the insulation resistance test (IET Regulation 643.3).

Assessment criteria 4.2

Identify the correct instrument for the test to be carried out

The electrical contractor is required by the IET Regulations to test all new installations and major extensions during erection and upon completion before being put into service. The contractor may also be called upon to test installations and equipment in order to identify and remove faults. These requirements imply the use of appropriate test instruments, and in order to take accurate readings consideration should be given to the following points:

* Is the instrument suitable for this test?
* Has the correct scale been selected?
* Is the test instrument correctly connected to the circuit?

Many commercial instruments are capable of making more than one test or have a range of scales to choose from. A range selector switch is usually used to choose the appropriate scale. A scale range should be chosen that suits the range of the current, voltage or resistance being measured. For example, when taking a reading in the 8 or 9 V range, the obvious scale choice would be one giving 10 V full-scale deflection. To make this reading on an instrument with 100 V full-scale deflection would lead to errors, because the deflection is too small. Ammeters must be connected in series with the load, and voltmeters in parallel across the load as shown in Fig. 4.8. The power in a resistive load may

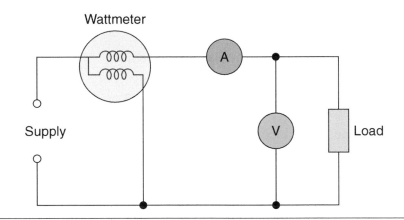

Figure 4.8 Wattmeter, ammeter and voltmeter correctly connected to a load.

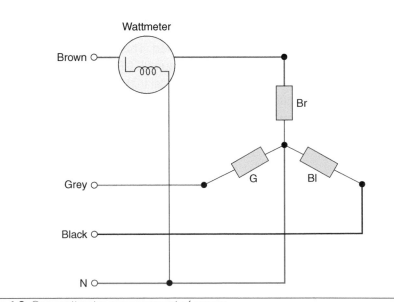

Figure 4.9 One-wattmeter measurement of power.

be calculated from the readings of voltage and current since $P = VI$. This will give accurate calculations on both a.c. and d.c. supplies, but when measuring the power of an a.c. circuit that contains inductance or capacitance, a watt meter must be used because the voltage and current will be out of phase.

Measurement of power in a three-phase circuit

One-wattmeter method

When three-phase loads are balanced, for example, in motor circuits, one wattmeter may be connected into any phase, as shown in Fig. 4.9. This wattmeter will indicate the power in that phase and, since the load is balanced, the total power in the three-phase circuit will be given by:

$$\text{Total power} = 3 \times \text{wattmeter reading}$$

Two-wattmeter method

This is the most commonly used method for measuring power in a three-phase, three-wire system since it can be used for both balanced and unbalanced loads connected in either star or delta. The current coils are connected to any two of the lines, and the voltage coils are connected to the other line, the one without a current coil connection, as shown in Fig. 4.10. Then,

$$\text{Total power} = W_1 + W_2$$

This equation is true for any three-phase load, balanced or unbalanced, star or delta connection, provided there is no fourth wire in the system.

Three-wattmeter method

If the installation is four-wire, and the load on each phase is unbalanced, then three-wattmeter readings are necessary, connected as shown in Fig. 4.11.

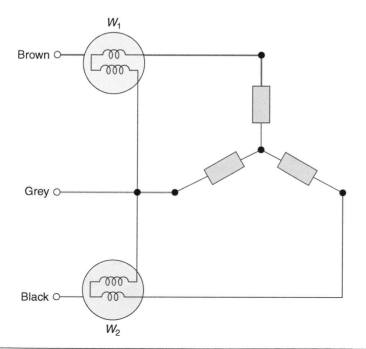

Figure 4.10 Two-wattmeter measurement of power.

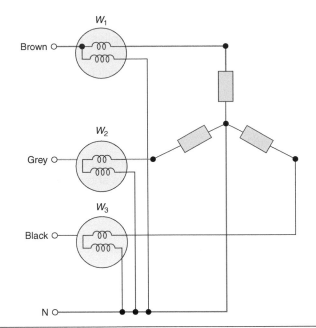

Figure 4.11 Three-wattmeter measurement of power.

Each wattmeter measures the power in one phase and the total power will be given by:

$$Total\ power = W_1 + W_2 + W_3$$

Tong tester

Figure 4.12 Tong tester or clip-on ammeter.

The tong tester, or clip-on ammeter, offers an easy method of measuring current without any connections being made. It works on the same principle as the transformer. The laminated core of the transformer can be opened and passed over the busbar or single-core cable. In this way a measurement of the current being carried can be made without disconnection of the supply. The construction is shown in Fig. 4.12.

Figure 4.13 Some of the electrical testing essentials: lock-offs, warning signs, padlock with key and a proving unit.

Assessment criteria 4.3

Specify the requirements for the safe and correct use of instruments to be used for testing and commissioning

Approved test instruments

The **test instruments and test leads** used by the electrician for testing an electrical installation must meet all the requirements of the relevant regulations.

The HSE has published guidance note GS 38 for test equipment used by electricians. The IET Regulations (BS 7671) and GN3 also specify the test voltage or current required to carry out particular tests satisfactorily. All test equipment must be chosen to comply with the relevant parts of BS EN 61557. All testing must, therefore, be carried out using an 'approved' test instrument if the test results are to be valid. The test instrument must also carry a **calibration certificate**, otherwise the recorded results may be void. *Calibration certificates* usually last for a year. Test instruments must, therefore, be tested and recalibrated each year by an approved supplier. This will maintain the accuracy of the instrument to an acceptable level, usually within 2% of the true value. Modern digital test instruments are reasonably robust, but to maintain them in good working order they must be treated with care. An approved test instrument costs as much as a good quality camera; it should, therefore, receive the same care and consideration. It therefore should go without saying that all test equipment should undergo a user check before and after testing to ensure that the equipment is serviceable.

Let us now look at the requirements of four often used test instruments.

Continuity tester

To measure accurately the resistance of the conductors in an electrical installation, we must use an instrument that is capable of producing an open circuit voltage of between 4 and 24 V a.c. or d.c., and delivering a short-circuit current of not less than 200 mA. The functions of continuity testing and insulation resistance testing are usually combined in one test instrument.

Insulation resistance tester

The test instrument must be capable of detecting insulation leakage between live conductors and between live conductors and earth. To do this and comply with IET Regulation 643.3 the test instrument must be capable of producing a test voltage of 250, 500 or 1,000 V and delivering an output current of not less than 1 mA at its normal voltage.

Earth fault loop impedance tester

The test instrument must be capable of delivering fault currents as high as 25 A for up to 40 ms using the supply voltage. During the test, the instrument does an Ohm's law calculation and displays the test result as a resistance reading.

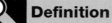
Definition

The *test instruments and test leads* used by the electricians for testing an electrical installation must meet all the requirements of the relevant regulations.

Definition

Calibration certificates usually last for a year. Test instruments must, therefore, be tested and recalibrated each year by an approved supplier.

Safety first

Not only should test equipment be in calibration but it must also be appropriate to the task.

Figure 4.14 Insulation tester.

RCD tester

Where circuits are protected by an RCD we must carry out a test to ensure that the device will operate very quickly under fault conditions and within the time limits set by the IET Regulations. The instrument must, therefore, simulate a fault and measure the time taken for the RCD to operate. The instrument is, therefore, calibrated to give a reading measured in milliseconds to an in-service accuracy of 10%.

If you purchase good-quality 'approved' test instruments and leads from specialist manufacturers they will meet all the regulations and standards and therefore give valid test results. However, to carry out all the tests required by the IET Regulations will require a number of test instruments and this will represent a major capital investment in the region of £1,000.

Let us now consider the individual tests.

Assessment criteria 4.6

State the reasons why it is necessary to verify the continuity of circuit protective conductors, earthing conductors, bonding conductors and ring final circuit conductors

Assessment criteria 4.7

Specify and apply the methods for verifying the continuity of circuit protective conductors and ring final circuit conductors and interpreting the obtained results

Testing for continuity of protective conductors, including main and supplementary equipotential bonding (IET Regulation 643.2)

Definition

A protective earthing conductor needs to form a low resistance path because it will enable any fault current to become high in value and isolate a circuit quickly.

The objective of the test is to ensure that the circuit protective conductor (CPC) is correctly connected, is electrically sound and has a total resistance that is low enough to permit the overcurrent protective device to operate within the disconnection time requirements of IET Regulation 411.4.201, should an earth fault occur. Every protective conductor must be separately tested from the main earthing terminal of the installation to verify that it is electrically sound and correctly connected, including the protective equipotential and supplementary bonding conductors as shown in Fig 4.15 of this chapter and Fig 3.106 of Chapter 3. The IET Regulations describe the need to consider additional protection by supplementary equipotential bonding in situations where there is a high risk of electric shock, such as kitchens and bathrooms (IET Regulation 415.2).

Key fact

Supplementary bonding is also used when circuit disconnection times cannot be achieved.

A d.c. test using an ohmmeter continuity tester is suitable where the protective conductors are of copper or aluminium up to 35 mm². The test is made with the supply disconnected, measuring from the main earthing terminal of the installation to the far end of each CPC, as shown in Fig. 4.16. The resistance of

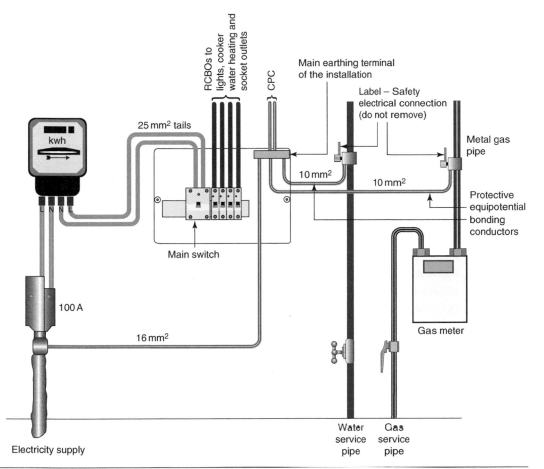

Figure 4.15 Cable Sheath Earth Supplies (TN-S System) showing earthing and main protective equipotential bonding arrangements.

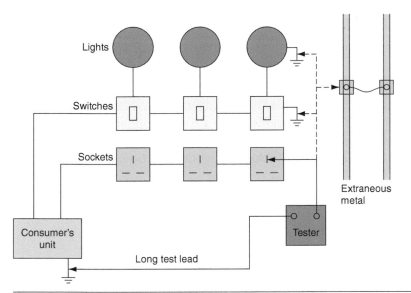

Figure 4.16 Testing continuity of protective conductors.

the long test lead is subtracted from these readings to give the resistance value of the CPC. The result is recorded on an installation schedule such as that given in Appendix 6 of the IET Regulations. A satisfactory test result for the bonding conductors will be in the order of 0.05 Ω or less (IET Guidance Note 3).

Where steel conduit or trunking forms the protective conductor, the standard test described above may be used, but additionally the enclosure must be visually checked along its length to verify the integrity of all the joints.

If the inspecting engineer has grounds to question the soundness and quality of these joints then the phase earth loop impedance test described later in this chapter should be carried out.

If, after carrying out this further test, the inspecting engineer still questions the quality and soundness of the protective conductor formed by the metallic conduit or trunking, then a further test can be done using an a.c. voltage not greater than 50 V at the frequency of the installation and a current approaching 1.5 times the design current of the circuit, but not greater than 25 A.

This test can be done using a low-voltage transformer and suitably connected ammeters and voltmeters, but a number of commercial instruments are available, such as the Clare tester, which give a direct reading in ohms. Because fault currents will flow around the earth fault loop path, the measured resistance values must be low enough to allow the overcurrent protective device to operate quickly. For a satisfactory test result, the resistance of the protective

Definition

The *circuit protective conductor* within all PVC flat profile cable with the exception of 1 mm² is smaller than the live conductors. This means that its resistance will be higher.

Table 4.1 Resistance values of some metallic containers

Metallic sheath	Size (mm)	Resistance at 20°C (mΩ/m)
Conduit	20	1.25
	25	1.14
	32	0.85
Trunking	50 × 50	0.949
	75 × 75	0.526
	100 × 100	0.337

Example

The CPC for a ring final circuit is formed by a 1.5 mm² copper conductor of 50 m approximate length. Determine a satisfactory continuity test value for the CPC using the value given in Table I1 of the *On-Site Guide*.

Table I1 gives resistance/metre for a 1.5 mm² copper conductor
$$= 12.10 \, \text{m}\Omega/\text{m}$$

Therefore, the resistance of 50m $= 50 \times 12.10 \times 10^{-3}$
$$= 0.605 \, \Omega$$

The protective conductor resistance values calculated by this method can only be an approximation since the length of the CPC can only be estimated. Therefore, in this case, a satisfactory test result would be obtained if the resistance of the protective conductor was about 0.6 Ω. A more precise result is indicated by the earth fault loop impedance test which is carried out later in the sequence of tests.

conductor should be consistent with those values calculated for a line conductor of similar length and cross-sectional area. Values of resistance per metre for copper and aluminium conductors are given in Table I1 of the *On-Site Guide*. The resistances of some other metallic containers are given in Table 4.1 of this book.

Testing for continuity of ring final circuit conductors (IET Regulation 643.2)

The objective of the test is to ensure that all ring circuit cables are continuous around the ring; that is, that there are no breaks and no interconnections in the ring, and that all connections are electrically and mechanically sound. This test also verifies the polarity of each socket outlet. The test is made with the supply disconnected, using an ohmmeter as follows. Disconnect and separate the conductors of both legs of the ring at the main fuse. There are three steps to this test:

Step 1

Measure the resistance of the line conductors (L1 and L2), the neutral conductors (N1 and N2) and the protective conductors (E1 and E2) at the mains position as shown in Fig. 4.17. End-to-end live and neutral conductor readings should be approximately the same (i.e. within 0.05 Ω) if the ring is continuous. The protective conductor reading will be 1.67 times as great as these readings if 2.5/1.5 mm cable is used. Record the results on a table such as that shown in Table 4.2.

Step 2

The live and neutral conductors should now be temporarily joined together as shown in Fig. 4.18. An ohmmeter reading should then be taken between live and neutral at *every* socket outlet on the ring circuit. The readings obtained should be substantially the same, provided that there are no breaks or multiple loops in the ring. Each reading should have a value of approximately half the live and neutral ohmmeter readings measured in Step 1 of this test. Sockets connected as a spur will have a slightly higher value of resistance because they are fed by only one cable, while each socket on the ring is fed by two cables. Record the results on a table such as that shown in Table 4.2.

Table 4.2 Table that may be used to record the readings taken when carrying out the continuity of ring final circuit conductors tests according to IET Regulation 643.2

Test	Ohmmeter connected to	Ohmmeter readings	This gives a value for
Step 1	L_1 and L_2		r_1
	N_1 and N_2		
	E_1 and E_2		r_2
Step 2	Live and neutral at each socket		
Step 3	Live and earth at each socket		$R_1 + R_2$

As a check $(R_1 + R_2)$ value should equal $(r_1 + r_2)/4$.

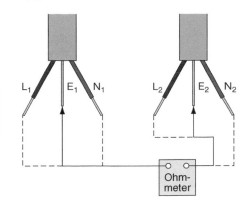
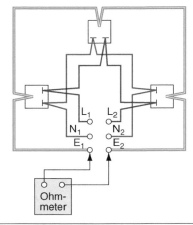

Figure 4.17 Step 1 test: measuring the resistance of phase, neutral and protective conductors.

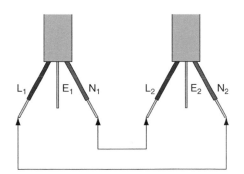
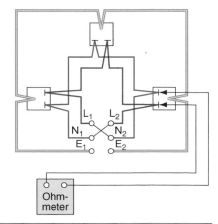

Figure 4.18 Step 2 test: connection of mains conductors and test circuit conditions.

Step 3

Where the CPC is wired as a ring, for example where twin and earth cables or plastic conduit is used to wire the ring, temporarily join the live and CPCs together as shown in Fig. 4.19. An ohmmeter reading should then be taken between live and earth at *every* socket outlet on the ring. The readings obtained should be substantially the same provided that there are no breaks or multiple loops in the ring. This value is equal to $R_1 + R_2$ for the circuit. Record the results on an installation schedule such as that given in Appendix 6 of the IET Regulations or a table such as that shown in Table 4.2. The Step 3 value of $R_1 + R_2$ should be equal to $(r_1 + r_2)/4$, where r_1 and r_2 are the ohmmeter readings from Step 1 of this test (see Table 4.2).

Definition

The reason why $R_1 + R_2$ is a quarter of $r_1 + r_2/4$ is because the length of the circuit conductors have been halved and the area doubled.

Key fact

Any socket reading greater than the ring measurement indicates the use of a spur outlet. Only one unfused spur is allowed per socket outlet.

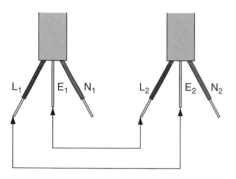

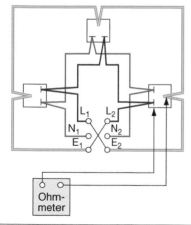

Figure 4.19 Step 3 test: connection of mains conductors and test circuit conditions.

Assessment criteria 4.8

State the effects that cables connected in parallel and variations in cable length can have on insulation resistance values

Assessment criteria 4.9

Interpret and apply the procedures for completing insulation resistance testing

Testing insulation resistance (IET Regulation 643.3)

The objective of the test is to verify that the quality of the insulation is satisfactory and has not deteriorated or short-circuited. The test should be made at the consumer's unit with the mains switch off, all fuses in place and all switches closed. Neon lamps, capacitors and electronic circuits should be disconnected, since they will respectively glow, charge up and be damaged by the test. There are two tests to be carried out using an insulation resistance tester, which must have a test voltage of 500 V d.c. for 230 V and 400 V installations. These are line and neutral conductors to earth and between line conductors. The procedures are:

Line and neutral conductors to earth:

1 Remove all lamps.
2 Close all switches and circuit-breakers.
3 Disconnect appliances.
4 Test separately between the line conductor and earth *and* between the neutral conductor and earth, for *every* distribution circuit at the consumer's unit as shown in Fig. 4.20(a). Record the results on a schedule of test results such as that given in Appendix 6 of the IET Regulations.

Between line conductors:

1 Remove all lamps.
2 Close all switches and circuit-breakers.
3 Disconnect appliances.
4 Test between line and neutral conductors of *every* distribution circuit at the consumer's unit as shown in Fig. 4.20(b) and record the result.

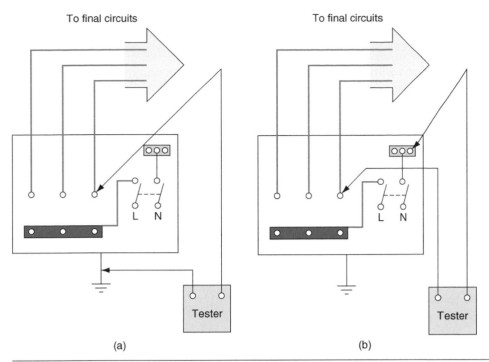

Figure 4.20 Insulation resistance test.

The insulation resistance readings for each test must be not less than 1.0 MΩ for a satisfactory result (IET Regulation 643.3.2).

Where the circuit includes electronic equipment that might be damaged by the insulation resistance test, a measurement between all live conductors (i.e. live and neutral conductors connected together) and the earthing arrangements may be made. The insulation resistance of these tests should be not less than 1.0 MΩ (IET Regulation 643.3.2).

Although an insulation resistance reading of 1.0 MΩ complies with the regulations, the IET guidance notes tell us that much higher values than this can be expected and that a reading of less than 2 MΩ might indicate a latent, but not yet visible, fault in the installation. In these cases each circuit should be separately tested to obtain a reading greater than 2 MΩ.

> **Definition**
>
> Whilst an insulation resistance reading of 1 MΩ is acceptable to BS 7671 further investigation is recommended due to the possibility of a latent fault.

> **Key fact**
>
> Doubling the length of a conductor will double its circuit resistance. However, it will halve its insulation resistance.

Assessment criteria 4.10

Explain why it is necessary to verify polarity

Assessment criteria 4.11

Interpret and apply the procedures for testing to identify correct polarity

Testing polarity (IET Regulation 643.6)

The object of this test is to verify that all fuses, circuit-breakers and switches are connected in the line or live conductor only, that all socket outlets are correctly wired and that Edison screw-type lampholders have the centre

4

contact connected to the live conductor. It is important to make a polarity test on the installation since a visual inspection will only indicate conductor identification.

The test is done with the supply disconnected using an ohmmeter or continuity tester as follows:

1 Switch off the supply at the main switch.
2 Remove all lamps and appliances.
3 Fix a temporary link between the line and earth connections on the consumer's side of the main switch.
4 Test between the 'common' terminal and earth at each switch position.
5 Test between the centre pin of any Edison screw lampholders and any convenient earth connection.
6 Test between the live pin (i.e. the pin to the right of earth) and earth at each socket outlet as shown in Fig. 4.21.

For a satisfactory test result the ohmmeter or continuity meter should read very close to zero for each test.

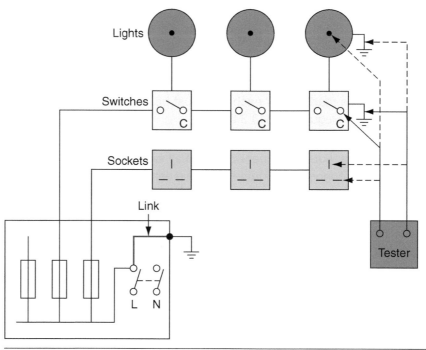

Figure 4.21 Polarity test.

Remove the test link and record the results on a schedule of test results such as that given in Appendix 6 of the IET Regulations.

Testing polarity – supply connected

Using an approved voltage indicator such as that shown at Fig. 4.2 or test lamp and probes that comply with the HSE Guidance Note GS 38, again carry out a polarity test to verify that all fuses, circuit-breakers and switches are connected in the live conductor. Test from the common terminal of switches to earth, the live pin of each socket outlet to earth and the centre pin of any Edison screw lamp holders to earth. In each case the voltmeter or test lamp should indicate the supply voltage for a satisfactory result.

Definition

Polarity is defined in two ways, ensuring that switching only occurs through the line conductor and that all protective devices are also installed in the line conductor.

Assessment criteria 4.12

Specify and apply the methods for measuring earth electrode resistance and correctly interpreting the results

Testing earth electrode resistance (IET Regulation 643.7.2)

When an earth electrode has been sunk into the general mass of earth, it is necessary to verify the resistance of the electrode. The general mass of earth can be considered as a large conductor that is at zero potential. Connection to this mass through earth electrodes provides a reference point from which all other voltage levels can be measured. This is a technique that has been used for a long time in power distribution systems.

The resistance to earth of an electrode will depend upon its shape, size and the resistance of the soil. Earth rods form the most efficient electrodes. A rod of about 1 m will have an earth electrode resistance of between 10 and 200 Ω. Even in bad earthing conditions, a rod of about 2 m will normally have an earth electrode resistance that is less than 500 Ω in the United Kingdom. In countries that experience long dry periods of weather the earth electrode resistance may be thousands of ohms. In the past, electrical engineers used the metal pipes of water mains as an earth electrode, but the recent increase in the use of PVC pipe for water mains now prevents the use of water pipes as the means of earthing in the United Kingdom, although this practice is still permitted in some countries. IET Regulation 542.2.2 recognizes the use of the following types of earth electrodes:

- earth rods or pipes;
- earth tapes or wires;
- earth plates;
- earth electrodes embedded in foundations;
- welded metallic reinforcement of concrete structures;
- other suitable underground metalwork;
- lead sheaths or other metallic coverings of cables.

The earth electrode is sunk into the ground, but the point of connection should remain accessible (IET Regulation 542.4.2). The connection of the earthing conductor to the earth electrode must be securely made with a copper conductor complying with Table 54.1 and IET Regulation 542.3.2 as shown in Fig. 4.22.

The installation site must be chosen so that the resistance of the earth electrode does not increase above the required value due to climatic conditions such as the soil drying out or freezing, or from the effects of corrosion (IET Regulations 542.2.1 and 4).

Under fault conditions the voltage appearing at the earth electrode will radiate away from the electrode like the ripples radiating away from a pebble thrown into a pond. The voltage will fall to a safe level in the first 2 or 3 m away from the point of the earth electrode.

The basic method of measuring earth electrode resistance is to pass a current into the soil through the electrode and to measure the voltage required to produce this current.

> **Definition**
>
> Using the centre contact to confirm polarity does not apply to Edison screw types E14 and E27.

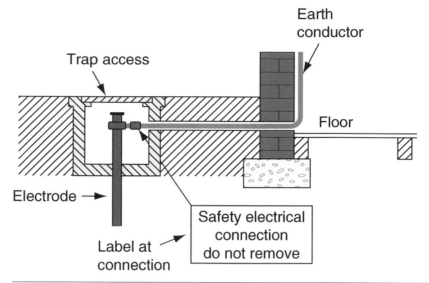

Figure 4.22 Termination of an earth electrode.

IET Regulation 643.7.2 demands that where earth electrodes are used they should be tested. If the electrode under test forms part of the earth return for a TT installation in conjunction with an RCD, Guidance Note 3 of the IET Regulations describes the following method:

1 Disconnect the installation protective equipotential bonding from the earth electrode to ensure that the test current passes only through the earth electrode.
2 Switch off the consumer's unit to isolate the installation.
3 Using a line earth loop impedance tester, test between the incoming line conductor and the earth electrode.
4 Reconnect the protective bonding conductors when the test is completed.

Record the result on a schedule of test results such as that given in Appendix 6 of the IET Regulations.

The IET Guidance Note 3 tells us that an acceptable value for the measurement of the earth electrode resistance would be less than 200 Ω. Providing the first five tests were satisfactory, the supply may now be switched on and the final tests completed with the supply connected.

Assessment criteria 4.13

Identify the earth fault loop paths for the following systems: TN-S, TN-C-S and TT

Assessment criteria 4.14

State the methods for verifying protection by automatic disconnection of the supply

Earth fault loop impedance Z_s

The three most common earthing systems are:

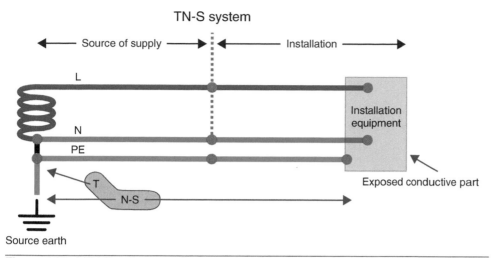

Figure 4.23 A TN-S system uses the armouring of the supply cable to provide a low resistance earth fault path.

- TN-S;
- TN-C-S;
- TT.

TN-S earthing systems use the supply armouring to create a low-resistance path to carry the fault current back to the supply transformer.

TN-C-S earthing systems make use of a protective earthing neutral (PEN) also known as a combined earthing neutral (CNE), which links the CPC and neutral. Line to CPC faults therefore become line to neutral faults but the loop impedance values are lower. Because of other installations sharing the same neutral there is a danger of faults being caused in other installations, for instance, through a break in the neutral conductor, which could bring about certain dangerous potential differences. To counter this the supply will be earthed at multiple points, which is why it is also known as protective multiple earthing. Because of the previous issue of sharing a neutral, PME earthing is not permitted in such places as: petrol stations, caravans and marinas.

TT earthing systems make use of an earth electrode commonly through an earth spike but not exclusively so. Again, the aim is to create a low resistance path but this time through the earth itself, with the supply transformer also being earthed in the same way. The problem with using the earth itself is that it is not a good conductor of electricity especially if the soil is dry. Therefore, because there is a high possibility that loop impedance values can be >200 Ω, BS 7671 states that TT systems must be protected by an RCD.

In order that an overcurrent protective device can operate successfully it must meet the required disconnection times of IET Regulation 411.3.2.2, that is, final circuits not exceeding 32 A shall have a disconnection time not exceeding 0.4 s.

To achieve this, the earth fault loop impedance value measured in ohms must be less than those values given in Appendix B of the *On-Site Guide* and Tables 41.2 and 41.3 of the IET Regulations. Bearing in mind, of course, that while values in the *On-Site Guide* and Guidance Note 3 can be taken directly, those in BS 7671 cannot. This is because values within BS 7671 have not been corrected regarding conductors attaining full operational temperature and therefore the working value is 80% of those listed.

Definition

TN-S earthing uses the supply cable armouring to carry any fault current.

Definition

TN-C-S makes use of a PEN conductor which changes line to CPC faults to line to neutral faults.

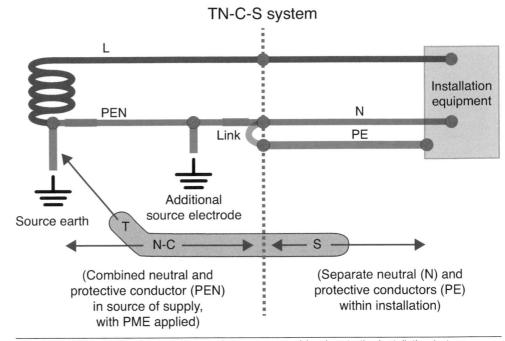

TN-C-S system

Figure 4.24 Neutral and protective conductors are combined up to the installation but separated within.

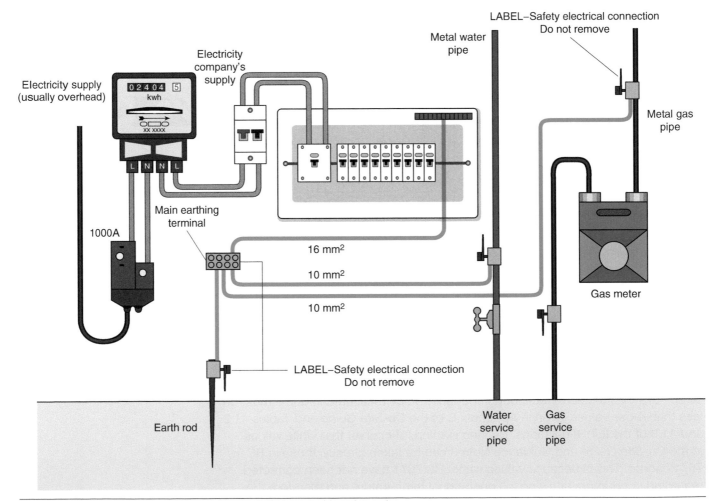

Figure 4.25 A TT earthing system operates by using an earthing electrode.

Values of the earth fault loop impedance may be verified by means of an earth fault loop impedance test, which is described in the next section. The formula is:

$$Z_S = Z_E + (R_1 + R_2) \ (\Omega)$$

Here Z_E is the impedance of the supply side of the earth fault loop. The actual value will depend upon many factors: the type of supply, the ground conditions, the distance from the transformer, etc. The value can be obtained from the area electricity companies, but typical values are 0.35 Ω for TN-C-S (protective multiple earthing, PME) supplies and 0.8 Ω for TN-S (cable sheath earth) supplies as described a little earlier in this chapter. Also in the above formula, R_1 is the resistance of the line conductor and R_2 is the resistance of the earth conductor. The complete earth fault loop path is shown in Fig. 4.26. Values of $R_1 + R_2$ have been calculated for copper and aluminium conductors and are given in Table I1 of the *On-Site Guide*.

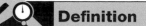

Definition

A TT earthing system uses an earthing electrode but if high earth loop impedance values are recorded then RCD protection must be in place.

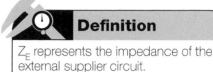

Definition

Z_E represents the impedance of the external supplier circuit.

Figure 4.26 Earth fault loop path for a TN-S system.

Testing earth fault loop impedance – supply connected (IET Regulation 643.7.3)

The objective of this test is to verify that the impedance of the whole earth fault current loop line to earth is low enough to allow the overcurrent protective device to operate within the disconnection time requirements of IET Regulations 411.3.2.2, 411.4.201 and 411.4.202, should a fault occur.

The whole earth fault current loop examined by this test comprises all the installation protective conductors, the main earthing terminal of the installation

Definition

Earth fault loop impedance testing must be carried out at the furthest point of the circuit. This will then gauge if sufficient fault current will flow in order to disconnect a circuit in the required time.

and protective bonding conductors, the earthed neutral point and the secondary winding of the supply transformer and the line conductor from the transformer to the point of the fault in the installation.

The test will, in most cases, be done with a purpose-made line earth loop impedance tester, which circulates a current in excess of 10 A around the loop for a very short time, so reducing the danger of a faulty circuit. The test is made with the supply switched on, and carried out from the furthest point of *every* final circuit, including lighting, socket outlets and any fixed appliances. Record the results on a schedule of test results.

Purpose-built testers give a readout in ohms and a satisfactory result is obtained when the loop impedance does not exceed the appropriate values given in Tables 41.2 and 41.3 of the IET Regulations. Table 41.3 gives a value of 2.73 ohm maximum for a circuit protected by a 16 A CB and 1.37 ohm maximum for a circuit protected by a 32 A CB.

Assessment criteria 4.15

Specify the methods for determining prospective fault current

The ways of determining values of prospective fault current are similar to obtaining earth loop impedance values in that they can be obtained by:

- enquiry directly from the service provider;
- measurement (earth loop impedance test instruments include a prospective fault current facility);
- calculation by Ohm's law: $I = V/R$.

The prospective fault current must be obtained at the supply origin to ensure that switchgear and protective devices can withstand this level of voltage during fault conditions.

Definition

Measurement of prospective fault current is normally carried out at the supply origin. This is because its value will be at its highest and switchgear and protective devices can be accessed if they can withstand such values.

Assessment criteria 4.16

Specify the methods for testing the correct operation of residual current devices (RCDs)

Additional protection: testing of RCD – supply connected (IET Regulation 643.8)

The object of the test is to verify the effectiveness of the RCD, that it is operating with the correct sensitivity and to prove the integrity of the electrical and mechanical elements. The test must simulate an appropriate fault condition and be independent of any test facility incorporated in the device. When carrying out the test, all loads normally supplied through the device are disconnected.

The testing of a ring circuit protected by a general-purpose RCD to BS EN 61008 in a split-board consumer unit is carried out as follows:

1 Using the standard lead supplied with the test instrument, disconnect all other loads and plug in the test lead to the socket at the centre of the ring (i.e. the socket at the furthest point from the source of supply).

2 Set the test instrument to the tripping current of the device and at a phase angle of 0°.

3 Press the test button – the RCD should trip and disconnect the supply within 200 ms.

4 Change the phase angle from 0° to 180° and press the test button once again. The RCD should again trip within 200 ms. Record the highest value of these two results on a schedule of test results such as that given in Appendix 6 of the IET Regulations.

5 Now set the test instrument to 50% of the rated tripping current of the RCD and press the test button. The RCD should *not trip* within two seconds. This test is testing the RCD for inconvenience or nuisance tripping.

6 Finally, the effective operation of the test button incorporated within the RCD should be tested to prove the integrity of the mechanical elements in the tripping device. This test should be repeated every three months.

If the RCD fails any of the above tests it should be changed for a new one. Where the RCD has a rated tripping current not exceeding 30 mA and has been installed to reduce the risk associated with 'basic' and/or 'fault' protection, as indicated in IET Regulation 411.1, a residual current of 150 mA should cause the circuit-breaker to open within 40 ms.

Whereever RCDs are installed, a label shall be fixed near to the RCD stating 'this device must be tested 6 monthly' (IET Regulation 514.12.2). Previously the test period was 3 monthly.

> **Key fact**
>
> An element involved in the testing of RCDs is to ensure that they are not the sole means of protection.

Assessment criteria 4.17

State the methods used to check for the correct phase sequence

Assessment criteria 4.18

Explain why having the correct phase sequence is important

Check for phase sequence (IET Regulation 643.9)

Phase sequence is the order in which each phase of a three-phase supply reaches its maximum value. The normal phase sequence for a three-phase supply is brown–black–grey, which means that first brown, then black and finally the grey phase reaches its maximum value. A phase sequence tester can be an indicator which is, in effect, a miniature induction motor, with three clearly colour-coded connection leads. A rotating disc with a pointed arrow shows the normal rotation for phase sequence brown–black–grey. If the sequence is reversed, the disc rotates in the opposite direction to the arrow.

Figure 4.27 Phase rotation indicator instrument.

Definition

Excessive lengths of cable can deny electrical loads some of its working voltage. This can be corrected by increasing the conductor size, which in turn lowers the circuit resistance.

Alternatively, an indicator lamp showing the phase rotation by L1–L2–L3 may be used as shown in Fig. 4.27.

Instruments are also available that contain both of the above indications. The phase sequence is first of all checked at the point of entry to the building of the three-phase supply and the phase rotation noted. The phase sequence is then checked at any other three-phase distribution boards within the building. The objective is to obtain confirmation that the phase rotation of every three-phase distribution board is the same as that at the origin of the installation. Phase rotation and therefore direction of rotation of any three-phase motors affected by this action can be reversed by swapping over any two phases.

Assessment criteria 4.19

State the need for functional testing and identify items that need to be checked

Functional testing (IET Regulation 643.10)

The objective of the functional test is to confirm that all switchgear, controls and switches work as they were intended to work and are properly installed, mounted and adjusted. The little test button on any RCDs must also be activated to verify that it will trip under fault conditions.

Assessment criteria 4.20

Specify the methods used for verification of voltage drop

Assessment criteria 4.21

State the cause of volt-drop in an electrical installation

IET Regulation 525 states that the drop in voltage from the supply terminals to the fixed current-using equipment must not exceed 3% for lighting circuits and 5% for other uses of the mains voltage. That is, a maximum of 6.9 V for lighting and 11.5 V for other uses on a 230 V installation. The volt drop for a particular cable may be found from:

$$VD = Factor \times Design\ current \times Length\ of\ run$$

The factor is given in the tables of Appendix 4 of the IET Regulations and Appendix 6 of the *On-Site Guide.* One of the main reasons behind ascertaining volt drop is that if the circuit length is excessive then voltage will be dropped across the conductor but in doing so it deprives the load of some working voltage. This means that light can appear dim or even electrical motors can be sluggish or slow in operation. Excessive length of cable and volt drop issues are corrected by increasing the size of cable, since increasing the cross sectional area of a conductor lowers the circuit resistance.

Assessment criteria 4.22

State the appropriate procedures for dealing with customers and clients during the commissioning and certification process

Commissioning electrical systems

The commissioning of the electrical and mechanical systems within a building is a part of the 'handing-over' process of the new building by the architect and main contractor to the client or customer in readiness for its occupation and intended use. To 'commission' means to give authority to someone to check that everything is in working order. If it is out of commission, it is not in working order. Following the completion, inspection and testing of the new electrical installation, the functional operation of all the electrical systems must be tested before they are handed over to the customer. It is during the commissioning period that any design or equipment failures become apparent, and this testing is one of the few quality controls possible on a building services installation.

This is the role of the commissioning engineer, who must assure him or herself that all the systems are in working order and that they work as they were designed to work. This engineer must also instruct the client's representative, or the staff who will use the equipment, in the correct operation of the systems as part of the handover arrangements.

The commissioning engineer must test the operation of all the electrical systems, including the motor controls, the fan and air conditioning systems, the fire alarm and emergency lighting systems. However, before testing the emergency systems, it is important to first notify everyone in the building of the intention to test the alarms so that they may be ignored during the period of testing. Commissioning has become one of the most important functions within the building project's completion sequence. The commissioning engineer will therefore have access to all relevant contract documents, including the building specifications and the electrical installation certificates as required by the IET Regulations (BS 7671), and have a knowledge of the requirements of the Electricity at Work Regulations and the Health and Safety at Work Act.

The building will only be handed over to the client if the commissioning engineer is satisfied that all the building services meet the design specification in the contract documents and all the safety requirements.

Assessment criteria 5.1

Explain the purpose of and relationship between different types of testing documentation

Assessment criteria 5.2

State the information that must be contained within test certificates

Assessment criteria 5.3

Describe the certification process for a completed installation and identify the responsibilities of different relevant personnel in relation to the completion of the certification process

Assessment criteria 5.4

Explain the procedures and requirements, in accordance with the IET Wiring Regulations, IET Guidance Note 3 and where appropriate customer/ client requirements for the recording and retention of certificates

Certification for Initial Verification (Regulation 644)

Following the completion of all new electrical work or additional work to an existing installation, the installation must be inspected and tested and an installation certificate issued and signed by a skilled or instructed person. The 'skilled or instructed person' must have a sound knowledge of the type of work undertaken, be fully versed in the inspection and testing procedures contained in the IET Regulations (BS 7671) and employ adequate testing equipment.

A certificate and test results shall be issued to those ordering the work in the format given in Appendix 6 of the IET Regulations. There are three model forms.

1 **Electrical Installation Certificate:** this is to be used when inspecting and testing a new installation or for alterations or additions to an existing installation. The documentation to be handed over to the client must include details of the extent of the installation covered by the certificate, a record of the inspection schedule and a copy of the test results plus a recommendation for the interval until the first periodic inspection (IET Regulation 644.1).

Figure 4.28 Testing and commissioning an electrical installation.

2 **Minor Electrical Installation Works Certificate:** this form is intended to be used for the addition of a socket outlet or lighting point to an existing circuit or for a repair or modification to the installation. The work does not extend to the addition of a new circuit. In this case, only the minor works certificate is given to the client upon completion (IET Regulation 644.4.201).

3 **Periodic Inspection and Testing:** all electrical installations must be periodically inspected and tested for the purpose of completing an electrical installation condition report (IET Regulation 651 and 653.1).

The documentation handed over to the client upon completion must include details of the extent of the installation covered by the report, a record of the inspection schedule, and a record of the test results plus a recommendation for the interval until the next periodic inspection (IET Regulation 653.2 and 4). Inspection schedules are discussed in more detail at the beginning of this chapter. In both cases the certificate must include the test values obtained from a 'calibrated' instrument which verify that the installation complies with the IET Regulations at the time of testing. The suggested frequency of periodic inspection intervals are given below:

- domestic installations – 10 years;
- commercial installations – 5 years;
- industrial installations – 3 years;
- agricultural installations – 3 years;
- caravan site installations – 1 year;
- caravans – 3 years but every year if used regularly;
- temporary installations on construction sites – 3 months.

Test your knowledge

When you have completed the questions check out the answers at the back of the book.

Note: More than one multiple-choice answer may be correct.

Learning outcome 1

1 All electrical test probes and leads used in applying safe isolation procedures must comply with certain standards. These standards are set by:
   a. BS EN 60898
   b   BS 7671
   c. HSE Guidance Note GS 38
   d. IET Regulations Part 2.

2 When carrying out safe isolation procedures for the purposes of safety, which of the following should be used?
   a. neon screwdriver
   b. a multimeter
   c. volt stick
   d. approved voltage indicator.

3 The Electricity at Work Regulations require that:
   a. only persons who have the appropriate level of knowledge and experience can work with electricity
   b. only duty holders can work with electricity
   c. only time served electricians can work with electricity
   d. only supervised apprentices can work with electricity.

4 Electrical safety is best achieved by:
   a. isolating and securing the supply
   b. switching the supply off
   c. working during quiet hours
   d. wearing electrician's gloves.

5 Unless it can be justified, there exists an expectation that an electrician should always work on:
   a. dead circuits
   b. live circuits
   c. RCD protected circuits
   d. SELV circuits.

6   The legislation that prohibits live working except in exceptional circumstances is:
   a.   Provision and Use of Work Equipment Regulations (1992)
   b.   Personal Protective Equipment Regulations (1992)
   c.   BS 7671 Requirements for Electrical Installations
   d.   Electricity at Work Regulations (1989).

7   If more than one person is working on a circuit, which of the following should be used when carrying out safety isolation procedure?
   a.   mortise clasp
   b.   multi-lock hasp
   c.   cable ties
   d.   multi-key lock.

8   Which statutory regulation ensures the safe inspection of all manufactured, purchased and installed electrical systems?
   a.   Electricity at Work Regulation 1989
   b.   Health and Safety at Work Act 1974
   c.   Management of Health and Safety
   d.   COSHH.

9   The most serious outcome following a failure in carrying out safe isolation procedures would be:
   a.   incarceration
   b.   imprisonment
   c.   quicker timescales
   d.   death.

Learning outcome 2

10   'To ensure that all the systems within a building work as they were intended to work' is one definition of the purpose of:
   a.   testing electrical equipment
   b.   inspecting electrical systems
   c.   commissioning electrical systems
   d.   isolating electrical systems.

11   The person carrying out the initial verification of an installation is confirming that the installation complies with:
   a.   BS 7671 in every regard
   b.   BS 7671 so far as reasonably practicable
   c.   Health and Safety Executive Requirements
   d.   Electricity at Work Regulation 1989.

12   The reason for initial verification is to ensure that installed electrical equipment is the:
   a.   correct type and complies to British Standards
   b.   correct size and complies to British Standards
   c.   correct rating and complies to British Standards
   d.   correct colour and complies to British Standards.

13   Initial verification is carried out to make sure that the fixed installation:
   a.   is correctly selected and erected
   b.   is correctly finished and sealed
   c.   is not defective
   d.   is not visibly damaged.

14   Which of the following is required before an initial inspection is able to be carried out?
   a.   equipment costing
   b.   earthing system
   c.   maximum demand
   d.   supply characteristics.

Learning outcome 3

15   Which of the following are the most appropriate human senses regarding electrical testing?
   a.   intuition and speech
   b.   taste and pain
   c.   touch and feel
   d.   touch and sight.

16   Inspection and testing should:
   a.   only be carried out when installation is complete
   b.   only be carried out midway during the installation
   c.   be carried out during the installation and on completion
   d.   be complete within one month of the installation erection.

17   Which of the following are most appropriate to verify initial verification?
   a.   basic protection
   b.   equipment suitability
   c.   termination of conductors and cable selection
   d.   designated storage areas.

18   Which of the following is most appropriate in establishing safety?
   a.   fault protection
   b.   basic protection
   c.   all systems comply with PELV
   d.   all systems have to be FELV.

19   Choose which option is correct:
   a.   inspection comes before testing
   b.   testing comes before inspection
   c.   testing is more important than inspection
   d.   inspection is more important than testing.

Learning outcome 4

20   Testing results should be compared to:
   a.   other installations
   b.   BS 7671
   c.   designer requirements
   d.   Electricity at Work Regulation 1989.

21 The first test to be carried out on initial verification is:
  a. polarity
  b. continuity of protective conductors
  c. functional testing
  d. earth loop impedance.

22 Establishing fault protection includes:
  a. earthing of class II equipment
  b. RCD protection if Ze in TT systems is high
  c. equipotential bonding of extraneous conductive parts
  d. earthing of class 1 equipment.

23 Why is continuity of protective conductors so important?
  a. due to their colour they are easier to see
  b. they are smaller than live conductors
  c. they are larger than live conductors
  d. they are safety critical.

24 What value of voltage should be used when calculating the resistance of supplementary bonding conductors?
  a. 110 V
  b. 50 V
  c. 400 V
  d. 230 V.

25 When testing a 230 V installation an insulation resistance tester must be set to a voltage of:
  a. less than 50 V
  b. 500 V
  c. less than 500 V
  d. greater than twice the supply voltage but less than 1000 V.

26 The value of a satisfactory insulation resistance test on each final circuit of a 230 V installation must be:
  a. less than 1 Ω
  b. less than 0.5 MΩ
  c. not less than 0.5 MΩ
  d. not less than 1 MΩ.

27 Instrument calibration certificates are usually valid for a period of:
  a. 3 months
  b. 1 year
  c. 3 years
  d. 5 years.

28 The correct order for the requirement regarding dead tests is:
  a. continuity of CPC, live polarity and earth loop impedance
  b. continuity of CPC, dead polarity and earth loop impedance
  c. continuity of CPC, insulation resistance and earth loop impedance
  d. continuity of CPC, insulation resistance, polarity.

29 To obtain R1 + R2 in a ring circuit the r1 + r2 measurement needs to be:
   a. divided by a factor of 4
   b. divided by a factor of 3
   c. divided by a ratio of 1.67
   d. multiplied by a ratio of 1.67.

30 Which of the following regarding testing ring final conductors is correct?
   a. one unfused spur allowed, readings will be higher than R1 + R2
   b. any number of unfused spurs allowed, readings will be higher than R1 + R2
   c. any number of fused spurs allowed, readings will be higher than R1 + R2
   d. one fused spur allowed, readings will be higher than R1 + R2.

31 For PVC flat profile cable aside from 1 mm², the resistance value of the live conductors with respect to the CPC is equivalent to:
   a. ratio of 1.67 – CPC readings are more than the neutral conductor
   b. ratio of 1.67 – CPC readings are less than the line conductor
   c. ratio of 1.67 – CPC readings are more than the line and neutral conductors
   d. ratio of 1.67 – CPC readings are less than the line and neutral conductors.

32 A tong test instrument is also known as:
   a. a continuity tester
   b. a clip/clamp-on ammeter
   c. an insulation resistance tester
   d. a voltage indicator.

33 One objective of dead polarity is to verify that:
   a. lampholder's outer thread is connected to the line conductor
   b. lampholder is fused
   c. lampholder's centre contact is connected to the neutral conductor
   d. lampholder's centre contact is connected to the line conductor.

34 Connecting circuits in parallel will:
   a. reduce the total resistance
   b. increase the total resistance
   c. halve the total resistance
   d. double the total resistance.

35 Increasing the length of a circuit will:
   a. reduce its insulation resistance
   b. increase its insulation resistance
   c. halve its insulation resistance
   d. double its insulation resistance.

36 Before carrying out an insulation resistance test, which of the following should be considered?
   a. all sensitive loads removed
   b. all loads and capacitors removed
   c. value of supply voltage
   d. switches off protective devices removed.

37 If a circuit returns an insulation resistance value of 1 MΩ, according to BS 7671:

a. the circuit does not comply

b. the circuit complies

c. no further investigation is required

d. further investigation is recommended.

38 The definition of polarity is:

a. functional switching takes place through all live conductors

b. functional switching takes place through the line conductor

c. functional switching takes place through the neutral conductor

d. functional switching takes place through the CPC.

39 Earth loop impedance values can be read directly from:

a. GN3

b. BS 7671

c. *On-Site Guide*

d. GS38.

40 Total earth loop impedance can be obtained by:

a. measuring Zs

b. inquiry of Ze and measuring R1 and R2

c. inquiry of Zs

d. measuring Ze and measuring R1 and R2.

41 Live polarity tests are carried out by connecting across which conductors?

a. line and line

b. neutral and line

c. phase and earth

d. line and earth.

42 To ensure that the breaking capacity of a CB is met, which of the following human senses needs to be used?

a. touch

b. sight

c. taste

d. sound.

Learning outcome 5

43 The recommended maximum inspection and re-test period for a domestic electrical installation is:

a. 3 months

b. 3 years

c. 5 years

d. 10 years.

44 What should be issued after the completion of two small additions to an existing circuit?

a. minor works certificate

b. two minor works certificates

c. electrical Installation Certificate to cover both circuits

d. two Electrical Installation Certificates.

45 Following the completion of a new installation, which of the following must be issued?
   a. test results schedules
   b. Electrical Installation Condition Report
   c. Electrical Installation Certificate
   d. inspection schedules.

46 Following completion of an installation the:
   a. inspector keeps a copy of the electrical installation certificate
   b. inspector keeps the original electrical installation certificate
   c. customer keeps a copy of the electrical installation certificate
   d. customer is given the original electrical installation certificate.

47 Use bullet points to state three reasons for testing a new electrical installation.

48 State five of the most important safety factors to be considered before electrical testing begins.

49 State the seven requirements of GS 38 when selecting probes, voltage indicators and measuring instruments.

50 Use bullet points to describe a safe isolation procedure of a final circuit fed from a CB in a distribution board.

51 IET Regulation 611.3 gives a checklist of about 20 items to be considered in the initial visual inspection of an electrical installation. Make a list of 10 of the most important items to be considered in the visual inspection process (perhaps by joining together similar items).

52 State three reasons why electricians must only use 'approved' test instruments.

53 State the first five tests to be carried out on a new electrical installation following a satisfactory 'inspection'. For each test:
   a. state the object (reason for) the test
   b. state a satisfactory test result.

Classroom activities

Match one of the numbered boxes to each lettered box on the left.

Dead test values

0.05–0.08Ω

High fault current will flow and lead to a protective device operating quickly

A

A =

Built-in test equipment facility that removes value of test leads from reading under test. Alternatively subtract meter leads from measured value

(1)

---

1 MΩ

Within limits but possible latent fault

B

B =

Acceptable range of values for continuity of circuit protective conductor

(2)

---

2 MΩ

C

C =

Acceptable value of insulation resistance but further investigation is recommended

(3)

---

500 V

D

D =

Acceptable value of insulation resistance

(4)

NULL

E

E =

Insulation resistance meter setting for a domestic single phase supply

(5)

Testing position

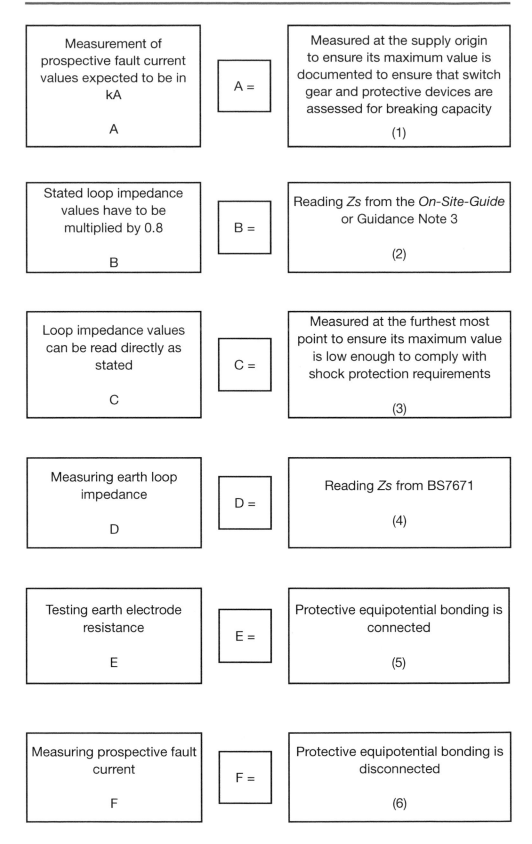

Measurement of prospective fault current values expected to be in kA A	A =	Measured at the supply origin to ensure its maximum value is documented to ensure that switch gear and protective devices are assessed for breaking capacity (1)
Stated loop impedance values have to be multiplied by 0.8 B	B =	Reading *Zs* from the *On-Site-Guide* or Guidance Note 3 (2)
Loop impedance values can be read directly as stated C	C =	Measured at the furthest most point to ensure its maximum value is low enough to comply with shock protection requirements (3)
Measuring earth loop impedance D	D =	Reading *Zs* from BS7671 (4)
Testing earth electrode resistance E	E =	Protective equipotential bonding is connected (5)
Measuring prospective fault current F	F =	Protective equipotential bonding is disconnected (6)

QELTK3/006 Understanding the principles, practices and legislation for the inspection, testing, commissioning and certification of electrotechnical systems and equipment in buildings, structures and the environment

Chapter checklist

Learning outcome	Assessment criteria	Page number
1. Understand the principles, regulatory requirements and procedures for completing the safe isolation of an electrical circuit and complete electrical installations in preparation for inspection, testing and commissioning.	1.1 State the requirements of the Electricity at Work Regulations 1989 for the safe inspection of electrical systems and equipment, in terms of those carrying out the work and those using the building during the inspection.	266
	1.2 Specify and undertake the correct procedure for completing safe isolation with regard to: • carrying out safe working practices • correct identification of circuit(s) to be isolated • identifying suitable points of isolation • selecting correct test and proving instruments in accordance with relevant industry guidance and standards • correct testing methods • selecting locking devices for securing isolation • correct warning notices • correct sequence for isolating circuits.	267
	1.3 State the implications of carrying out safe isolations to: • other personnel • customers/clients • public • building systems (loss of supply).	270
	1.4 State the implications of not carrying out safe isolations to: • self • other personnel • customers/clients • public • building systems (presence of supply).	270
	1.5 Identify all health and safety requirements that apply when inspecting, testing and commissioning electrical installations and circuits including those which cover: • working in accordance with risk assessments / permits to work / method statements • safe use of tools and equipment • safe and correct use of measuring instruments • provision and use of PPE • reporting of unsafe situations.	270

Learning outcome	Assessment criteria	Page number
2. Understand the principles and regulatory requirements for inspecting, testing and commissioning electrical systems, equipment and components.	2.1 State the purpose of and requirements for initial verification and periodic inspection of electrical installations	271
	2.2 Identify and interpret the requirements of the relevant documents associated with the inspection, testing and commissioning of an electrical installation, including: • Electricity at Work Regulations 1989 • IET wiring Regulations • IET Guidance Note 3.	271
	2.3 Specify the information that is required to correctly conduct the initial verification of an electrical installation in accordance with the IET Wiring Regulations and IET Guidance Note 3.	273
3. Understand the regulatory requirements and procedures for completing the inspection of electrical installations.	3.1 Identify the items to be checked during the inspection process for given electrotechnical systems and equipment, and their locations as detailed in the IET Wiring Regulations.	274
	3.2 State how human senses (sight, touch etc.) can be used during the inspection process.	275
	3.3 State the items of an electrical installation that should be inspected in accordance with IET Guidance Note 3.	274

	3.4 Specify the requirements for the inspection of the following:
	• earthing conductors
	• circuit protective conductors
	• protective bonding conductors
	  • main bonding conductors
	  • supplementary bonding conductors
	• isolation
	• type and rating of overcurrent protective devices.
		275

4. Understand the regulatory requirements and procedures for the safe testing and commissioning of electrical installations.	4.1 State the tests to be carried out on an electrical installation in accordance with the IET Wiring Regulations and IET Guidance Note 3.	276
	4.2 Identify the correct instrument for the test to be carried out in terms of: • the instrument is fit for purpose • identifying the right scale/settings of the instrument appropriate to the test to be carried out.	277
	4.3 Specify the requirements for the safe and correct use of instruments to be used for testing and commissioning, including: • checks required to prove that test instruments and leads are safe and functioning correctly • the need for instruments to be regularly checked and calibrated and that this be done in accordance with the requirements of the IET Wiring Regulations and other relevant guidance documents (HSE guidance document GS38).	281

Learning outcome	Assessment criteria	Page number
	4.4 Explain why it is necessary for test results to comply with standard values and state the actions to take in the event of unsatisfactory results being obtained.	276
	4.5 Explain why testing is carried out in the exact order as specified in the IET Wiring Regulations and IET Guidance Note 3	276
	4.6 State the reasons why it is necessary to verify the continuity of circuit protective conductors, earthing conductors, bonding conductors and ring final circuit conductors.	282
	4.7 Specify and apply the methods for verifying the continuity of circuit protective conductors and ring final circuit conductors and interpreting the obtained results.	282
	4.8 State the effects that: • cables connected in parallel • variations in cable length can have on insulation resistance values.	286
	4.9 Interpret and apply the procedures for completing insulation resistance testing, including: • precautions to be taken before conducting insulation resistance tests • methods of testing insulation resistance • the required test voltages and minimum insulation resistance values for circuits operating at various voltages.	286
	4.10 Explain why it is necessary to verify polarity.	287
	4.11 Interpret and apply the procedures for testing to identify correct polarity.	287
	4.12 Specify and apply the methods for measuring earth electrode resistance and correctly interpreting the results.	289
	4.13 Identify the earth fault loop paths for the following systems: • TN-S • TN-C-S • TT.	290
	4.14 State the methods for verifying protection by automatic disconnection of the supply, including: • the measurement of the earth fault loop impedance (Zs) and external impedance (Ze) • establishing Ze from enquiry • calculate the value of Zs from given information • comparing Zs and the maximum tabulated figures as specified in the IET Wiring Regulations.	290
	4.15 Specify the methods for determining prospective fault current.	294
	4.16 Specify the methods for testing the correct operation of residual current devices (RCDs).	294
	4.17 State the methods used to check for the correct phase sequence	295

Learning outcome	Assessment criteria	Page number
	4.18 Explain why having the correct phase sequence is important.	295
	4.19 State the need for functional testing and identify items that need to be checked.	296
	4.20 Specify the methods used for verification of voltage drop.	296
	4.21 State the cause of volt-drop in an electrical installation.	296
	4.22 State the appropriate procedures for dealing with customers and clients during the commissioning and certification process, including: • Ensuring the safety of customers and clients during the completion of work activities • Keeping customers and clients informed during the process • Labelling electrical circuits, systems and equipment that is yet to be commissioned • Providing customers and clients with all appropriate documentation upon work completion.	297
5. Understand the procedures and requirements for the completion of electrical installation certificates and related documentation.	5.1 Explain the purpose of and relationship between: • an Electrical Installation Certificate • a Minor Electrical Installation Works Certificate • schedule of Inspections • schedule of Test results.	297
	5.2 State the information that must be contained within; • an Electrical Installation Certificate • a Minor Electrical Installation Works Certificate • Schedule of Inspections • Schedule of Test Results.	297
	5.3 Describe the certification process for a completed installation and identify the responsibilities of different relevant personnel in relation to the completion of the certification process	298
	5.4 Explain the procedures and requirements, in accordance with the IET Wiring Regulations, IET Guidance Note 3 and where appropriate customer/client requirements for the recording and retention of completed: • Electrical Installation Certificates • Minor Electrical Installation Works Certificates • Schedules of Inspections • Schedules of Test Results.	298

CHAPTER 5

Unit QELTK3/007

Understanding the principles, practices and legislation for diagnosing and correcting electrical faults in electro technical systems and equipment in buildings, structures and the environment

Learning outcomes

When you have completed this chapter you should:

1. Understand the principles, regulatory requirements and procedures for completing the safe isolation of electrical circuits and complete electrical installations.

2. Understand how to complete the reporting and recording of electrical fault diagnosis and correction work.

3. Understand how to complete the preparatory work prior to fault diagnosis and correction work.

4. Understand the procedures and techniques for diagnosing electrical faults.

5. Understand the procedures and techniques for correcting electrical faults.

   Note that Learning Outcomes 1.1–1.4 have been covered in QELTK3/006 but are again summarized above.

EAL Electrical Installation Work – Level 3, 2nd Edition 978 0 367 19564 9
© 2019 T. Linsley. Published by Taylor & Francis. All rights reserved.
https://www.routledge.com/9780367195649

Requirements for safe working procedures

Assessment criteria 1.1 to 1.4 have already been covered in Chapter 4 but should be embraced alongside the following requirements. The following five safe working procedures must be applied before undertaking fault diagnosis:

1   The circuits must be isolated using a 'safe isolation procedure', such as that described in Fig 5.7.
2   All test equipment must be 'approved' and calibrated and connected to the test circuits by recommended test probes as described by the Health and Safety Executive (HSE) Guidance Note GS 38. The test equipment used must also be 'proved' on a known supply or by means of a proving unit such as that shown in Figure 4.2 and 4.4.
3   Isolation devices must be 'secured' in the 'off' position as shown in Figure 4.13. The key is retained by the person working on the isolated equipment.
4   Warning notices must be posted.
5   All relevant safety and functional tests must be completed before restoring the supply.

Safety first

Undertaking fault diagnosis is inherently dangerous because there is a need to work on live supplies.

Assessment criteria 2.1

State the procedures for reporting and recording information on electrical fault diagnosis and correction work

Assessment criteria 2.2

State the procedures for informing relevant persons about information on electrical fault diagnosis and correction work and the completion of relevant documentation

Assessment criteria 2.3

Explain why it is important to provide relevant persons with information on fault diagnosis and correction work clearly, courteously and accurately

---

Definition

The *absence of circuit protective conductors* would attract a C1 code, since it poses an immediate danger if a fault were to develop.

Recording faults

If an electrical installation fault is identified as a result of a periodic inspection for the purpose of completing an Electrical Installation Condition Report, then it should be documented as detailed in Appendix 6 of the IET Regulations.

The 18th Edition of the IET Regulations (2018) has introduced changes to the Periodic Inspection Report. This has now become the Electrical Installation Condition Report and incorporates changes to the classification coding system and a new inspection schedule that you will find in Appendix 6 of the IET Regulations. The inspection schedule for a domestic installation includes over 60 items that must be inspected and the electrician carrying out the inspection must comment on the condition of each item using the coding system described below. Commenting on the condition of items in relation to the code can also

Figure 5.1 Undertaking fault finding is potentially hazardous since there is an expectation that you have to work with live supplies.

inform on the corrective process, for example whether an item can be repaired or justification on why it must be replaced and why further cost is necessary.

The objective in producing this new Electrical Installation Condition Report inspection schedule is to provide the electrician carrying out the inspection with guidelines, so that the inspection is done in a structured and consistent way and for the client to better understand the result of the inspection. The classification codes to be used on the condition report inspection schedule are shown below where green means 'Good to Go'. Red means 'Stop' there is a problem here. Think of the colour code like traffic lights, and further investigation at a later date is a proceed with caution action.

✓ meaning an acceptable condition. The item inspected has been classified as acceptable.
C1 meaning an unacceptable condition. The item inspected has been classified as unacceptable. Immediate danger is present and the safety of those using the installation is at risk. For example, live parts are directly accessible. *Immediate remedial action is required.*
C2 meaning an unacceptable condition. The item inspected has been classified as unacceptable. Potential danger is present and the safety of those using the installation may be at risk; for example, the absence of the main protective bonding conductors. *Urgent remedial action is required.*
C3 meaning improvement is recommended. The item inspected is not dangerous for continued use but the inspector recommends that improvements be made. For example, the installation does not have RCDs installed for additional protection. *Improvements are recommended.*
FI (further investigation) has uncovered a deficiency that was not possible to identify or uncover due to certain limitations or the actual extent of the inspection being carried out.

Let us assume that while carrying out a visual inspection, the electrician observes a badly damaged socket outlet. On close inspection he realizes that it is possible to touch a live part through the cracked plate. This fault will attract a C1 code because there is immediate danger. Alternatively, if on close inspection the electrician realizes it is possible to touch an earth connection, then it will attract a C2 code: there is potential danger here because the earth wire may become live under fault conditions.

The presence of a C1 or C2 code will result in the overall condition of the electrical installation being reported as unsatisfactory – it has failed to meet the standard. The inspector should report the findings to the duty holder or person responsible for the installation immediately, both verbally and in writing, that a risk of injury exists. If possible, immediately dangerous situations, given a C1 code, should be made safe, isolated or rectified on discovery.

A code C3 in itself would not warrant an unsatisfactory result. It is simply stating that while the installation is not compliant with the latest edition of the IET Regulations, it is compliant with a previous edition, and the item is not unsafe and does not necessarily require upgrading. This differs from when an inspection has revealed a deficiency or non-compliance that could not be fully identified due to the extent or limitations of the inspection. An FI should be recorded if any further investigation is liable to uncover and lead to a classification code C1 (danger present) or C2 (potentially dangerous) but should not be recorded if it is thought that any investigation would lead, at worst, to a C3 classification.

The **Electrical Installation Condition Report** should only be used for an existing building. The report should include:

- schedules of both the inspection and test results in section A;
- the reasons for producing the report such as a change of occupancy or landlord's maintenance should be identified in section B;
- the extent and limitations of the report should be stated in section D;
- the inspector producing the final report should give a summary of the condition of the installation in terms of the safety of the installation in section E, stating how any C1, C2, C3 and F1 codes given for items in the schedule, may affect the overall condition of the installation being classified as unsatisfactory;
- the recommended date of the next Electrical Installation Condition Report should be given in section F.

Further information on the new coding system can be downloaded free from the Electricity Safety Council's Best Practice Guide 4.

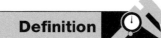

Definition

The *absence of equipotential bonding conductors* would attract a C2 code, since there is potential danger of dangerous voltage developing.

Definition

Electrical Installation Condition Reports are used to report on the condition of existing installations.

- N/V meaning not verified. Although a particular item on the schedule is relevant to the installation, the inspector was unable to verify its condition.
- LIM meaning limitation. A particular item on the schedule is relevant to the installation but there were certain limitations in being able to check the condition.
- N/A meaning not applicable. The particular item on the schedule is not relevant to the installation being inspected.

Assessment criteria 3.1

Specify safe working procedures that should be adopted for completion of fault diagnosis and correction work

Carrying out a safe isolation procedure is only one element in ensuring that any work activities involving fault diagnosis are sound, logical and safe. For instance, given that it might be necessary for the supply to be live when tracing down faults, priority should be given to the positioning of notices and barriers to physically restrict access to unauthorized personnel. The notices involved need to be positioned in view of all routes to the point of isolation and work. Not only must signage contribute to effective communication but all technical personnel involved must be informed of the fault-finding process taking place, and equally important, informing those who might be affected by such activities.

Assessment criteria 3.2

Interpret and apply the logical stages of fault diagnosis and correction work that should be followed

Assessment criteria 3.3

Identify and describe common symptoms of electrical faults

Assessment criteria 3.4

State the causes of different types of fault

Fault diagnosis

To diagnose and find faults in electrical installations and equipment is probably one of the most difficult tasks undertaken by an electrician. The knowledge of fault finding and the diagnosis of faults can never be completely 'learned' because no two fault situations are exactly the same. As the systems we install become more complex, then the faults developed on these systems become more complicated to solve. To be successful the individual must have a thorough knowledge of the installation or piece of equipment and have a broad range of the skills and competences associated with the electrical industries. The ideal person will tackle the problem using a reasoned and logical approach, recognize his own limitations and seek help and guidance where necessary.

The tests recommended by the IET Regulations can be used as a diagnostic tool but the safe working practices described by the Electricity at Work Regulations and elsewhere must always be observed during the fault-finding procedures.

If possible, fault finding should be planned ahead to avoid inconvenience to other workers and to avoid disruption of the normal working routine. However, a faulty piece of equipment or a fault in the installation is not normally a planned event and usually occurs at the most inconvenient time. The diagnosis and rectification of a fault is therefore often carried out in very stressful circumstances.

Symptoms of an electrical fault

The basic symptoms of an electrical fault may be described in one or a combination of the following ways:

1 There is a complete loss of power.
2 There is partial or localized loss of power.
3 The installation or piece of equipment is failing because of the following:
   – an individual component is failing;
   – the whole plant or piece of equipment is failing;

Key fact

Fault finding involves understanding the systems in use, using a logical and reasoned approach and communicating when supplies are to be shut off or re-applied.

Top tip

Fault finding should be a systematic process aligning specific indicators and a circuit's behaviour in order to determine a cause of action. A circuit breaker tripping is normally an indication of excess current (short circuit conditions).

Figure 5.2 Symptoms aren't always as obvious as this.

– the insulation resistance is low;
– the overload or protective devices operate frequently;
– electromagnetic relays will not latch, giving an indication of undervoltage.

Causes of electrical faults

A fault is not a natural occurrence; it is an unplanned event that occurs unexpectedly. The fault in an electrical installation or piece of equipment may be caused by:

* negligence – that is, lack of proper care and attention;
* misuse – that is, not using the equipment properly or correctly;
* abuse – that is, deliberate ill treatment of the equipment.

If the installation was properly designed in the first instance to perform the tasks required of it by the user, then the *negligence, misuse or abuse* must be the fault of the user. However, if the installation does not perform the tasks required of it by the user, then the negligence is due to the electrical contractor not designing the installation to meet the needs of the user.

Negligence on the part of the user may be due to insufficient maintenance or lack of general care and attention, such as not repairing broken equipment or removing covers or enclosures that were designed to prevent the ingress of dust or moisture.

Misuse of an installation or piece of equipment may occur because the installation is being asked to do more than it was originally designed to do, because of the expansion of a company, for example. Circuits are sometimes overloaded because a company grows and a greater demand is placed on the existing installation by the introduction of new or additional machinery and equipment that is otherwise healthy. All electrical connections should be both electrically and mechanically sound, but often bad workmanship or inappropriate or non-precision tooling can cause high resistance joints, which can lead to not only volt drop issues but more seriously in electrical fires. Other faults such as transient voltages can be generated by relays and contactors

Definition

A *fault* is not a natural occurrence; it is an unplanned event that occurs unexpectedly.

which cause transient spikes which can be specially damaging to electronic equipment.

The 18th Edition of the IET Regulations at 532.6 and 421.1.7 has introduced a new topic, Arc Fault Detection Devices. (AFDD) These are compulsory in the EU Countries but only recommended in this country by the IET and the British standards. They are recognised as giving additional protection against fires caused by arc faults. Arc faults can be formed by cable insulation defects, damage to cables by impact or penetration by nails and screws, and loose terminal connections. AFDDs detect faults which MCBs and RCDs cannot detect. They are installed at the origin of the final circuit to be protected.

Before an electrician can begin to diagnose the cause of a fault he must:

- have a thorough knowledge and understanding of the electrical installation or electrical equipment;
- collect information about the fault and the events occurring at or about the time of the fault from the people who were in the area at the time;
- begin to predict the probable cause of the fault using his own and other people's skills and expertise;
- test some of the predictions using a logical approach to identify the cause of the fault.

Most importantly, electricians must use their detailed knowledge of electrical circuits and equipment learned through training and experience and then apply this knowledge to look for a solution to the fault.

Requirements for successful electrical fault finding

The steps involved in successfully finding a fault can be summarized as follows:

1 Gather *information* by talking to people and looking at relevant sources of information such as manufacturer's data, circuit diagrams, charts and schedules.
2 *Analyse* the evidence and use standard tests and a visual inspection to predict the cause of the fault.
3 *Interpret* test results and diagnose the cause of the fault.
4 *Rectify* the fault.
5 *Carry out* functional tests to verify that the installation or piece of equipment is working correctly and that the fault has been rectified.

Key fact

Fault finding can be broken down into: who, where, when and why?

Designing out faults

The designer of the installation cannot entirely design out the possibility of a fault occurring but he can design in 'damage limitation' should a fault occur. For example, designing in two, three or four lighting and power circuits will reduce the damaging effect of any one circuit failing because not all lighting and power will be lost as a result of a fault. Limiting faults to only one of many circuits is good practice because it limits the disruption caused by a fault. IET Regulation 314 tells us to divide an installation into circuits as necessary so as to:

1 avoid danger and minimize inconvenience in the event of a fault occurring;
2 facilitate safe operation, inspection, testing and maintenance;
3 reduce unwanted RCD tripping, sometimes called 'nuisance tripping'.

Figure 5.3 Circuits are divided to avoid danger, inconvenience and reduce unwanted tripping of RCDs.

> **Assessment criteria 3.5**
>
> **Specify the types of faults and their likely locations**

Where do electrical faults occur?

Definition

All terminations and connections need to be both mechanically and electrically secure. Badly formed joints are a prime cause of electrical fires.

1 Faults occur in wiring systems, but not usually along the length of the cable, unless it has been damaged by a recent event such as an object being driven through it or a JCB digger pulling up an underground cable. Cable faults usually occur at each end, where the human hand has been at work at the point of cable interconnections. This might result in broken conductors, trapped conductors or loose connections in joint boxes, accessories or luminaires. All cable connections must be made mechanically and electrically secure. They must also remain accessible for future inspection, testing and maintenance (IET Regulation 526.3). The only exceptions to this rule are when:

- underground cables are connected in a compound-filled or encapsulated joint;
- floor-warming or ceiling-warming heating systems are connected to a cold tail;
- a joint is made by welding, brazing, soldering or a compression tool;
- a joint is made in a maintenance free accessory complying with BS 5733 and marked with the symbol MF, as shown in Fig. 3.85 of Chapter 3.

Since they are accessible, cable interconnections are an obvious point of investigation when searching out the cause of a fault.

2   Faults also occur at cable terminations. The IET Regulations require that a cable termination of any kind must securely anchor all conductors to reduce mechanical stresses on the terminal connections. All conductors of flexible cords must be terminated within the terminal connection, otherwise the current-carrying capacity of the conductor is reduced, which may cause local heating. Flexible cords and fine multiwire conductors are delicate – has the terminal screw been over-tightened, thus breaking the connection as the conductors flex or vibrate? Cables and flexible cords must be suitable for the temperature to be encountered at the point of termination or must be provided with additional insulation sleeves to make them suitable for the surrounding temperatures (IET Regulations 522.2 and 526.9). Even mineral insulated cable, which is robust in design, is sometimes used with a PVC over-sheath to protect from corrosion.

3   Faults also occur at accessories such as switches, sockets, control gear, motor contactors either when they enter or at the point of connection with electronic equipment. The source of a possible fault is again at the point of human contact with the electrical system and again the connections must be checked as described in the first two points above as well as ensuring grommets are fitted where needed. Contacts that make and break a circuit are another source of wear and possible failure, so switches and motor contactors may fail after extensive use. For instance, relay or contactor contacts can weld shut, thus causing a short circuit, or can partially close, causing a high resistance, or even fail to close altogether, causing an open circuit fault.

Socket outlets that have been used extensively and loaded to capacity, in say kitchens, are another source of fault due to overheating or loose connections. Electronic equipment can be damaged by the standard tests described in the IET Regulations and must, therefore, be disconnected before testing begins.

4   Faults occur on instrumentation panels either as a result of a faulty instrument or as a result of a faulty monitoring probe connected to the instrument. Many panel instruments are standard sizes connected to CTs or VTs and this is another source of possible faults of the types described in points 1–3.

5   Faults occur in protective devices for the reasons given in points 1–3 above but also because they may have been badly selected for the job in hand and do not offer adequate protection, discrimination or selectivity as described in Chapter 3 of this book.

6   Faults often occur in luminaires (light fittings) because the lamp has expired. Discharge lighting (fluorescent fittings) also requires a 'starter' to be in good condition, although many fluorescent luminaires these days use starter-less electronic control gear. The points made in 1–3 about cable and flexible cord connections are also relevant to luminaire faults.

7   Faults occur when terminating flexible cords and fine multiwire conductors as a result of the flexible cable being of a smaller cross-section than the load demands, because it is not adequately anchored to reduce mechanical stresses on the connection or because the flexible cord is not suitable for the ambient temperature to be encountered at the point of connection. When terminating flexible cords, the insulation should be carefully removed without cutting out any flexible cord strands of wire because this effectively reduces the cross-section of the conductor. The conductor strands should be twisted together and then doubled over, if possible, and terminated in the appropriate connection as shown in Fig. 3.84. The connection screws should

Key fact

A prime cause of electrical fires stems from badly formed electrical connections.

Key fact

Equipment that is not designed or appropriate to their environment can suffer premature failure.

Key fact

If conductor strands are damaged during the termination process the conductor has effectively become smaller, reducing its current-carrying capacity.

Key fact

Doubling over conductor strands increases the contact area of the conductor termination which lowers its contact resistance.

Figure 5.4 It is essential that any test equipment used in fault finding is appropriate, calibrated and maintained in a good condition.

be opened fully so that they will not snag the flexible cord as it is eased into the connection. The insulation should go up to, but not into, the termination. The terminal screws should then be tightened. When terminating very fine conductors, see also IET Regulation 526.9.

8 Faults occur in electrical components, equipment and accessories such as motors, starters, switch gear, control gear, distribution panels, switches, sockets and luminaires because these all have points at which electrical connections are made. It is unusual for an electrical component to become faulty when it is relatively new because it will have been manufactured and tested to comply with the appropriate British Standard, however it is not unknown.

Through overuse or misuse, components and equipment do become faulty, but most faults are caused by poor installation techniques or design. Ingress of moisture into an enclosure will be largely due to inadequate IP requirements. When re-terminating a connection care must be taken to use the same size and type of termination. This is because using an aluminium crimp with a copper conductor can actually introduce dissimilar metal corrosion.

Overall, however, modern electrical installations using new materials can now last longer than 50 years. Therefore, they must be properly installed. Good design, good workmanship and the use of proper materials are essential if the installation is to comply with the relevant regulations (IET Regulations 133.1 to 134.1).

Assessment criteria 3.6

State the special precautions that should be taken

Special situations or hazardous electrical locations

Figure 5.5 Excessive lengths of cable or high resistance connections can generate volt drop issues, which in turn can result in lights appearing dim.

All electrical installations and installed equipment must be safe to use and free from the dangers of electric shock, but some installations or locations require special consideration because of the inherent dangers of the installed conditions. The danger may arise because of the corrosive or explosive nature of the atmosphere, because the installation must be used in damp or low-temperature conditions or because there is a need to provide additional mechanical protection for the electrical system. Part 7 of the IET Regulations deals with these special installations or locations. In this section we will consider some of the installations that require special consideration.

Working alone

Some working situations are so potentially hazardous that not only must PPE be worn, but you must also never work alone and safe working procedures must be in place before your work begins to reduce the risk. It is unsafe to work in isolation in the following situations:

* when working above ground;
* when working below ground;
* when working in confined spaces;
* when working close to unguarded machinery;
* when a fire risk exists;
* when working close to toxic or corrosive substances such as battery acid.

Permit-to-work system

The **permit-to-work procedure** is a type of 'safe system to work' procedure used in specialized and potentially dangerous plant process situations. The procedure was developed for the chemical industry, but the principle is equally applicable to the management of complex risk in other industries or situations. For example:

* working on part of an assembly line process where goods move through a complex, continuous process from one machine to another (e.g. the food industry);
* repairs to railway tracks, tippers and conveyors;
* working in confined spaces (e.g. vats and storage containers);

Safety first

Working alone
Never work in:
* confined spaces
* storage tanks
* enclosed ductwork.

Safety first

When working within hazardous environments one way of incorporating a safe system of work is to use a permit to work, which will control access.

Definition

Permit to work is a type of 'safe system to work' procedure used in specialized and potentially dangerous plant process situations.

Figure 5.6 Some places are more dangerous than others, but if there is always the danger of an electric shock.

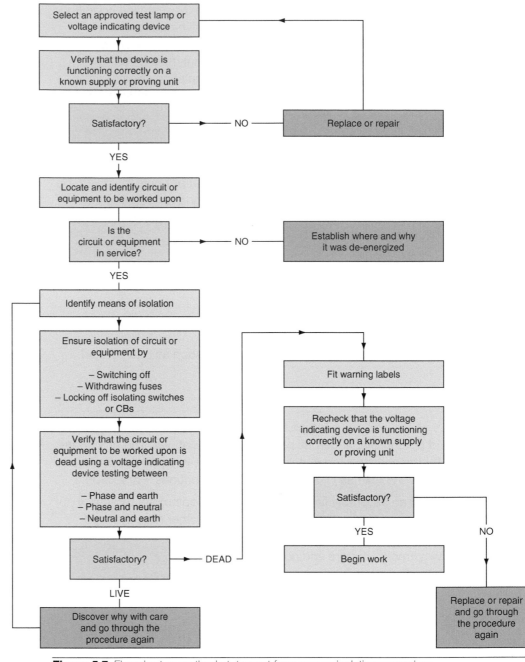

Figure 5.7 Flowchart or method statement for a secure isolation procedure.

- working on or near overhead crane tracks;
- working underground or in deep trenches;
- working on pipelines;
- working near live equipment or unguarded machinery;
- roof work;
- working in hazardous atmospheres (e.g. the petroleum industry);
- working near or with corrosive or toxic substances.

All the above situations are high-risk working situations that should be avoided unless you have received special training and will probably require the completion of a permit to work. Permits to work must adhere to the following eight principles:

1 Wherever possible the hazard should be eliminated so that the work can be done safely without a permit to work.

2   The site manager has overall responsibility for the permit to work even though he may delegate the responsibility for its issue.

3   The permit must be recognized as the master instruction, which, until it is cancelled, overrides all other instructions.

4   The permit applies to everyone on-site, other trades and subcontractors.

5   The permit must give detailed information, for example: (i) which piece of plant has been isolated and the steps by which this has been achieved; (ii) what work is to be carried out; (iii) the time at which the permit comes into effect.

6   The permit remains in force until the work is completed and is cancelled by the person who issued it.

7   No other work is authorized. If the planned work must be changed, the existing permit must be cancelled and a new one issued.

8   Responsibility for the plant must be clearly defined at all stages because the equipment that is taken out of service is released to those who are to carry out the work.

The people doing the work, the people to whom the permit is given, take on the responsibility of following and maintaining the safeguards set out in the permit, which will define what is to be done (no other work is permitted) and the time scale in which it is to be carried out. The permit-to-work system must help communication between everyone involved in the process or type of work. Employers must train staff in the use of such permits and, ideally, training should be designed by the company issuing the permit, so that sufficient emphasis can be given to particular hazards present and the precautions which will be required to be taken. For further details see Permit to Work @www.hse. gov.uk.

Working in hazardous areas

The British Standards concerned with hazardous areas were first published in the 1920s and were concerned with the connection of electrical apparatus in the mining industry. Since those early days many national and international standards, as well as codes of practice, have been published to inform the manufacture, installation and maintenance of electrical equipment in all hazardous areas. The relevant British Standards for Electrical Apparatus for Potentially Explosive Atmospheres are BS 5345, BS EN 60079 and BS EN 50014: 1998. They define a **hazardous area** as 'any place in which an explosive atmosphere may occur in such quantity as to require special precautions to protect the safety of workers'. Clearly these regulations affect the petroleum industry, but they also apply to petrol filling stations.

Most flammable liquids only form an explosive mixture between certain concentration limits. Above and below this level of concentration the mix will not explode. The lowest temperature at which sufficient vapour is given off from a flammable substance to form an explosive gas–air mixture is called the **flashpoint**. A liquid that is safe at normal temperatures will require special consideration if heated to flashpoint. An area in which an explosive gas–air mixture is present is called a hazardous area, as defined by the British Standards, and any electrical apparatus or equipment within a hazardous area must be classified as flameproof.

Flameproof electrical equipment is constructed so that it can withstand an internal explosion of the gas for which it is certified, and prevent any spark or

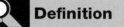
Definition

The British Standards define a *hazardous area* as 'any place in which an explosive atmosphere may occur in such quantity as to require special precautions to protect the safety of workers'.

Definition

The lowest temperature at which sufficient vapour is given off from a flammable substance to form an explosive gas-air mixture is called the flashpoint.

flame resulting from that explosion leaking out and igniting the surrounding atmosphere. This is achieved by manufacturing flameproof equipment to a robust standard of construction. All access and connection points have wide machined flanges that damp the flame in its passage across the flange. Flanged surfaces are firmly bolted together with many recessed bolts, as shown in Fig. 5.8. Wiring systems within a hazardous area must comply with flameproof fittings using an appropriate method, such as:

- PVC cables encased in solid drawn heavy-gauge screwed steel conduit terminated at approved enclosures having wide flanges and bolted covers.
- Mineral insulated cables terminated into accessories with approved flameproof glands. These have a longer gland thread than normal MICC glands. Where the cable is laid underground, it must be protected by a PVC sheath and laid at a depth of not less than 500 mm.
- PVC armoured cables terminated into accessories with approved flameproof glands or any other wiring system that is approved by the British Standard.

All certified flameproof enclosures will be marked Ex, indicating that they are suitable for potentially explosive situations, or EEx, where equipment is certified to the harmonized European Standard. All the equipment used in a flameproof installation must carry the appropriate markings, as shown in Fig. 5.9, if the integrity of the wiring system is to be maintained.

Flammable and explosive installations are to be found in the petroleum and chemical industries, which are classified as group II industries. Mining is classified as group I and receives special consideration from the Mining Regulations because of the extreme hazards of working underground. Petrol filling pumps must be wired and controlled by flameproof equipment to meet the requirements of the Petroleum Regulation Act 1928 and 1936 and any local licensing laws concerning the keeping and dispensing of petroleum spirit.

Hazardous area classification

The British Standard divides the risk associated with inflammable gases and vapours into three classes or zones.

- Zone 0 is the most hazardous, and is defined as a zone or area in which an explosive gas–air mixture is *continuously present* or present for long periods. ('Long periods' is usually taken to mean that the gas–air mixture will be present for longer than 1,000 hours per year.)
- Zone 1 is an area in which an explosive gas–air mixture is *likely to occur* in normal operation. (This is usually taken to mean that the gas–air mixture will be present for up to 1,000 hours per year.)
- Zone 2 is an area in which an explosive gas–air mixture is *not likely* to occur in normal operation and if it does occur it will exist for a very short time. (This is usually taken to mean that the gas–air mixture will be present for fewer than 10 hours per year.)

If an area is not classified as zone 0, 1 or 2, then it is deemed to be non-hazardous, so that normal industrial electrical equipment may be used.

The electrical equipment used in zone 2 will contain a minimum amount of protection. For example, normal sockets and switches cannot be installed in a zone 2 area, but oil-filled radiators may be installed if they are directly connected and controlled from outside the area. Electrical equipment in this area should be

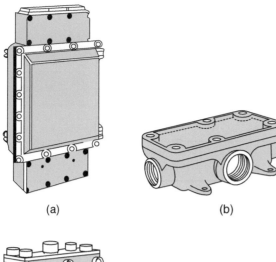

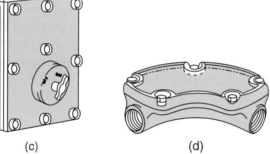

Figure 5.8 Flameproof fittings: (a) flameproof distribution board; (b) flameproof rectangular junction box; (c) double-pole switch; (d) flameproof inspection bend.

Figure 5.9 Flameproof equipment markings.

marked Ex'o' for oil-immersed or Ex'p' for powder-filled. In zone 1 all electrical equipment must be flameproof, as shown in Fig. 5.8, and marked Ex'd' to indicate a flameproof enclosure.

Ordinary electrical equipment cannot be installed in zone 0, even when it is flameproof protected. However, many chemical and oil-processing plants are entirely dependent upon instrumentation and data transmission for their safe operation. Therefore, very low-power instrumentation and data-transmission circuits can be used in special circumstances, but the equipment must be *intrinsically safe*, and used in conjunction with a 'safety barrier' installed outside the hazardous area. Intrinsically safe equipment must be marked Ex'ia' or Ex's', specially certified for use in zone 0.

Intrinsic safety

By definition, an intrinsically safe circuit is one in which no spark or thermal effect is capable of causing ignition of a given explosive atmosphere. The intrinsic safety of the equipment in a hazardous area is assured by incorporating

Definition

An *intrinsically safe circuit* is one in which no spark or thermal effect is capable of causing ignition of a given explosive atmosphere.

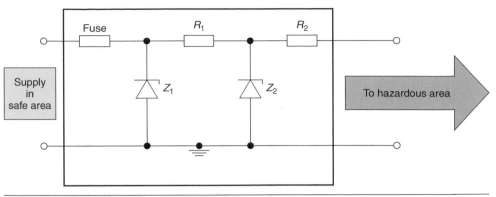

Figure 5.10 Zener safety barrier.

a Zener diode safety barrier into the control circuit such as that shown in Fig. 5.10. In normal operation, the voltage across a Zener diode is too low for it to conduct, but if a fault occurs, the voltage across Z_1 and Z_2 will rise, switching them on and blowing the protective fuse. Z_2 is included in the circuit as a 'backup' in case the first Zener diode fails. An intrinsically safe system, suitable for use in zone 0, is one in which *all* the equipment, apparatus and interconnecting wires and circuits are intrinsically safe.

Optical fibre cables

Top tip

Although fibre optics are inherently safer given their use of light as a medium, faults can occur, for instance, if they are not handled with care and their maximum bend radius is exceeded.

The introduction of fibre-optic cable systems and digital transmissions will undoubtedly affect future cabling arrangements and the work of the electrician. Networks based on the digital technology currently being used so successfully by the telecommunications industry are very likely to become the long-term standard for computer systems. Fibre-optic systems dramatically reduce the number of cables required for control and communications systems, and this will in turn reduce the physical room required for these systems. Fibre-optic cables are also immune to electrical noise when run parallel to mains cables and, therefore, the present rules of segregation and screening may change in the future. There is no spark risk if the cable is accidentally cut and, therefore, such circuits are intrinsically safe. That said there is a danger of blinding a user, should they peer directly at an open-ended termination when the cable is live.

Figure 5.11 Fibre-optic cable.

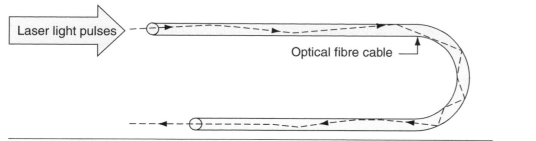

Figure 5.12 Digital pulses of laser light down an optical fibre cable.

The testing of fibre-optic cables requires that special instruments be used to measure the light attenuation (i.e. light loss) down the cable. Finally, when working with fibre-optic cables, electricians should avoid direct eye contact with the low-energy laser light transmitted down the conductors.

Electrostatic discharge

Static electricity is a voltage charge that builds up to many thousands of volts between two surfaces when they rub together. A dangerous situation occurs when the static charge has built up to a potential capable of striking an arc through the air gap separating the two surfaces.

Static charges build up in a thunderstorm. A lightning strike is the discharge of the thunder cloud, which might have built up to a voltage of 100 MV, to the general mass of earth which is at 0 V. Lightning discharge currents are of the order of 20 kA, hence the need for lightning conductors on vulnerable buildings in order to discharge the energy safely.

Static charge builds up between any two insulating surfaces or between an insulating surface and a conducting surface, but it is not apparent between two conducting surfaces.

A motor car moving through the air builds up a static charge, which sometimes gives the occupants a minor shock as they step out and touch the door handle. Static electricity also builds up in modern offices and similar carpeted areas. The combination of synthetic carpets, man-made footwear materials and dry air conditioned buildings contribute to the creation of static electrical charges building up on people moving about these buildings. Individuals only become aware of the charge if they touch earthed metalwork, such as a stair banister rail, before the static electricity has been dissipated. The effect is a sensation of momentary shock.

The precautions against this problem include using floor coverings that have been 'treated' to increase their conductivity or that contain a proportion of natural fibres that have the same effect. The wearing of leather-soled footwear also reduces the likelihood of a static charge persisting, as does increasing the humidity of the air in the building.

A nylon overall and nylon bed sheets build up static charge which is the cause of the 'crackle' when you shake them. Many flammable liquids have the same properties as insulators, and therefore liquids, gases, powders and paints moving through pipes build up a static charge. Petrol pumps, operating theatre oxygen masks and car spray booths are particularly at risk because a spark in these situations may ignite the flammable liquid, powder or gas.

So how do we protect ourselves against the risks associated with static electrical charges? I said earlier that a build-up of static charge is not apparent between

Definition

Static electricity is a voltage change that builds up to many thousands of volts between two surfaces when they run together.

Safety first

Some equipment that is designated as a static sensitive device could be severely damaged through electrostatic discharge.

Definition

Static charge builds up between any two insulating surfaces or between an insulating surface and a conducting surface, but it is not apparent between two conducting surfaces.

two conducting surfaces, and this gives a clue to the solution. Bonding surfaces together with protective equipotential bonding conductors prevents a build-up of static electricity between the surfaces. If we use large-diameter pipes, we reduce the flow rates of liquids and powders and, therefore, we reduce the build-up of static charge. Hospitals use cotton sheets and uniforms, and use protective equipotential bonding extensively in operating theatres. Rubber, which contains a proportion of graphite, is used to manufacture anti-static trolley wheels and surgeons' boots. Rubber constructed in this manner enables any build-up of static charge to 'leak' away. Increasing humidity also reduces static charge because the water droplets carry away the static charge, thus removing the hazard.

IT equipment

Every modern office now contains computers, and many systems are linked together or networked. Most computer systems are sensitive to variations or distortions in the mains supply and many computers incorporate filters that produce high-protective conductor currents of around 2 or 3 mA. This is clearly not a fault current, but is typical of the current that flows in the circuit protective conductor of IT equipment under normal operating conditions. IET Regulation 543.7 deals with the earthing requirements for the installation of equipment having high protective conductor currents. The IET Guidance Note 7 recommends that IT equipment should be connected to double socket outlets as shown in Figure 5.13.

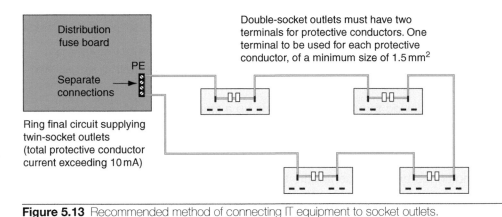

Figure 5.13 Recommended method of connecting IT equipment to socket outlets.

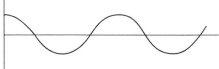

Clean supply

Spikes, caused by an over-voltage transient surging through the mains

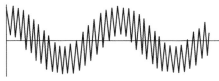

'Noise': unwanted electrical signals picked up by power lines or supply cords

Figure 5.14 Distortions in the a.c. mains supply.

Surge protection

A transient overvoltage or surge is a voltage spike of very short duration. It may be caused by a lightning strike or a switching action on the system. It sends a large voltage spike for a few microseconds down the mains supply, which is sufficient to damage sensitive electronic equipment. Supplies to computer circuits must be 'clean' and 'secure'. Mainframe computers and computer networks are sensitive to mains distortion or interference, which is referred to as 'noise'. Noise is mostly caused by switching an inductive circuit, which causes a transient spike, or by brush gear making contact with the commutator segments of an electric motor. These distortions in the mains supply can cause computers to 'crash' or provoke errors and are shown in Fig. 5.14.

To avoid this, a 'clean' supply is required for the computer network. This can be provided by taking the ring or radial circuits for the computer supplies from a point as close as possible to the intake position of the electrical supply to the building. A clean earth can also be taken from this point, which is usually one core of the cable and not the armour of an SWA cable, and distributed around the final wiring circuit. Alternatively, the computer supply can be cleaned by means of a filter such as that shown in Fig. 5.15 or by installing surge protection devices. A surge protection device is a device intended to limit transient overvoltages, and to divert damaging surge currents away from sensitive equipment. Section 534 of the IET Regulations contains the requirements for the installation of surge protective devices (SPDs) to limit transient overvoltages where required by Section 443 of the IET Regulations or where specified by the designer.

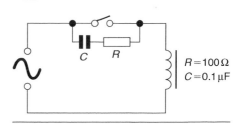

$R = 100\,\Omega$
$C = 0.1\,\mu F$

Figure 5.15 A simple noise suppressor.

Damage to electronic devices by 'overvoltage'

The use of electronic circuits in all types of electrical equipment has increased considerably over recent years. Electronic circuits and components can now be found in leisure goods, domestic appliances, motor starting and control circuits, discharge lighting, emergency lighting, alarm circuits and special-effects lighting systems. All electronic circuits are low-voltage circuits carrying very small currents.

Electrical installation circuits usually carry in excess of 1A and often carry hundreds of amperes. Electronic circuits operate in the milliampere or even microampere range. The test instruments used on electronic circuits must have a *high impedance* so that they do not damage the circuit when connected to take readings.

The use of an insulation resistance test as described by the IET Regulations (described in Chapter 4 of this book), must be avoided with any electronic equipment. The working voltage of this instrument can cause total devastation to modern electronic equipment. When carrying out an insulation resistance test as part of the prescribed series of tests for an electrical installation, all electronic equipment must first be disconnected or damage will result.

Any resistance measurements made on electronic circuits must be achieved with a battery-operated ohmmeter, with a high impedance to avoid damaging electronic components which is explained further on page 341.

Key fact

All sensitive loads as well as items that can store electrical energy such as capacitors must be disconnected before insulation resistance testing is carried out.

Risks associated with high-frequency or large capacitive circuits

Induction heating processes use high-frequency power to provide focused heating in industrial processes. The induction heater consists of a coil of large cross-section. The work-piece or object to be heated is usually made of ferrous metal and is placed inside the coil. When the supply is switched on, eddy currents are induced into the work-piece and it heats up very quickly so that little heat is lost to conduction and convection.

The frequency and size of the current in the coil determines where the heat is concentrated in the work-piece:

- the higher the current, the greater is the surface penetration;
- the longer the current is applied, the deeper the penetration;
- the higher the frequency, the less is the depth of heat penetration.

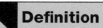

Definition

Induction heating processes use high-frequency power to provide focused heating in industrial processes.

5

Figure 5.16 You may be more familiar with the principle of induction heating than you think – this cooker uses it to cook food.

For shallow penetration, high frequency, high current, short time application is typically used for tool tempering. Other applications are brazing and soldering industrial and domestic gas boiler parts.

When these machines are not working they look very harmless but when they are working they operate very quietly and there is no indication of the intense heat that they are capable of producing. Domestic and commercial microwave ovens operate at high frequency. The combination of risks of high frequency and intense heating means that before any maintenance, repair work or testing is carried out, the machine must first be securely isolated and no one should work on these machines unless they have received additional training to enable them to do so safely.

Industrial wiring systems are very inductive because they contain many inductive machines and circuits, such as electric motors, transformers, welding plants and discharge lighting. The inductive nature of the industrial load causes the current to lag behind the voltage and creates a bad power factor. Power factor is the percentage of current in an alternating current circuit that can be used as energy for the intended purpose. A power factor of say 0.7 indicates that 70% of the current supplied is usefully employed by the industrial equipment.

An inductive circuit, such as that produced by an electric motor, induces an electromagnetic force that opposes the applied voltage and causes the current waveform to lag the voltage waveform. Magnetic energy is stored up in the load during one half cycle and returned to the circuit in the next half cycle. If a capacitive circuit is employed, the current leads the voltage since the capacitor stores energy as the current rises and discharges it as the current falls. So here we have the idea of a solution to the problem of a bad power factor created by inductive industrial loads. Power factor and power factor improvement is discussed below.

The **power factor** at which consumers take their electricity from the local electricity supply authority is outside the control of the supply authority. The power factor of the consumer is governed entirely by the electrical plant and equipment that is installed and operated within the consumer's buildings. Domestic consumers do not have a bad power factor because they use very little inductive equipment. Most of the domestic load is neutral and at unity power factor.

Electricity supply authorities discourage the use of equipment and installations with a low power factor because they absorb part of the capacity of the

Definition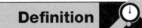

The *power factor* of the consumer is governed entirely by the electrical plant and equipment that is installed and operated within the consumer's buildings.

generating plant and the distribution network to no useful effect. They, therefore, penalize industrial consumers with a bad power factor through a maximum demand tariff, metered at the consumer's intake position. If the power factor falls below a datum level of between 0.85 and 0.9 then extra charges are incurred. In this way industrial consumers are encouraged to improve their power factor.

Power factor improvement of most industrial loads is achieved by connecting capacitors to either:

* individual items of equipment; or
* banks of capacitors may be connected to the main busbars of the installation at the intake position.

The method used will depend upon the utilization of the installed equipment by the industrial or commercial consumer. If the load is constant then banks of capacitors at the mains intake position would be indicated. If the load is variable then power factor correction equipment could be installed adjacent to the machine or piece of equipment concerned. Power factor correction by capacitors is the most popular method because of the following:

* They require no maintenance.
* Capacitors are flexible and additional units may be installed as an installation or system is extended.
* Capacitors may be installed adjacent to individual pieces of equipment or at the mains intake position. Equipment may be placed on the floor or fixed high up and out of the way.

Capacitors store charge and must be disconnected before the installation or equipment is tested in accordance with Section 6 of the IET Regulations BS 7671.

Small power factor correction capacitors, as used in discharge lighting, often incorporate a high-value resistor connected across the mains terminals. This discharges the capacitor safely when not in use. Banks of larger capacity capacitors may require discharging to make them safe when not in use. To discharge a capacitor safely and responsibly it must be discharged slowly over a period in excess of five 'time-constants' through a suitable discharge resistor. Capacitors are discussed in this book in Chapter 6.

Presence of storage batteries

Since an emergency occurring in a building may cause the mains supply to fail, the emergency lighting should be supplied from a source that is independent from the main supply. A battery's ability to provide its output instantly makes it a very satisfactory source of standby power. In most commercial, industrial and public service buildings housing essential services, the alternative power supply would be from batteries, but generators may also be used. Generators can have a large capacity and duration, but a major disadvantage is the delay of time while the generator runs up to speed and takes over the load. In some premises a delay of more than five seconds is considered unacceptable, and in these cases a battery supply is required to supply the load until the generator can take over. The emergency lighting supply must have an adequate capacity and rating for the specified duration of time (IET Regulation 313.2). BS 5266 and BS EN 1838 states that after a battery is discharged by being called into operation for its specified duration of time, it should be capable of once again operating for the specified duration of time following a recharge period of not longer than 24

Definition

Since an emergency occurring in a building may cause the mains supply to fail, the emergency lighting should be supplied from a source that is independent from the main supply.

Figure 5.17 Batteries come in all shapes and sizes – these deep charge industrial batteries can store power for long periods.

hours. The duration of time for which the emergency lighting should operate will be specified by a statutory authority but is normally one to three hours. The British Standard states that escape lighting should operate for a minimum of one hour. Standby lighting operation time will depend upon financial considerations and the importance of continuing the process or activity within the premises after the mains supply has failed.

The contractor installing the emergency lighting should provide a test facility that is simple to operate and secure against unauthorized interference. The emergency lighting installation must be segregated completely from any other wiring, so that a fault on the main electrical installation cannot damage the emergency lighting installation (IET Regulation 528.1). The batteries used for the emergency supply should be suitable for this purpose. Motor vehicle batteries are not suitable for emergency lighting applications, except in the starter system of motor-driven generators. The fuel supply to a motor-driven generator should be checked. The battery room of a central battery system such as that shown in Fig 5.17 must be well ventilated and, in the case of a motor-driven generator, adequately heated to ensure rapid starting in cold weather. The British Standard recommends that the full load should be carried by the emergency supply for at least one hour in every six months. After testing, the emergency system must be carefully restored to its normal operative state. A record should be kept of each item of equipment and the date of each test by a qualified or responsible person. It may be necessary to produce the record as evidence of satisfactory compliance with statutory legislation to a duly authorized person.

Self-contained units are suitable for small installations of up to about 12 units. The batteries contained within these units should be replaced about every five years, or as recommended by the manufacturer.

Storage batteries are secondary cells. A secondary cell has the advantage of being rechargeable. If the cell is connected to a suitable electrical supply, electrical energy is stored on the plates of the cell as chemical energy. When the cell is connected to a load, the chemical energy is converted to electrical

Definition

Storage batteries are secondary cells. A secondary cell has the advantage of being rechargeable. If the cell is connected to a suitable electrical supply, electrical energy is stored on the plates of the cell as chemical energy. When the cell is connected to a load, the chemical energy is converted to electrical energy.

energy. A lead-acid cell is a secondary cell. Each cell delivers about 2 V, and when six cells are connected in series a 12 V battery is formed. A lead-acid battery is constructed of lead plates which are deeply ribbed to give maximum surface area for a given weight of plate. The plates are assembled in groups, with insulating separators between them. The separators are made of a porous insulating material, such as wood or ebonite, and the whole assembly is immersed in a dilute sulphuric acid solution in a plastic container. The capacity of a cell to store charge is a measure of the total quantity of electricity that it can cause to be displaced around a circuit after being fully charged. It is stated in ampere-hours, abbreviation Ah, and calculated at the 10-hour rate, which is the steady load current that would completely discharge the battery in 10 hours. Therefore, a 50 Ah battery will provide a steady current of 5A for 10 hours.

Maintenance of lead-acid batteries

The plates of the battery must always be covered by dilute sulphuric acid. If the level falls, it must be topped up with distilled water.

- Battery connections must always be tight and should be covered with a thin coat of petroleum jelly.
- The specific gravity or relative density of the battery gives the best indication of its state of charge. A discharged cell will have a specific gravity of 1.150, which will rise to 1.280 when fully charged. The specific gravity of a cell can be tested with a hydrometer.
- To maintain a battery in good condition it should be regularly trickle-charged. A rapid charge or discharge encourages the plates to buckle, and may cause permanent damage. Most batteries used for standby supplies today are equipped with constant voltage chargers. The principle of these is that after the battery has been discharged by it being called into operation, the terminal voltage will be depressed and this enables a relatively large current (1–5 A) to flow from the charger to recharge the battery. As the battery becomes more fully charged its voltage will rise until it reaches the constant voltage level where the current output from the charger will drop until it is just sufficient to balance the battery's internal losses. The main advantage of this system is that the battery controls the amount of charge it receives and is therefore automatically maintained in a fully charged condition without human intervention and without the use of any elaborate control circuitry.
- The room used to charge the emergency supply storage batteries must be well ventilated because the charged cell gives off hydrogen and oxygen, which are explosive in the correct proportions.

Assessment criteria 4.1

State the dangers of electricity in relation to the nature of fault diagnosis work

An electrical fault is not a natural occurrence. It is an unplanned event which occurs unexpectedly. The fault may cause a piece of equipment to fail, a production process to stop, or plunge the whole installation into darkness causing normal activities to stop. The electrician trying to find the cause of the fault will be placed in a very stressful environment. That person must have a thorough knowledge and understanding of the electrical systems and equipment, and resolve the problem using a reasoned and logical approach which may involve some live testing.

The tests recommended by the Regulations can be used as a diagnostic tool but the safe working practices described by the Electricity at Work Regulations, discussed elsewhere in this book, must be observed during the fault finding and diagnosis process.

Assessment criteria 4.2

Describe how to identify supply voltages

Construction and Demolition Site Installations (section 704)

Construction and demolition sites are potentially dangerous in many ways. The risk of electric shock is high because of the following factors:

- a construction site or demolition site is, by definition a temporary state. Upon completion there will be either a building with all the necessary safety features, or a brown field site where the building previously stood.
- there is the possibility of damage to cables and equipment as a consequence of the temporary nature of the site and because the site is not always sealed in the early stages from the weather.
- mobile equipment such as electrical tools and hand lamps with trailing leads may be in use.
- there will be many extraneous conductive parts on site which cannot practically be bonded because of the changing nature of the construction process.

Temporary electrical supplies provided on construction sites can save many man hours of labour by providing the energy required for fixed and portable tools and lighting, which speeds up the completion of a project. However, construction sites are dangerous places and the temporary electrical supply that is installed to assist the construction process must comply with all of the relevant wiring regulations for permanent installations (IET Regulation 110.1). All equipment must be of a robust construction in order to fulfil the on-site electrical requirements while being exposed to rough handling, vehicular nudging, the wind, rain and sun. All equipment socket outlets, plugs and couplers must be of the industrial type to BS EN 60439 and BS EN 60309 and specified by IET Regulation 704.511.1 as shown in Fig. 5.18.

IET Regulations 704.313, 704.410.3.10 and 411.8 tell us that reduced low voltage is strongly *preferred* for portable hand lamps and tools used on construction and demolition sites. The distribution of electrical supplies on a construction site would typically be as follows:

- 400 V three phase for supplies to major items of plant having a rating above 3.75 kW such as cranes and lifts. These supplies must be wired in armoured cables.
- 230 V single phase for supplies to items of equipment that are robustly installed such as floodlighting towers, small hoists and site offices. These supplies must be wired in armoured cable unless run inside the site offices.

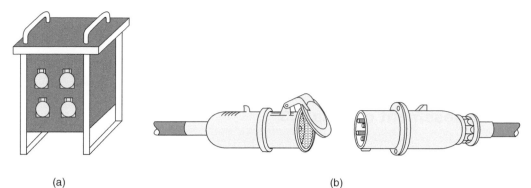

(a)   (b)

Figure 5.18 110 V distribution unit and cable connector, suitable for construction site electrical supplies: (a) reduced-voltage distribution unit incorporating industrial sockets to BS EN 60309 and (b) industrial plug and connector.

Figure 5.19 Temporary electrical supply.

110 V single phase for supplies to all mobile hand tools and all mobile lighting equipment. The supply is usually provided by a reduced voltage distribution unit that incorporates splashproof sockets fed from a centre-tapped 110 V transformer. This arrangement limits the voltage to earth to 55 V, which is recognized as safe in most locations. A 110 V distribution unit is shown in Fig. 5.18. Edison screw lamps are used for 110 V lighting supplies so that they are not interchangeable with 230 V site office lamps.

There are occasions when even a 110 V supply from a centre-tapped transformer is too high; for example, supplies to inspection lamps for use inside damp or confined places. In these circumstances a safety extra-low voltage (SELV) supply would be required.

Industrial plugs have a keyway that prevents a tool from one voltage being connected to the socket outlet of a different voltage. They are also colour coded for easy identification as follows:

* 400 V – red
* 230 V – blue

Safety first

Construction sites
Low voltage or battery tools must be used on construction sites.

- 110 V – yellow
- 50 V – white
- 25 V – violet.

Assessment criteria 4.3

Select the correct test instruments (in accordance with HSE guidance document GS 38) for fault diagnosis work

Assessment criteria 4.4

Describe how to confirm test instruments are fit for purpose, functioning correctly and are correctly calibrated

Selecting test equipment

The HSE has published a guidance note (GS 38) that advises electricians and other electrically competent people on the selection of suitable test probes, voltage-indicating devices and measuring instruments. This is because they consider suitably constructed test equipment to be as vital for personal safety as the training and practical skills of the electrician. In the past, unsatisfactory test probes and voltage indicators have frequently been the cause of accidents, and therefore all test probes must now incorporate the following features:

1 The probes must have finger barriers or be shaped so that the hand or fingers cannot make contact with the live conductors under test.
2 The probe tip must not protrude more than 2 mm, and preferably only 1 mm, be spring-loaded and screened.
3 The lead must be adequately insulated and coloured so that one lead is readily distinguished from the other.
4 The lead must be flexible and sufficiently robust.
5 The lead must be long enough to serve its purpose but not too long.
6 The lead must not have accessible exposed conductors even if it becomes detached from the probe or from the instrument.

GS 38 also tells us that where the test is being made simply to establish the presence or absence of a voltage, the preferred method is to use a proprietary test lamp or voltage indicator that is suitable for the working voltage, rather than a multimeter. Accident history has shown that incorrectly set multimeters or makeshift devices for voltage detection have frequently caused accidents.

Figure 4.2 shows a suitable voltage indicator. Test lamps and voltage indicators are not fail-safe, and therefore GS 38 recommends that they should be regularly proved, preferably before and after use, as described previously in the flowchart for a safe isolation procedure. The IET Regulations (BS 7671) also specify the test voltage or current required to carry out particular tests satisfactorily. All testing must, therefore, be carried out using an 'approved' test instrument if the test results are to be valid.

The test instrument must also carry a calibration certificate, otherwise the recorded results may be void. Calibration certificates usually last for a year. Test instruments must, therefore, be tested and recalibrated each year by an approved supplier. This will maintain the accuracy of the instrument to an acceptable level, usually within 2% of the true value. Modern digital test instruments are reasonably robust, but to maintain them in good working order they must be treated with care. An approved test instrument costs as much as a good-quality camera; it should, therefore, receive the same care and consideration.

Continuity tester

To measure accurately the resistance of the conductors in an electrical installation, we must use an instrument that is capable of producing an open circuit voltage of between 4 and 24 V a.c. or d.c., and delivering a short-circuit current of not less than 200 mA (IET Regulation 643.2.1). The functions of continuity testing and insulation resistance testing are usually combined in one test instrument.

Insulation resistance tester

The test instrument must be capable of detecting insulation leakage between live conductors and between live conductors and earth. To do this and comply with IET Regulation 643.3 the test instrument must be capable of producing a test voltage of 250, 500 or 1,000 V and delivering an output current of not less than 1 mA at its normal voltage.

Earth fault loop impedance tester

The test instrument must be capable of delivering fault currents as high as 25 A for up to 40 ms using the supply voltage. During the test, the instrument does an Ohm's law calculation and displays the test result as a resistance reading.

RCD tester

Where circuits are protected by an RCD we must carry out a test to ensure that the device will operate very quickly under fault conditions and within the time limits set by the IET Regulations. The instrument must, therefore, simulate a fault and measure the time taken for the RCD to operate. The instrument is, therefore, calibrated to give a reading measured in milliseconds to an in-service accuracy of 10%. If you purchase good-quality 'approved' test instruments and leads from specialist manufacturers they will meet all the regulations and standards and therefore give valid test results. However, to carry out all the tests required by the IET Regulations will require a number of test instruments and this will represent a major capital investment in the region of £1,000. The specific tests required by the IET Regulations: BS 7671 are described in detail in Chapter 4 of this book under the sub-heading 'Inspection and testing techniques'.

Electrical installation circuits usually carry in excess of 1 A and often carry hundreds of amperes. Electronic circuits operate in the milliampere or even microampere range. The test instruments used on electronic circuits must have a circuit because they consume some power in order to provide the torque required to move the pointer. In power applications these small disturbances seldom give

Figure 5.20 Digital multimeter suitable for testing electrical and electronic circuits.

rise to obvious errors, but in electronic circuits a small disturbance can completely invalidate any readings taken. We must, therefore, choose our electronic test equipment with great care as described in Chapter 4 of this publication.

So far in this chapter, I have been considering standard electrical installation circuits wired in conductors and cables using standard wiring systems. However, you may be asked to diagnose and repair a fault on a system that is unfamiliar to you or outside your experience and training. If this happens to you, I would suggest that you immediately tell the person ordering the work or your supervisor that it is beyond your knowledge and experience. I have said earlier that fault diagnosis can only be carried out successfully by someone with a broad range of experience and a thorough knowledge of the installation or equipment that is malfunctioning. The person ordering the work will not think you a fool for saying straightaway that the work is outside your experience. It is better to be respected for your honesty than to attempt something that is beyond you at the present time and which could create bigger problems and waste valuable repair time. Let us now consider some situations where special precautions or additional skills and knowledge may need to be applied.

Assessment criteria 4.5

State the appropriate documentation that is required for fault diagnosis work and explain how and when it should be completed

Assessment criteria 4.8

Identify whether test results are acceptable and state the actions to take where unsatisfactory results are obtained

Assessment criteria 5.3

State the appropriate documentation that is required for fault correction work and explain how and when it should be completed

Various types of information are required when engaging in fault-finding. This can and should include the IET Wiring Regulations BS 7671, which contain essential data such as earth loop impedance and insulation resistance values. This is especially important regarding insulation resistance since BS 7671 stipulates that a reading of 1 MΩ complies with BS 7671 but does warrant further investigation, which could mean further expenditure for the customer. Moreover, BS 7671 can be used to obtain the appropriate level of index protection that equipment requires, size of circuit and protective conductors, as well as rating and classification of protective devices. Maintenance records can be accessed in order to determine why certain components were replaced including any relevant or re-occurring symptoms or sharp degradation of insulation resistance values. Manufacturers' information such as manuals and user guides can also assist in defining if equipment is faulty or malfunctioning or simply not being operated in the correct mode of operation as designed. This is a very important part of commissioning an installation whereby the hand over process would include, where necessary, a demonstration of how certain equipment operates.

Schedules of previous testing can compare current results against previous readings, such as insulation values, whereby slow deterioration is indicative of a gradual decline due to ageing of the system. This is in comparison with a sharp decline, which is more likely to have been caused by inadvertent component failure or work activity.

Assessment criteria 4.6

Explain why carrying out fault diagnosis work can have implications for customers and clients

Assessment criteria 5.4

Explain how and why relevant people need to be kept informed during completion of fault correction work

Live testing

The Electricity at Work Regulations tell us that it is 'preferable' that supplies be made dead before work commences (Regulation 4(3)). However, they do acknowledge that some work, such as fault finding and testing, may require the electrical equipment to remain energized. Therefore, if the fault finding and testing can only be successfully carried out 'live', then the person carrying out the fault diagnosis must:

- be trained so that he understands the equipment and the potential hazards of working live and can, therefore, be deemed to be 'competent' to carry out the activity;
- only use approved test equipment;
- set up barriers and warning notices so that the work activity does not create a situation dangerous to others.

Note that while live testing may be required in order to find the fault, live repair work must not be carried out. The individual circuit or item of equipment must first be isolated.

Inevitably, both testing and fault finding will affect the supply of an installation irrespective of what environment you are working in. Although in certain circumstances, if it can be justified, work around live supplies may be permitted subject to certain safety measures, undertaking fault finding in an industrial, commercial or domestic environment can disrupt any meaningful activity or production that requires electricity. It is vital therefore that the communication stream is maintained in order to give people fair warning of disruption, when this is necessary. Planning of some maintenance events can alleviate disruption by being implemented during quiet hours either during the night, weekend and even in some cases over bank holidays. Failure to incorporate such measures is effectively a serious breakdown of communication, causing not only tension when information is not cascaded down, but in certain situations can also impinge on the safety of the persons engaged in or affected by the work activities. A further consideration is the detrimental effect it may have on customer relations or even souring the working relationship with another company engaged in similar operations in the same vicinity.

Assessment criteria 5.1

Identify and explain factors that can affect fault correction, repair or replacement

Faulty equipment: to repair or replace?

Having successfully diagnosed the cause of the fault we have to decide if we are to repair or replace the faulty component or piece of equipment. In many cases the answer will be straightforward and obvious, but in some circumstances the solution will need to be discussed with the customer. Some of the issues that may be discussed are as follows:

- What is the cost of replacement? Will the replacement cost be prohibitive? Is it possible to replace only some of the components? Will the labour costs of the repair be more expensive than a replacement? Do you have the skills necessary to carry out the repair? Would the repaired piece of equipment be as reliable as a replacement?
- Is a suitable replacement available within an acceptable time? These days, manufacturers carry small stocks to keep costs down.
- Can the circuit or system be shut down to facilitate a repair or replacement?
- Can alternative or temporary supplies and services be provided while replacements or repairs are carried out?

Figure 5.21 Some things have to be replaced for health and safety reasons, like the cable shown above.

Assessment criteria 5.2

Specify the procedures for functional testing and identify tests that can verify fault correction

Assessment criteria 4.7

Specify the procedures for functional testing and identify tests that verify fault correction

Restoring systems to working order

When the fault has been identified and repaired as described in this chapter, the circuit, system or equipment must be inspected, tested and functional checks carried out as required by IET Regulations Chapter 61. The purpose of inspecting and testing the repaired circuit, system or equipment is to confirm the electrical integrity of the system before it is re-energized.

The tests recommended by Part 6 of the IET Regulations are:

1 Test the continuity of the protective conductors including the protective equipotential and supplementary bonding conductors.
2 Test the continuity of all ring final circuit conductors.
3 Test the insulation resistance between live conductors and earth.
4 Test the polarity to verify that single pole control and protective devices are connected in the line conductor only.
5 Test the earth electrode resistance where the installation incorporates an earth electrode as a part of the earthing system.

The supply may now be connected and the following tests carried out:

6 Test the polarity using an approved test lamp or voltage indicator.
7 Test the earth fault loop impedance where the protective measures used require a knowledge of earth fault loop impedance.

These tests *where relevant* must be carried out in the order given above to comply with IET Regulation 643.1.

If any test indicates a failure, that test and any preceding test must be repeated after the fault has been rectified. This is because the earlier tests may have been influenced by the fault.

The above tests are described in Chapter 4 of this book, in Part 6 of the IET Regulations and in Guidance Note 3 published by the IET.

Following the relevant tests described above we must carry out functional testing to ensure that:

- the circuit, system or equipment works correctly;
- it works as it did before the fault occurred;
- it continues to comply with the original specification;
- it is electrically safe;
- it is mechanically safe;
- it meets all the relevant regulations, in particular the IET Regulations (BS 7671).

IET Regulation 643.10 tells us to check the effectiveness of the following assemblies to show that they are properly mounted, adjusted and installed:

- residual current devices (RCDs);
- switchgear;
- control gear;
- controls and interlocks.

> **Key fact**
>
> GN3 specifies in Chapter 1 general requirements that when an addition or alteration has been carried out then the minimum tests required are protective conductor continuity and earth fault loop impedance.

Assessment criteria 5.5

Specify the methods for restoring the condition of building fabric

Restoration of the building structure

If the structure, or we sometimes call it the fabric, of the building has been damaged as a result of your electrical repair work, it must be made good before you hand the installation, system or equipment back to the client.

> **Definition**
>
> A major consideration of making good is to use non-combustible material.

Where a wiring system passes through elements of the building construction such as floors, walls, roofs, ceilings, partitions or cavity barriers, the openings remaining after the passage of the wiring system must be sealed according to the degree of fire resistance demonstrated by the original building material (IET Regulation 527.2).

You should always make good the structure of the building using appropriate materials *before* you leave the job so that the general building structural performance and fire safety are not reduced. If, additionally, there is a little cosmetic plastering and decorating to be done, then who will actually carry out this work is a matter of negotiation between the client and the electrical contractor.

Assessment criteria 5.6

State the methods to ensure the safe disposal of any waste and that the work area is left in a safe and clean condition

Disposal of waste

Having successfully diagnosed the electrical fault and carried out the necessary repairs or having completed any work in the electrical industry, we come to the final practical task, leaving the site in a safe and clean condition and the removal of any waste material. This is an important part of your company's 'good customer relationships' with the client. We know that we have a 'duty of care' for the waste that we produce as an electrical company.

We have also said many times in this book that having a good attitude to health and safety, working conscientiously and neatly, keeping passageways clear and regularly tidying up the workplace are signs of a good and competent craftsman. But what do you do with the rubbish that the working environment produces? Well, all the packaging material for electrical fittings and accessories usually goes into either your employer's skip or the skip on site designated for that purpose.

However, some substances require special consideration and disposal. We will now look at asbestos and large quantities of used fluorescent tubes, which are classified as 'special waste' or 'hazardous waste'.

Asbestos is a mineral found in many rock formations. When separated it becomes a fluffy, fibrous material with many uses. It was used extensively in the construction industry during the 1960s and 1970s for roofing material, ceiling and floor tiles, fire resistant board for doors and partitions, for thermal insulation and commercial and industrial pipe lagging.

In the buildings where it was installed some 40 years ago, when left alone, it does not represent a health hazard, but those buildings are increasingly becoming in need of renovation and modernization. It is in the dismantling and breaking up of these asbestos materials that the health hazard increases. Asbestos is a serious health hazard if the dust is inhaled. The tiny asbestos particles find their way into delicate lung tissue and remain embedded for life, causing constant irritation and, eventually, serious lung disease.

Working with asbestos materials is not a job for anyone in the electrical industry. If asbestos is present in situations or buildings where you are

---

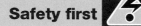

Safety first

If your electrical activities cause damage to the fabric of the building then:

- the openings remaining must be sealed according to the degree of fire resistance demonstrated by the original building material.

IET Regulation 527.2

All the offcuts of conduit, trunking and tray also go into the skip.

- In fact, most of the general site debris will probably go into the skip and the waste disposal company will take the skip contents to a designated local council landfill area for safe disposal.

- The part coils of cable and any other reusable leftover lengths of conduit, trunking or tray will be taken back to your employer's stores area. Here it will be stored for future use and the returned quantities deducted from the costs allocated to that job.

- What goes into the skip for normal disposal into a landfill site is usually a matter of common sense.

Figure 5.22 Disposing of waste in this manner can be harmful to wildlife.

expected to work, it should be removed by a specialist contractor before your work commences. Specialist contractors, who will wear fully protective suits and use breathing apparatus, are the only people who can safely and responsibly carry out the removal of asbestos. They will wrap the asbestos in thick plastic bags and store them temporarily in a covered and locked skip. This material is then disposed of in a special landfill site with other toxic industrial waste materials and the site monitored by the Local Authority for the foreseeable future.

There is a lot of work for electrical contractors in many parts of the country, updating and improving the lighting in government buildings and schools. This work often involves removing the old fluorescent fittings hanging on chains or fixed to beams and installing a suspended ceiling and an appropriate number of recessed modular LED fittings. So what do we do with the old fittings? Well, the fittings are made of sheet steel, a couple of plastic lamp holders, a little cable, a starter and ballast. All of these materials can go into the ordinary skip. However, the fluorescent tubes contain a little mercury and fluorescent powder with toxic elements, which cannot be disposed of in the normal land-fill sites. Hazardous Waste Regulations were introduced in July 2005 and under these regulations lamps and tubes are classified as hazardous. While each lamp contains only a small amount of mercury, vast numbers of lamps and tubes are disposed of in the United Kingdom every year resulting in a significant environmental threat. The environmentally responsible way to dispose of fluorescent lamps and tubes is to recycle them.

Test your knowledge

When you have completed the questions check out the answers at the back of the book.

Note: more than one multiple-choice answer may be correct.

Learning outcome 1

1　Which of the following is correct regarding safe isolation and investigating an electrical fault?
   a.　begin investigation, verify circuit is dead, isolate and lock off, re-check voltage device
   b.　obtain permission, check voltage device, verify circuit is dead, isolate and lock off, re-check voltage device, begin investigation
   c.　begin work, lock off, verify circuit is dead, re-check voltage device, begin investigation
   d.　obtain permission, isolate and lock off, check voltage device, verify circuit is dead, re-check voltage device, begin investigation.

2　The very first action involved in carrying out safe isolation is to:
   a.　test the circuit
   b.　shut off, lock off then assess all hazards
   c.　assess all hazards through a risk assessment
   d.　carry out a full functional test.

3　When applying safe isolation protocols you:
   a.　can work on live circuits
   b.　can use the circuit identification as found
   c.　need to correctly identify the circuit
   d.　can use the mains to check test equipment.

4　The HSE recommends which of the following in order to embrace safety?
   a.　approved volt stick
   b.　equipment listed and endorsed by BS 7671
   c.　equipment listed and endorsed by GS38
   d.　approved neon screwdriver.

5　The legislation that prohibits live working except in exceptional circumstances is:
   a.　Provision and Use of Work Equipment Regulations (1992)
   b.　Personal Protective Equipment Regulations (1992)
   c.　BS 7671 Requirements for Electrical Installations
   d.　Electricity at Work Regulations (1989).

Learning outcome 2

6   Whilst carrying out fault diagnosis, a serious issue has come to light. You actions should be to:
    a.   leave the circuit as it is and go and inform the client in a clear and courteous manner
    b.   ensure the circuit is safe then go and inform the client in a clear and courteous manner
    c.   complete the investigation and then inform the client in a clear and courteous manner
    d.   carry on with the investigation and then inform the client when you see them next.

7   Keeping the customer abreast of development and any correction work in a clear, courteous and accurate manner:
    a.   makes good business sense
    b.   ensures that safety is maintained
    c.   can boost your earnings
    d.   justifies any extra expense.

8   Which of the following will physically reduce the possibility of interference from unauthorized personnel?
    a.   barriers
    b.   polite notices
    c.   secure locking devices
    d.   multiple number of keys.

9   Notices can only be effective if:
    a.   they are well positioned
    b.   they inform management only
    c.   they are clear and accurate
    d.   they are popular.

10   When fault finding, which of the following apply to customer relations?
    a.   the electrician needs to be flexible and sensitive regarding when supply can be isolated
    b.   the electrician needs to be discourteous if they encounter a customer
    c.   the electrician does not have to worry about the impact of noise
    d.   the electrician does not have to worry about any damage incurred.

11   If an electrician informs a client of the need to isolate the supply:
    a.   there is no need for any more control measures
    b.   there is no further health and safety obligation required
    c.   it will remind the client of the importance of turning the supply back on
    d.   the implication and timing when supply is turned off can be discussed.

Learning outcome 3

12   Which of the following has priority?
    a.   establish the events that led up to the problem
    b.   check if supply is present
    c.   check if any protective devices have operated
    d.   check rating of protective devices.

13   Isolating a power supply includes:
   a.   switching off secondary supplies only
   b.   switching off primary supplies only
   c.   discharging only stored energy
   d.   isolating and discharging all forms of energy.

14   The symptom of a high resistance joint could include:
   a.   burning smell
   b.   protective device operating
   c.   collection of moisture
   d.   corrosion.

15   High resistance joints (terminations) are associated with:
   a.   unsheathed cables
   b.   loose connection
   c.   over-tightened connection
   d.   damaged conductors.

16   Typical symptoms of excessive circuit current include:
   a.   an undersized conductor
   b.   protective device operating
   c.   melting of conductor insulation
   d.   an oversized conductor.

17   If the solenoid contacts of a relay became open circuit, what would happen when power was applied?
   a.   the relay would operate as advertised and pull in, but very quickly get hot
   b.   the relay would fail to operate
   c.   the relay would chatter
   d.   the relay would get hot.

18   If the contacts of a relay were to be short circuited, what would happen when power was applied?
   a.   the relay would operate without being energized
   b.   the relay would fail to operate
   c.   the relay would chatter
   d.   the relay would get hot.

19   Symptoms associated with a high resistance termination and excessive volt drop are:
   a.   light appearing dim
   b.   tripping of protective devices
   c.   motors slow to operate
   d.   cables becoming hot.

20   Which of the following apply to IT equipment?
   a.   static sensitive devices
   b.   clean earths
   c.   uninterruptable supply
   d.   sensitivity regarding IR testing.

21 Discovering moisture within an enclosure is symptomatic of issues in relation to:

    a.   insulation resistance

    b.   fault protection

    c.   index protection

    d.   basic protection.

22 Discovering a copper cable that has been terminated with an aluminium crimp will require the cable to be re-terminated or replaced. This is because of:

    a.   dissimilar metal corrosion

    b.   protective device will operate

    c.   volt drop issues

    d.   termination contact resistance will be low.

Learning outcome 4

23 One of the dangers of electricity in relation to testing and fault finding is that

    a.   the supply may need to be dead

    b.   the supply may need to be live

    c.   manufacturing may be affected

    d.   heavy fines are possible.

24 When conducting voltage measurements, which of the following will more likely stop inadvertent tripping of RCDs?

    a.   equipment with a valid calibration

    b.   approved equipment listed within GS38

    c.   expensive equipment

    d.   extremely sensitive equipment.

25 When diagnosing faults, which of the following can aid diagnosis?

    a.   questioning customers

    b.   maintenance records

    c.   previous test results

    d.   manufacturer's warranty.

26 Following replacement of a faulty piece of equipment, completion of this particular task would include:

    a.   full functional test

    b.   completion of job sheet

    c.   replace any protective devices that operated

    d.   full set of dead tests.

27 Looking at the circuit diagram below, if the voltmeter across the load measures 12V, what should the other voltmeter read?

    a.   0 V

    b.   12 V

    c.   36 V

    d.   6 V.

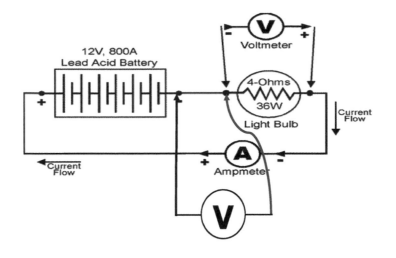

28   Looking at the circuit diagram below, if the voltmeter across the load measures 12 V, what should the other voltmeter read?

    a.   0 V

    b.   12 V

    c.   36 V

    d.   6 V.

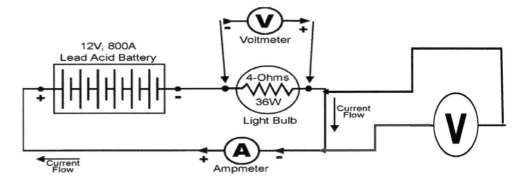

29   With an open circuit as shown in the diagram below, if the voltmeter across the load measures 12 V, what should the other voltmeter read?

    a.   0 V

    b.   12 V

    c.   36 V

    d.   6 V.

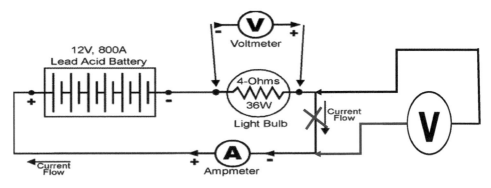

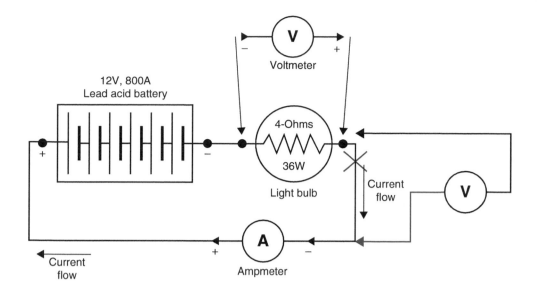

30 A circuit breaker tripping is symptomatic of:
  a. an open circuit
  b. a short circuit
  c. a high resistance
  d. a faulty circuit breaker.

31 Lights appearing dim is symptomatic of:
  a. excessive length of cable
  b. short circuit
  c. high resistance
  d. volt drop issues.

32 Melting insulation is symptomatic of:
  a. excessive length of cable
  b. a short circuit
  c. high resistance
  d. undersized conductor.

33 Excessive volt drop means that the voltage at the supply end:
  a. is less than the voltage at the consumer load
  b. is more than the voltage at the consumer load
  c. is the same as the voltage at the consumer load
  d. is double the voltage at the consumer load.

34 A burnt out relay is symptomatic of:
  a. excessive circuit volt drop
  b. excessive circuit resistance
  c. excessive circuit current
  d. incorrect rating.

35 An electrical termination should:
  a. be accessible
  b. be mechanically sound
  c. be electrically sound
  d. be undersized if possible.

36 If the circuit cable supplying the solenoid element of a heavy duty contactor became open circuit, which of the following would correctly match the symptom?

   a. the contactor would energize but instantly drop out

   b. the contactor would fail to energize

   c. the contactor would energize but start to chatter

   d. the contactor would work as normal but would soon overheat.

37 If the normally open switch that supplies the solenoid element of a heavy duty contactor shorted out, which of the following would correctly match the symptom?

   a. the contactor would energize but instantly drop out

   b. the contactor would fail to operate

   c. the contactor would energize without being functionally switched

   d. the contactor would de-energize when the normally closed switched is operated but then re-energize.

38 Which of the following tests can ascertain the serviceability of a solenoid?

   a. insulation resistance

   b. resistance measurement

   c. power rating measurement

   d. power factor measurement.

39 Increased circuit resistance could be symptomatic of:

   a. undersized conductor

   b. oversized conductor

   c. corrosion

   d. stripped or damaged conductor strands.

Learning outcome 5

40 Which of the following would influence the decision if a replacement part is repaired or replaced?

   a. cost

   b. effect on production

   c. availability of replacement parts

   d. access.

41 Which of the following is a major consideration of making good?

   a. colour

   b. cost

   c. nonflammable material

   d. soundproof.

42 The correct procedure for disposing of fluorescent tubes is:

   a. in black plastic bags having smashed them to bits

   b. in cardboard boxes to protect against the glass shards

   c. through landfill

   d. using a hazardous waste management company.

43 Keeping relevant people informed during fault correction work will:
   a. speed up fault correction
   b. keep everyone happy
   c. keep everyone safe
   d. bring about agreement on the best possible solution.

Classroom activities

Fault symptoms

Match one of the numbered boxes to each lettered box on the left.

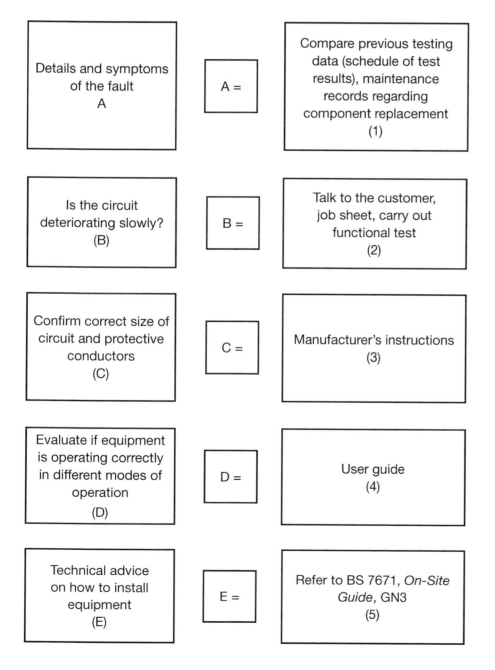

| Details and symptoms of the fault A | A = | Compare previous testing data (schedule of test results), maintenance records regarding component replacement (1) |

| Is the circuit deteriorating slowly? (B) | B = | Talk to the customer, job sheet, carry out functional test (2) |

| Confirm correct size of circuit and protective conductors (C) | C = | Manufacturer's instructions (3) |

| Evaluate if equipment is operating correctly in different modes of operation (D) | D = | User guide (4) |

| Technical advice on how to install equipment (E) | E = | Refer to BS 7671, *On-Site Guide*, GN3 (5) |

Symptoms and corrective action

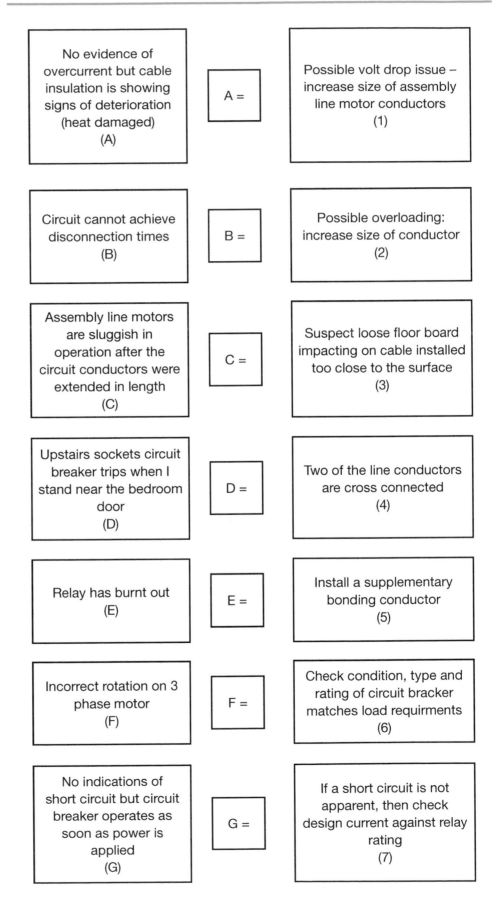

Symptom		Corrective action
No evidence of overcurrent but cable insulation is showing signs of deterioration (heat damaged) (A)	A =	Possible volt drop issue – increase size of assembly line motor conductors (1)
Circuit cannot achieve disconnection times (B)	B =	Possible overloading: increase size of conductor (2)
Assembly line motors are sluggish in operation after the circuit conductors were extended in length (C)	C =	Suspect loose floor board impacting on cable installed too close to the surface (3)
Upstairs sockets circuit breaker trips when I stand near the bedroom door (D)	D =	Two of the line conductors are cross connected (4)
Relay has burnt out (E)	E =	Install a supplementary bonding conductor (5)
Incorrect rotation on 3 phase motor (F)	F =	Check condition, type and rating of circuit bracker matches load requirments (6)
No indications of short circuit but circuit breaker operates as soon as power is applied (G)	G =	If a short circuit is not apparent, then check design current against relay rating (7)

QELTK3/007 Understanding the principles, practices and legislation for the diagnosing and correcting of electrical faults in electrotechnical systems and equipment in buildings, structures and the environment

Chapter checklist

Learning outcome	Assessment criteria		Page number
1. Understand the principles, regulatory requirements and procedures for completing the safe isolation of electrical circuits and complete electrical installations.	1.1	Specify and undertake the correct procedure for completing the safe isolation of an electrical circuit with regard to: • assessment of safe working practices • correct identification of circuits to be isolated • the selection of suitable points of isolation • the selection of correct test and proving instruments in accordance with relevant industry guidance and standards • the use of correct testing methods • the selection of locking devices for securing isolation • the use of correct warning notices • the correct sequence for isolating circuits.	266
	1.2	State the implications of carrying out safe isolations to: • other personnel • customers/clients • public • building systems (loss of supply).	267
	1.3	State the implications of not carrying out safe isolations to: • self • other personnel • customers/clients • public • building systems (presence of supply).	270
	1.4	Identify all health and safety requirements that apply when diagnosing and correcting electrical faults in electrotechnical systems and equipment including those that cover: • working in accordance with risk assessments / permits to work/ method statements • safe use of tools and equipment • safe and correct use of measuring instruments • provision and use of PPE • reporting of unsafe situations.	270
2. Understand how to complete the reporting and recording of electrical fault diagnosis and correction work.	2.1	State the procedures for reporting and recording information on electrical fault diagnosis and correction work	314

Learning outcome	Assessment criteria		Page number
	2.2	State the procedures for informing relevant persons about information on electrical fault diagnosis and correction work and the completion of relevant documentation	314
	2.3	Explain why it is important to provide relevant persons with information on fault diagnosis and correction work clearly, courteously and accurately	314
3. Understand how to complete the preparatory work prior to fault diagnosis and correction work.	3.1	Specify safe working procedures that should be adopted for completion of fault diagnosis and correction work, including: • effective communication with others in the work area • use of barriers • positioning of notices • safe isolation.	316
	3.2	Interpret and apply the logical stages of fault diagnosis and correction work that should be followed: • identification of symptoms • collection and analysis of data • use of sources/types of information such as the IET Wiring Regulations, installation certificates • installation specifications, drawings/diagrams, manufacturer's information and operating instructions • maintenance records • experience (personal and of others) • checking and testing (e.g. supply, protective devices) • interpreting results/information • fault correction • functional testing • restoration.	317
	3.3	Identify and describe common symptoms of electrical faults, including: • loss of supply • low voltage • operation of overload or fault current devices • component/equipment malfunction/failure • arcing.	317
	3.4	State the causes of the following types of fault: • high resistance • transient voltages • insulation failure • excess current • short-circuit • open circuit.	317

Learning outcome	Assessment criteria		Page number
	3.5	Specify the types of faults and their likely locations in: • wiring systems • terminations and connections • equipment/accessories (switches, luminaires, switchgear and control equipment • instrumentation/metering.	320
	3.6	State the special precautions that should be taken with regard to the following: • lone working • hazardous areas • fibre-optic cabling • electro-static discharge (friction, induction, separation) • electronic devices (damage by over voltage) • IT equipment (e.g. shutdown, damage) • high-frequency or capacitive circuits • presence of batteries (e.g. lead acid cells, connecting cells).	322
4. Understand the procedures and techniques for diagnosing electrical faults.	4.1	State the dangers of electricity in relation to the nature of fault diagnosis work.	335
	4.2	Describe how to identify supply voltages.	336
	4.3	Select the correct test instruments (in accordance with HSE guidance document GS 38) for fault diagnosis work, including: • voltage indicator • low resistance ohmmeter • Insulation resistance testers • EFLI and PFC tester • RCD tester • tong tester/clamp on ammeter • phase sequence tester.	338
	4.4	Describe how to confirm test instruments are fit for purpose, functioning correctly and are correctly calibrated.	338
	4.5	State the appropriate documentation that is required for fault diagnosis work and explain how and when it should be completed.	340
	4.6	Explain why carrying out fault diagnosis work can have implications for customers and clients.	341
	4.7	Specify the procedures for functional testing and identify tests that verify fault correction • continuity • insulation resistance • polarity • earth fault loop impedance • RCD operation • current and voltage measurement • phase sequence.	342
	4.8	Identify whether test results are acceptable and state the actions to take where unsatisfactory results are obtained.	340

Learning outcome	Assessment criteria		Page number
5. Understand the procedures and techniques for correcting electrical faults.	5.1	Identify and explain factors that can affect fault correction, repair or replacement, including: • cost • availability of replacement parts, resources and staff • down time (planning) • legal and personal responsibility (e.g. contracts, warranties, relevant personnel) • access to systems and equipment • provision of emergency or stand-by supplies • client demand (continuous supply, out of hours working).	342
	5.2	Specify the procedures for functional testing and identify tests that can verify fault correction, including: • continuity • insulation resistance • polarity • earth fault loop impedance • RCD operation • values of current and voltage • phase sequencing.	342
	5.3	State the appropriate documentation that is required for fault correction work and explain how and when it should be completed	340
	5.4	Explain how and why relevant people need to be kept informed during completion of fault correction work including; • other workers/colleagues • customers/clients • representatives of other services.	341
	5.5	Specify the methods for restoring the condition of building fabric including: • brickwork • plastering • decorative finishings • supporting structures.	343
	5.6	State the methods to ensure the safe disposal of any waste and that the work area is left in a safe and clean condition.	344

CHAPTER 6

Unit ELEC3/008

Principles of electrical science

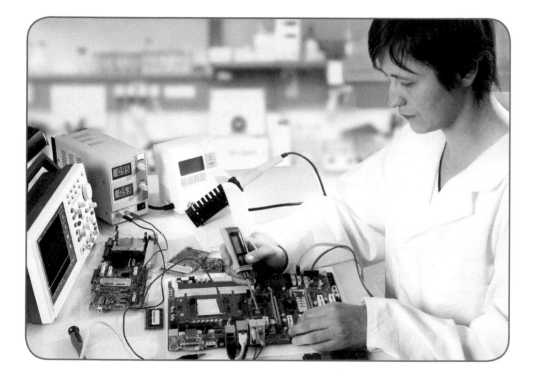

Learning outcomes

When you have completed this chapter you should:

1. Understand the mathematical methods that are appropriate to electrical science and principles.
2. Understand the principles of electronic components and devices in electrotechnical systems.
3. Understand the principles of alternating current circuits.
4. Understand the principles and applications of luminaires.
5. Understand the principles and applications of direct current machines and alternating current motors.
6. Understand the operating principles of electrical components.
7. Understand the principles of electric heating.

EAL Electrical Installation Work – Level 3, 2nd Edition 978 0 367 19564 9
© 2019 T. Linsley. Published by Taylor & Francis. All rights reserved.
https://www.routledge.com/9780367195649

Assessment criteria 1.1

Apply mathematical methods relevant to electrical science and principles

Mathematics is used in many different engineering applications but are underpinned through certain conventions, which are not always signposted. For instance, all numbers tend to be positive unless otherwise stated. The number 5 for example is actually positive otherwise it would be shown as –5. In fact, it can also be represented as 5/1 but we tend not to indicate the numerator in normal convention. There are other mathematical conventions used when adding or subtracting numbers. For example:

like signs add: unlike signs subtract

$3 + (+7) = 3 + 7 = 10$ (positive meets a positive = like signs add)
$3 + (-7) = 3 - 7 = -4$ (positive meets a negative = unlike signs subtract)

The same applies when multiplying and dividing, like signs show a positive results, unlike signs a negative one. This applies when you multiply two negative numbers.

$+4 \times +5 = 20$ (like signs give a positive result)
$-4 \times +5 = -20$ (unlike signs give negative results)
$-4 \times -5 = 20$ (like signs give positive results)

BODMAS

BODMAS is a classic mathematical convention and very important especially since modern day calculators automatically apply the rules surrounding BODMAS, which means that its reasoning is sometimes lost.

In my experience this is shown when applicants are given a basic mathematical test when taking part in an interview process and candidates appear horrified that calculators are not permitted.

BODMAS stands for:

B: Brackets
O: Other operation such as indices
D: Division
M: Multiplication
A: Addition
S: Subtraction

Because the other function is mostly indices, it is sometimes referred to as BIDMAS.

Try to solve the following example:

$6 + 4 \times 2 =$

Some readers might conclude that the answer is 20, since they would apply the logic of $6 + 4 = 10 \times 2 = 20$.

Unfortunately, BODMAS dictates that multiplying comes before adding, which means that the answer is $4 \times 2 = 8 + 6 = 14$.

BODMAS is applied to equations by working out brackets inside initially, before multiplying them out. Other means applying indices such as 3^3, which remember actually works out as 3 × 3 × 3 = 27. The remainder would follow as directed through their function of: dividing, multiplying, adding, and subtracting. However, be aware that when only addition and subtraction remain then the equation should be processed from left to right.

Let's take the following example:

$2 \times 4 + 5 \times 4 (10 - 8)^2$

Work out the brackets inside first $(10 - 8)^2$ becomes

$2 \times 4 + 5 \times 4 (2)^2$

Next, multiply the brackets out, which includes indices: $4 \times 2^2 = 4 \times 4 = 16$

Which leaves us with:

$2 \times 4 + 5 \times 16$

Next, any multiplication $2 \times 4 = 8$ and $5 \times 16 = 80$

now addition

$8 + 80 = 88$

Try the next example:

$20 \times 4 - 10 + 2$

If your answer is 68, then you have forgotten that when only + and − remain, then the equation should be processed from left to right.

The answer should be 72.

Percentages

Percentages are based on a fraction but are then multiplied by 100 to show it as a percentage.

For example, if a student gets 42 correct answers out of a total of 48 questions in an exam how can we work this out as a percentage (%)?

First of all written as a fraction it becomes $\dfrac{42}{48}$

Dividing 42 into 48 will change this into a decimal $\dfrac{42}{48} = 0.875$

If we now multiply by 100, we will change the answer into a percentage (%).

$0.875 \times 100 = 87.5\%$.

A very important stage in designing an electrical circuit is to ensure that it is not too long, and this is known as confirming volt drop requirements. Both power and lighting circuits are permitted a certain percentage of voltage drop which is lost across the resistance of its conductors, but excessive amounts robs electrical loads of working voltage. This can then manifest in problems such as electrical motors operating slowly or lights appearing dim. The wiring regulations BS 7671 specify that lighting and power circuits must operate within 3% and 5% respectively. If these percentages are not met, then the cross sectional area of the cable must be increased.

When applying these volt drop tolerances to a domestic single phase supply the maximum voltage that can be dropped across a lighting circuit is:

3% of 230 V

$\dfrac{3 \times 230}{100}$

Volt drop = 6.9 V

Power circuits are permitted

5% of 230 V

$$\frac{5 \times 230}{100}$$

Volt drop = 11.5 V

Powers of 10

Engineering make great use of applied mathematics such as powers of 10, a system of numbers that is arrived at by multiplying the number 10 by itself a number of times. Such numbers include:

10, 100, 1000, 10,000, 100,000, 1000000.

It is very easy to multiply and divide any multiple of 10 by following a simple rule.

Table 6.1 Basic SI units

Quantity	Measure of	Basic unit	Symbol	Notes
Area	Length × length	Metre squared	m^2	
Current	Electric current	Ampere	A	
Energy	Ability to do work	Joule	J	Joule is a very small unit 3.6×10^6 J × 1 kWh
Force	The effect on a body	Newton	N	
Frequency	Number of cycles	Hertz	Hz	Mains frequency is 50 Hz
Length	Distance	Metre	m	
Mass	Amount of material	Kilogram	kg	One metric tonne = 1000 kg
Magnetic flux Φ	Magnetic energy	Weber	Wb	
Magnetic flux density B	Number of lines of magnetic flux	Tesla	T	
Potential or pressure	Voltage	Volt	V	
Period	Time taken to complete one cycle	Second	s	The 50 Hz mains supply has a period of 20 ms
Power	Rate of doing work	Watt	W	
Resistance	Opposition to direct current flow	Ohm	Ω	
Resistivity	Resistance of a sample piece of material	Ohm metre	ρ	Resistivity of copper is $17.5 = 10^{-9}$ Ωm
Temperature	Hotness or coldness	Kelvin	K	0°C = 273 K. A change of 1 K is the same as 1°C
Time	Time	Second	s	60 s = 1 min 60 min = 1 h
Weight	Force exerted by a mass	Kilogram	kg	1000 kg = 1 tonne
Electric charge	Charge transported by a constant current of one ampere in one second	Coulomb	C	Charge of approximately 6.241×1018 electrons

Note: A more detailed description may be found in this chapter.

For example:

1000 × 1000 = 1000000 (when multiplying, simply combine the total number of zeros)

When dividing, cancel out the number of zeros in the denominator with respect to the numerator

$$\frac{1\ 000\ 000}{1000}$$

$$\frac{1\ 000\ 000}{1\cancel{0}\cancel{0}\cancel{0}}$$

$$= 1000$$

We also relate powers of 10 to what is known as scientific notation, therefore 3000 can be expressed as: 3×10^3 and 3000000 as 3×10^6 accordingly.

We can also express these types of numbers through some common prefixes as can be seen from the table below. For instance 3000 V or 3 000 000 V would be written as 3 kV and 3 MV respectively.

Table 6.2 Symbols and multiples for use with SI units

Prefix	Symbol	Multiplication factor		
Terra	T	$\times 10^{12}$	or	X 1000,000,000,000
Giga	G	$\times 10^{9}$	or	X 1000,000,000
Mega	M	$\times 10^{6}$	or	X 1000,000
Kilo	k	$\times 10^{3}$	or	X 1000
milli	m	$\times 10^{-3}$	or	÷ 1000
micro	μ	$\times 10^{-6}$	or	÷ 1000,000
nano	n	$\times 10^{-9}$	or	÷ 1000,000,000
pico	p	$\times 10^{-12}$	or	÷ 1000,000,000,000

Fractions

To consider fractions think of a cake.

Cut into two, each part is a half

Cake fractions

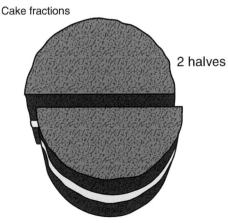

2 halves

Figure 6.1 Fractions can be considered as breaking or dividing something into equal parts. The simplest is two halves.

Adding fractions

When adding fractions we form what is called a common denominator: a number that is divisible (go into) to both sides.

For example

$$\frac{1}{4} + \frac{1}{2}$$

$$\frac{\frac{1}{4} + \frac{1}{2}}{4}$$

$$\frac{1+2}{4}$$ 4 is the common denominator, because both 4 goes into itself once and 2 goes into 4 twice

Then simplify by adding both numerators.

Subtraction follows the same procedure, simplifying by taking away both elements at the end.

Cake fractions

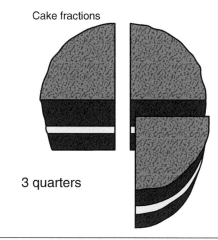

3 quarters

Figure 6.2 If you eat one quarter (1/4) of the cake, then you are left with three quarters (3/4) or 3 slices.

Multiplying fractions

Simply multiply all the top numbers and all the bottom numbers.

$$\frac{1}{4} \times \frac{4}{8} \quad \text{thus becomes} \quad \frac{1 \times 4}{4 \times 8} = \frac{4}{32}$$

We can simplify the answer by working out how many times 4 goes into 32.

$$\frac{1}{8}$$

Dividing fractions

A very useful rule when dividing fractions is to invert the fraction on the right.

Therefore

$$\frac{1}{4} \div \frac{3}{7} \quad \text{becomes} \quad \frac{1}{4} \times \frac{7}{3}$$

Then treat the fraction the same as you would when multiplying.

$$\frac{1 \times 7}{4 \times 3} = \frac{7}{12}$$

Algebra

In algebra another mathematical convention exists in that the multiplication sign is not always shown but must be applied to the letters or values involved. For instance, F = B I L is a formula regarding electrical motors but the relationship between B I L is actually B × I × L. The rule to remember here is that a lack of a + or − sign in a cluster of elements signifies that they should be multiplied.

The letters represent values which if known can be substituted to find the amount of force that the motor produces.

We also carry over some previous BODMAS rules, such as multiplication before addition or subtraction.

For example

2PR − 1PR if P = 4 and R = 2, then

$2 \times P \times R - 1 \times P \times R$

$2 \times 4 \times 2 - 1 \times 4 \times 2$

$16 - 8 = 6$

Laws of indices

Large numbers can be represented by indices.

For instance: 2^3 is actually $2 \times 2 \times 2 = 8$

The number 2 is known as the base and the number 3 the index and certain rules can be applied but only to numbers using the same base.

To multiply indices with the same base, add the indices.

$2^9 \times 2^3$

$= 2^{11}$

When dividing indices with the same base, subtract the indices.

$2^9 \div 2^3$

$= 2^6$

Remember, these rules only apply to indices that share the same base number.

Another rule that can help to factorize is an equation which is the reverse of the Lowest Common Factor used in fractions and is known as the Highest Common Factor (HCF) of the numbers involved. For example, in order to factorize the following equation: 27a + 18

The highest HCF that applies to both numbers is 9 because it applies to 27 $(3 \times 9 = 27)$ and 18 $(9 \times 2 = 18)$.

This means that our equation becomes

$9(3a + 2)$

Trigonometry

A Greek mathematician called Pythagoras found that if we know the length of two of sides, we can work out the length of the other.

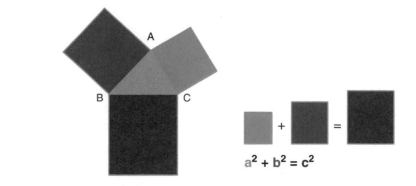

Figure 6.3 The area of a + b = c.

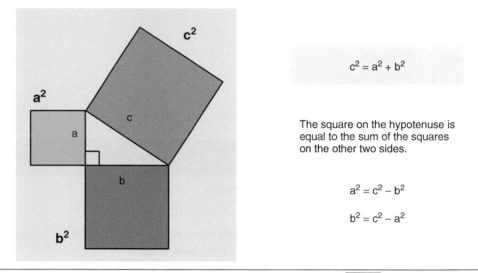

$$c^2 = a^2 + b^2$$

The square on the hypotenuse is equal to the sum of the squares on the other two sides.

$$a^2 = c^2 - b^2$$

$$b^2 = c^2 - a^2$$

Figure 6.4 By using transposition of formula we can determine $b = \sqrt{c^2 - a^2}$.

Using the rules of transposition any of the mathematical relationships can be found if at least two sides are known.

The sides are labelled hypotenuse, opposite and adjacent and, while students are exposed to this theorem in school, in electrical installation it is used to indicate and correct power factor issues which are brought about when inductors cause a circuit's current to lag the voltage.

The example given below shows how the length of the hypotenuse (C) can be found because we know the values of the other two sides.

To calculate the angle, however, we need to use trigonometry principles and the relationship between sine, cosine and tangent.

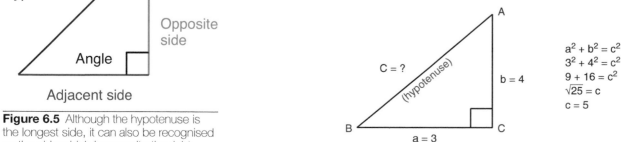

Figure 6.5 Although the hypotenuse is the longest side, it can also be recognised as the side which is opposite the right angle.

$$a^2 + b^2 = c^2$$
$$3^2 + 4^2 = c^2$$
$$9 + 16 = c^2$$
$$\sqrt{25} = c$$
$$c = 5$$

Figure 6.6 An example of a right sided triangle whose sides form a ratio of 3:4:5.

A famous acronym used to remember these relationships in the form of a formula is: SOH CAH TOA. In other words:

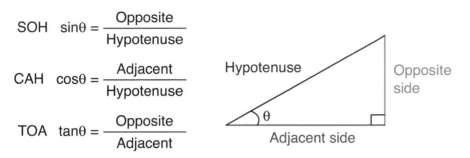

$$\text{SOH} \quad \sin\theta = \frac{\text{Opposite}}{\text{Hypotenuse}}$$

$$\text{CAH} \quad \cos\theta = \frac{\text{Adjacent}}{\text{Hypotenuse}}$$

$$\text{TOA} \quad \tan\theta = \frac{\text{Opposite}}{\text{Adjacent}}$$

Figure 6.7 If we know two sides of a right angled triangle, we can calculate the unknown side as well as the angle between them.

Referring to the aircraft, we can find the angle of its elevation because again we know two of its sides, adjacent (400) and opposite (300). The trigonometry relationships that involves the adjacent and the opposite is tangent (TOA). Our formula therefore becomes $\tan = \dfrac{300}{400} = 0.75$

To obtain the actual angle in degrees we use the inverse function of tan: 0.75 $\tan^{-1} = 36.869°$ (36.7°)

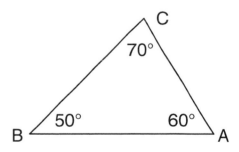

Figure 6.8 Opposite and adjacent are known, therefore we use tangent the function, TOA = opposite/adjacent.

Triangles

The right-sided triangle has already been covered through Pythagoras' theorem but others exist such as shown below:

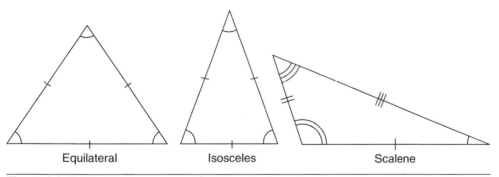

Equilateral   Isosceles   Scalene

Figure 6.9 Equilateral, Isosceles and Scalene triangles.

Figure 6.10 The three interior angles of a triangle always add up to 180°.

Their respective properties are as follows:

- Equilateral: all angles are 60° and sides are equal in sides.
- Isosceles: has two equal sides and the angles opposite these sides are also equal.
- Scalene: a triangle with no equal sides or equal angles.

All angles within a triangle add up to 180°.

Work out the unknown angle from the example below:

60°+ 40°= 100°. 180° − 100°= 80°

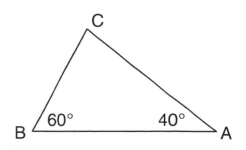

Figure 6.11 60° + 40° − 180° = 80°.

Transposion of formula

There are many different ways of transposing a formula but the simplest makes use of the OPPOSITE RULE. For instance:

1   The opposite of Addition is Subtraction

$$+ \qquad -$$

2   The opposite of Multiply is Divide

$$\times \qquad \div$$

3   The opposite of Squaring is Square root

$$x^2 \qquad \sqrt{}$$

By way of an example we can see this below:

Rule number 1

$5 + 5 = 10$

$10 - 5 = 5$

(The opposite of Addition is Subtraction.)

Rule number 2

$5 \times 5 = 25$

$25 \div 5 = 5$

(The opposite of Multiply is Divide.)

In electrical installation and indeed engineering we tend not to use the symbol $\div$ when we are showing a divide sign. Therefore $1 \div 2$, is effectively ½ but written in an equation it would be: $\dfrac{1}{2}$

Rule number 3

Revision: (The square root of a number is a value that, when multiplied by itself, gives that number. For example: $4 \times 4 = 16$, so the square root of 16 is 4.)

5^2 means 5×5, which equals 25.

The square root of 25 is 5.

(The opposite of Squaring is Square Root.)

The reason why the opposite rule works is that we can keep an equation or formula balanced by changing the opposite function.

Remember to keep an equation balanced, whatever you do to one side you must do the same or the opposite to the other side.

Let us work through a few examples in order to solve certain values.

Example 1

$a + c = d$

However, what if I need to re-arrange the formula to find c?

Figure 6.12 To keep the balance correct within an equation changes on one side must also be made to the other.

We do this by following three simple steps. Remember, the aim is to get c by itself.

Step 1: focus on the letter you want to find and write this down.

   c =

Step 2: Look which letter or value is already on its own, or in other words which letter or value is already on the other side of the = sign. In this case, d.

   c = d

Step 3: Move the letter or value that is left away from the one you want to find.

In this case the letter remaining is +a, which becomes –a as it moves across.

   c = d – a

Example 2

Ohm's law states that $V = I \times R$.

However, what if I need to re-arrange the formula to find I?

Remember the aim is to get I by itself.

Step 1: focus on the letter you want to find and write this down.

   I =

Step 2: Look which letter or value is already on its own, or in other words which letter or value that is already on the other side of the = sign. In this case V.

   I – V

Step 3: Move the letter or value that is left away from the one you want to find.

In this case the letter remaining is R, which is shown on top, which we therefore consider as multiply. The opposite of multiply is divide therefore simply move the R from the top of the equation to the bottom (underneath).

   $$I = \frac{V}{R}$$

Example 3

   $a^2 + b^2 = c$

Re-arrange the formula by transposing to find b.

Step 1: focus on the letter you want to find and write this down. Just for now we write b^2 as we would if it was b.

   b² =

Step 2: Look which letter or value is already on its own, or in other words which letter or value that is already on the other side of the = sign. In this case c.

   b² = c

Step 3: Move the letter or value that is left away from the one you want to find. In this example a^2 becomes $-a^2$. Therefore our equation becomes:

   b² = c – a²

Last part coming up, we need to find b not b^2. Therefore, using the opposite rule remove the ^{2} from b, but keep the formula balanced by square-rooting the other side.

   $b = \sqrt{c - a^2}$.

If we now work back in order to find c, the first thing we need to do is get rid of the square root sign, and we do this by squaring b.

$b^2 = c - a^2$.

We then follow the three steps.

c =
$c = b^2$
$c = b^2 + a^2$

Statistics

When calculating statistics the mean (or average) is the most popular and well known measurement. In the UK we know this value more as the average, but this can be confusing because the term median is also a measure of average but is calculated in a slightly different way.

For instance, in electrical installation if we used different values of insulation resistance obtained by testing several circuits then we can work out what the mean or median values are, but the usefulness or transparency of what these values represent largely depends on the actual data being measured or compared. Examine the following values:

2 MΩ, 1.6 MΩ, 2 MΩ, 1 MΩ, 0.5 MΩ, 100 MΩ, 300 MΩ, 5 MΩ, 10 MΩ.

To work out the mean I add up all the values and divide by the number of values, which is 9 in this case.

$$= \frac{422.1 \text{ M}\Omega}{9}$$

This gives me a mean value of = 46.9 MΩ

To work out the median I first have to put them in order:

Therefore:

2 MΩ, 1.6 MΩ, 2 MΩ, 1 MΩ, 0.5 MΩ, 100 MΩ, 300 MΩ, 5 MΩ, 10 MΩ.

Becomes:

0.5 MΩ, 1 MΩ, 1.6 MΩ, 2 MΩ, 2 MΩ, 5 MΩ, 10 MΩ, 100 MΩ, 300 MΩ.

Given that the sequence of numbers is odd (9) then the middle number will be the median, which is the fifth number = 2 MΩ.

But the question is which average is more accurate for representing the data being compared. On reflection, the data used were not symmetrical with a wide spectrum of values ranging from 0.5 to 300 MΩ. This means that the median gives a more accurate reflection of the true average because it sits in the middle of the data.

By the way, had the sequence of numbers been even as shown below with 12 values, then again having put them in order I would work out the median by taking the middle two values and dividing by two.

300 MΩ, 0.5 MΩ, 1 MΩ, 1.6 MΩ, 2 MΩ, 2 MΩ, 5 MΩ, 10 MΩ, 5 MΩ, 100 MΩ, 300 MΩ, 300 MΩ

Becomes:

0.5 MΩ, 1 MΩ, 1.6 MΩ, 2 MΩ, 2 MΩ, 5 MΩ, 5 MΩ, 10 MΩ, 100 MΩ, 300 MΩ, 300 MΩ, 300 MΩ

Definition

Averages
The mean should be used to calculate average if the numbers are similar in size. If not, use the median.

$$\frac{2M\Omega \text{ and } 5\ M\Omega}{2} = 3.5\ M\Omega$$

The other average that is sometimes used is the mode, which shows the most re-occurring value. In the above example this would be 300 MΩ since it showed up three times.

0.5 MΩ, 1 MΩ, 1.6 MΩ, 2 MΩ, 2 MΩ, 5 MΩ, 5 MΩ, 10 MΩ, 100 MΩ, 300MΩ, 300MΩ, 300MΩ

Key fact

To be able to make calculations accurately and confidently are an important skill at this level in the electrical installation training course. I always recommend a good math book to my students called 'Electrical Installation Calculations': Advanced, Level 3. By Christopher Kitcher and A.J.Watkins.

Assessment criteria 2.1

Outline the operating principles of components and devices

Basic electronics

There are numerous types of electronic components – diodes, transistors, thyristors and integrated circuits (ICs) – each with its own limitations, characteristics and designed application. When repairing electronic circuits it is important to replace a damaged component with an identical or equivalent component. Manufacturers issue comprehensive catalogues with details of working voltage, current, power dissipation, etc., and the reference numbers of equivalent components. These catalogues of information, together with a high impedance multimeter, should form a part of the extended toolkit for anyone in the electrical industries proposing to repair electronic circuits.

Electronic circuit symbols

The British Standard BS EN 60617 recommends that particular graphical symbols should be used to represent a range of electronic components on circuit diagrams. The same British Standard recommends a range of symbols suitable for electrical installation circuits with which electricians will already be familiar. Figure 6.15 shows a selection of electronic symbols.

Resistors

All materials have some resistance to the flow of an electric current but, in general, the term resistor describes a conductor specially chosen for its resistive properties.

Resistors are the most commonly used electronic component and they are made in a variety of ways to suit the particular type of application. They are usually manufactured as either carbon composition or carbon film. In both cases the base resistive material is carbon and the general appearance is of a small cylinder with leads protruding from each end, as shown in Fig. 6.16(a). If subjected to overload, carbon resistors usually decrease in resistance since carbon has a negative temperature coefficient. This causes more current to flow through the resistor, so that the temperature rises and failure occurs, usually by fracturing. Carbon resistors have a power rating between 0.1 and 2 W, which should not be exceeded.

Definition

Resistors are used to limit the current in a circuit.

All materials have some resistance to the flow of an electric current, but, in general, the term resistor describes a conductor specially chosen for its resistive properties.

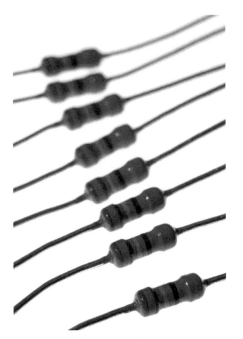

Figure 6.13 Resistors used in electrical circuits.

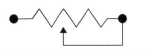

Figure 6.14 A rheostat is a resistor whose resistance can be changed in order to control circuit current.

When a resistor of a larger power rating is required a wire-wound resistor should be chosen. This consists of a resistance wire of known value wound on a small ceramic cylinder which is encapsulated in a vitreous enamel coating, as shown in Fig. 6.16(b). Wire-wound resistors are designed to run hot and have a power rating of up to 20 W. Care should be taken when mounting wire-wound resistors to prevent the high operating temperature affecting any surrounding components.

A variable resistor is one that can be varied continuously from a very low value to the full rated resistance. This characteristic is required in tuning circuits to adjust the signal or voltage level for brightness, volume or tone. The most common type used in electronic work has a circular carbon track contacted by a metal wiper arm. The wiper arm can be adjusted by means of an adjusting shaft (rotary type) or by placing a screwdriver in a slot (preset type), as shown in Fig. 6.17. Variable resistors are also known as potentiometers because they can be used to adjust the potential difference (voltage) in a circuit. The variation in resistance can be either a logarithmic or a linear scale.

The value of the resistor and the tolerance may be marked on the body of the component either by direct numerical indication or by using a standard colour code. The method used will depend upon the type, physical size and manufacturer's preference, but in general the larger components have values marked directly on the body and the smaller components use the standard resistor colour code.

A rheostat, on the other hand, is a similar device but is used to control the current in a circuit.

Abbreviations used in electronics

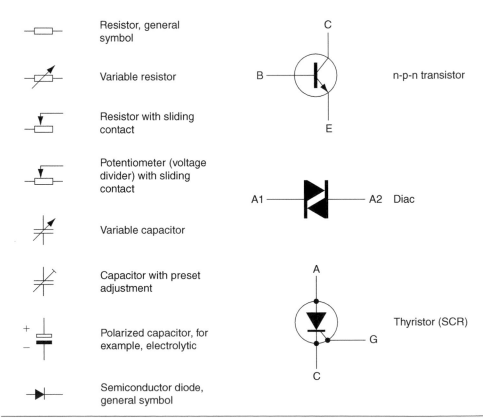

Figure 6.15 Some BS EN 60617 graphical symbols used in electronics.

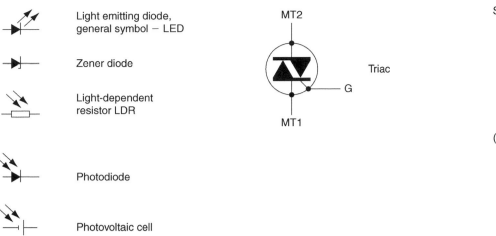

Light emitting diode, general symbol – LED

Zener diode

Light-dependent resistor LDR

Photodiode

Photovoltaic cell

MT2

Triac

G

MT1

Figure 6.15 (Continued)

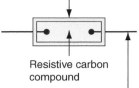

Silicon lacquer or paint coating

Resistive carbon compound

Embedded connection leads

(a) Carbon-composition resistor

End cap

Vitreous enamel coating

Resistance winding wound on ceramic former

(b) Wire-wound resistor

Figure 6.16 Construction of resistors.

Where the numerical value of a component includes a decimal point, it is standard practice to include the prefix for the multiplication factor in place of the decimal point, to avoid accidental marks being mistaken for decimal points. Multiplication factors and especially factors of 10 have already been dealt with at the beginning of this chapter.

The abbreviation   R means × 1

k means × 1000

M means × 1,000,000

Therefore, a 4.7 kΩ resistor would be abbreviated to 4 k7, a 5.6Ω resistor to 5R6 and a 6.8 MΩ resistor to 6 M8.

Tolerances may be indicated by adding a letter at the end of the printed code.

The abbreviation F means ±1%, G means ±2%, J means ±5%, K means ±10% and M means ±20%. Therefore a 4.7 kΩ resistor with a tolerance of 2% would be abbreviated to 4 k7G. A 5.6 Ω resistor with a tolerance of 5% would be abbreviated to 5R6J. A 6.8 MΩ resistor with a 10% tolerance would be abbreviated to 6 M8K.

This is the British Standard BS 1852 code, which is recommended for indicating the values of resistors on circuit diagrams and components when their physical size permits.

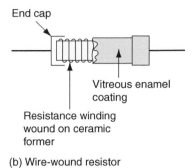

Figure 6.17 Types of variable resistors.

The standard colour code

Small resistors are marked with a series of coloured bands, as shown in Table 6.3. These are read according to the standard colour code to determine the resistance. The bands are located on the component towards one end. If the resistor is turned so that this end is towards the left, the bands are then read from left to right. Band (a) gives the first number of the component value, band (b) the second number, band (c) the number of zeros to be added after the first two numbers and band (d) the resistor tolerance. If the bands are not clearly oriented towards one end, first identify the tolerance band and turn the resistor so that this is towards the right before commencing to read the colour code as described.

The tolerance band indicates the maximum tolerance variation in the declared value of resistance. Thus a 100 Ω resistor with a 5% tolerance will have a value somewhere between 95 and 105 Ω, since 5% of 100 Ω is 5 Ω.

Table 6.3 The resistor colour code

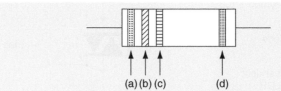

Colour	Band (a) first number	Band (b) second number	Band (c) number of zeros	Band (d) tolerance band (%)
Black	0	0	None	–
Brown	1	1	1	1
Red	2	2	2	2
Orange	3	3	3	–
Yellow	4	4	4	–
Green	5	5	5	–
Blue	6	6	6	–
Violet	7	7	7	–
Grey	8	8	–	–
White	9	9	–	–
Gold	–	–	÷10	5
Silver	–	–	÷100	10
None	–	–	–	20

Example 1

A resistor is colour coded yellow, violet, red, gold. Determine the value of the resistor.

Band (a) – yellow has a value of 4
Band (b) – violet has a value of 7
Band (c) – red has a value of 2
Band (d) – gold indicates a tolerance of 5%

The value is therefore 4700 ± 5%
This could be written as 4.7 kΩ ± 5% or 4k7J.

Example 2

A resistor is colour coded green, blue, brown, silver. Determine the value of the resistor.

Band (a) – green has a value of 5
Band (b) – blue has a value of 6
Band (c) – brown has a value of 1
Band (d) – silver indicates a tolerance of 10%

The value is therefore 560 ± 10% and could be written as 560 Ω ± 10% or 560RK.

Example 3

A resistor is colour coded blue, grey, green, gold. Determine the value of the resistor.

Band (a) – blue has a value of 6
Band (b) – grey has a value of 8
Band (c) – green has a value of 5
Band (d) – gold indicates a tolerance of 5%

The value is therefore 6,800,000 ± 5% and could be written as 6.8 M Ω ± 5% or 6 M8J.

Example 4

A resistor is colour coded orange, white, silver, silver. Determine the value of the resistor.

Band (a) – orange has a value of 3
Band (b) – white has a value of 9
Band (c) – silver indicates divide by 100 in this band
Band (d) – silver indicates a tolerance of 10%

The value is therefore 0.39 ± 10% and could be written as 0.39 Ω ± 10% or R39 K.

Try this

Electronics

Electricians are increasingly coming across electronic components and equipment. Make a list in the margin of some of the electronic components that you have come across at work.

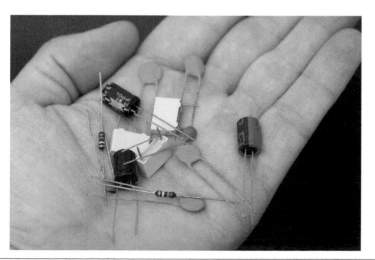

Figure 6.18 A handful of electronic resistors and capacitors.

Preferred values

It is difficult to manufacture small electronic resistors to exact values by mass production methods. This is not a disadvantage as in most electronic circuits the value of the resistors is not critical. Manufacturers produce a limited range of *preferred* resistance values rather than an overwhelming number of individual resistance values. Therefore, in electronics, we use the preferred value closest to the actual value required.

A resistor with a preferred value of 100 Ω and a 10% tolerance could have any value between 90 and 110 Ω. The next larger preferred value which would give the maximum possible range of resistance values without too much overlap would be 120 Ω. This could have any value between 108 and 132 Ω. Therefore, these two preferred value resistors cover all possible resistance values between 90 and 132 Ω. The next preferred value would be 150 Ω, then 180 Ω, 220 Ω and so on.

There is a series of preferred values for each tolerance level, as shown in Table 6.3, so that every possible numerical value is covered. Table 6.4 indicates the values between 10 and 100, but larger values can be obtained by multiplying

Table 6.4 Preferred values

E6 series 20% tolerance	E12 series 10% tolerance	E24 series 5% tolerance
10	10	10
		11
	12	12
		13
15	15	15
		16
	18	18
		20
22	22	22
		24
	27	27
		30
33	33	33
		36
	39	39
		43
47	47	47
		51
	56	56
		62
68	68	68
		75
	82	82
		91

these preferred values by some multiplication factor. Resistance values of 47 Ω, 470 Ω, 4.7 kΩ, 470 kΩ, 4.7 MΩ, etc., are available in this way.

Testing resistors

The resistor being tested should have a value close to the preferred value and within the tolerance stated by the manufacturer. To measure the resistance of a resistor which is not connected into a circuit, the leads of a suitable ohmmeter should be connected to each resistor connection lead and a reading obtained. If the resistor to be tested is connected into an electronic circuit it is *always necessary* to disconnect one lead from the circuit before the test leads are connected, otherwise the components in the circuit will provide parallel paths, and an incorrect reading will result.

Capacitors

In this section we shall consider the practical aspects associated with capacitors in electronic circuits.

A capacitor stores a small amount of electric charge; it can be thought of as a small rechargeable battery that can be quickly recharged. In electronics we are not only concerned with the amount of charge stored by the capacitor but in the way the value of the capacitor determines the performance of timers and oscillators by varying the time constant of a simple capacitor–resistor circuit.

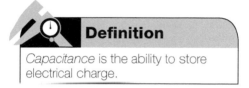

Definition

Capacitance is the ability to store electrical charge.

Capacitors in action

If a test circuit is assembled as shown in Fig. 6.20 and the changeover switch connected to d.c., the signal lamp will only illuminate for a very short pulse as the capacitor charges. The charged capacitor then blocks any further d.c. current flow. If the changeover switch is then connected to a.c. the lamp will illuminate at full brilliance because the capacitor will charge and discharge continuously at the supply frequency. Current is *apparently* flowing through the capacitor because electrons are moving to and fro in the wires joining the capacitor plates to the a.c. supply.

Coupling and decoupling capacitors

Capacitors can be used to separate a.c. and d.c. in an electronic circuit. If the output from circuit A, shown in Fig. 6.21(a), contains both a.c. and d.c. but only an a.c. input is required for circuit B then a *coupling* capacitor is connected between them. This blocks the d.c. while offering a low reactance to the a.c.

Figure 6.19 Capacitors come in several shapes and sizes.

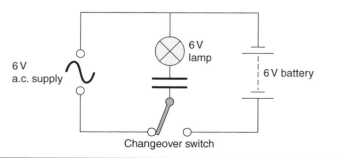

Figure 6.20 Test circuit showing capacitors in action.

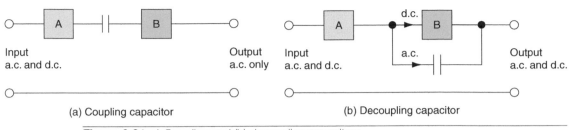

Input
a.c. and d.c.

Output
a.c. only

(a) Coupling capacitor

d.c.

Input
a.c. and d.c.

a.c.

Output
a.c. and d.c.

(b) Decoupling capacitor

Figure 6.21 a) Coupling and (b) decoupling capacitors.

Figure 6.22 Capacitors, resistors and other components mounted on a circuit board.

component. Alternatively, if it is required that only d.c. be connected to circuit B, shown in Fig. 6.21(b), a *decoupling* capacitor can be connected in parallel with circuit B. This will provide a low reactance path for the a.c. component of the supply and only d.c. will be presented to the input of B. This technique is used to *filter out* unwanted a.c. in, for example, d.c. power supplies.

Types of capacitor

There are two broad categories of capacitor, the non-polarized and polarized type. The non-polarized type can be connected either way round, but polarized capacitors *must* be connected to the polarity indicated otherwise a short circuit and consequent destruction of the capacitor will result. There are many different types of capacitor, each one being distinguished by the type of dielectric used in its construction. Figure 6.23 shows some of the capacitors used in electronics.

Polyester capacitors

Polyester capacitors are an example of the plastic film capacitor. Polypropylene, polycarbonate and polystyrene capacitors are other types of plastic film capacitor. The capacitor value may be marked on the plastic film, or the capacitor colour code given in Table 6.5 may be used. This dielectric material gives a compact capacitor with good electrical and temperature characteristics. They are used in many electronic circuits, but are not suitable for high-frequency use.

Mica capacitors

Mica capacitors have excellent stability and are accurate to +1% of the marked value. Since costs usually increase with increased accuracy, they tend to be more expensive than plastic film capacitors. They are used where high stability is required, for example in tuned circuits and filters.

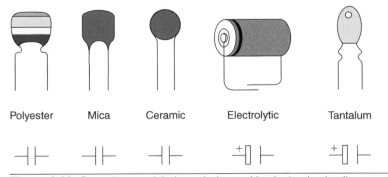

Polyester   Mica   Ceramic   Electrolytic   Tantalum

Figure 6.23 Capacitors and their symbols used in electronic circuits.

Table 6.5 Colour code for plastic film capacitors (values in picofarads)

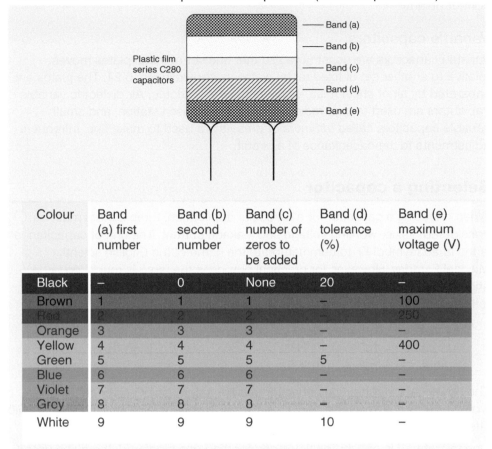

Colour	Band (a) first number	Band (b) second number	Band (c) number of zeros to be added	Band (d) tolerance (%)	Band (e) maximum voltage (V)
Black	–	0	None	20	–
Brown	1	1	1	–	100
Red	2	2	2	–	250
Orange	3	3	3	–	–
Yellow	4	4	4	–	400
Green	5	5	5	5	–
Blue	6	6	6	–	–
Violet	7	7	7	–	–
Grey	8	8	8	–	–
White	9	9	9	10	–

Ceramic capacitors

Ceramic capacitors are mainly used in high-frequency circuits subjected to wide temperature variations. They have high stability and low loss.

Electrolytic capacitors

Electrolytic capacitors are used where a large value of capacitance coupled with a small physical size is required. They are constructed on the 'Swiss roll' principle as are the paper dielectric capacitors used for power-factor correction in electrical installation circuits. The electrolytic capacitors' high capacitance for very small volume is derived from the extreme thinness of the dielectric coupled with a high dielectric strength. Electrolytic capacitors have a size gain of approximately 100 times over the equivalent non-electrolytic type. Their main disadvantage is that they are polarized and must be connected to the correct polarity in a circuit. Their large capacity makes them ideal as smoothing capacitors in power supplies.

Tantalum capacitors

Tantalum capacitors are a new type of electrolytic capacitor using tantalum and tantalum oxide to give a further capacitance/size advantage. They look like a 'raindrop' or 'blob' with two leads protruding from the bottom. The polarity and values may be marked on the capacitor, or a colour code may be used. The voltage ratings available tend to be low, as with all electrolytic capacitors. They are also extremely vulnerable to reverse voltages in excess of 0.3 V. This means

Safety first

Electrolytic capacitors are polarized, which means that one leg should be connected to the positive supply whilst the other to the negative and must never be cross connected.

Capacitance increase

Movable vanes

(a) Variable type

(b) Trimmer or preset type

Figure 6.24 Variable capacitors and their symbols: (a) variable type; (b) trimmer or preset type.

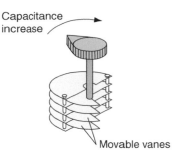

Safety first

You should never exceed a capacitor's working voltage.

that even when testing with an ohmmeter, extreme care must be taken to ensure correct polarity.

Variable capacitors

Variable capacitors are constructed so that one set of metal plates moves relative to another set of fixed metal plates as shown in Fig. 6.24. The plates are separated by air or sheet mica, which acts as a dielectric. Air dielectric variable capacitors are used to tune radio receivers to a chosen station, and small variable capacitors called *trimmers* or *presets* are used to make fine, infrequent adjustments to the capacitance of a circuit.

Selecting a capacitor

When choosing a capacitor for a particular application, three factors must be considered: value, working voltage and leakage current. The unit of capacitance is the *farad* (symbol F), to commemorate the name of the English scientist Michael Faraday. However, for practical purposes the farad is much too large and in electrical installation work and electronics we use fractions of a farad as follows:

$$1 \text{ microfarad} = 1\mu F = 1 \times 10^{-6} \text{ F}$$
$$1 \text{ nanofarad} = 1nF = 1 \times 10^{-9} \text{ F}$$
$$1 \text{ picofarad} = 1pF = 1 \times 10^{-12} \text{ F}$$

The power-factor correction capacitor used in a domestic fluorescent luminaire would typically have a value of 8μF at a working voltage of 400 V. In an electronic filter circuit a typical capacitor value might be 100 pF at 63 V.

One microfarad is one million times greater than one picofarad. It may be useful to remember that:

$$1,000 \text{ pF} = 1nF$$
$$1,000 \text{ nF} = 1\mu F$$

The working voltage of a capacitor is the *maximum* voltage that can be applied between the plates of the capacitor without breaking down the dielectric insulating material. This is a d.c. rating and, therefore, a capacitor with a 200 V rating must only be connected across a maximum of 200 V d.c. Since a.c. voltages are usually given as r.m.s. values, a 200 V a.c. supply would have a maximum value of about 283 V, which would damage the 200 V capacitor. When connecting a capacitor to the 230 V mains supply we must choose a working voltage of about 400 V because 230 V r.m.s. is approximately 325 V maximum. The 'factor of safety' is small and, therefore, the working voltage of the capacitor must not be exceeded.

An ideal capacitor that is isolated will remain charged forever, but in practice no dielectric insulating material is perfect, and the charge will slowly *leak* between the plates, gradually discharging the capacitor. The loss of charge by leakage through it should be very small for a practical capacitor.

Capacitor colour code

The actual value of a capacitor can be identified by using the colour codes given in Table 6.4 in the same way that the resistor colour code was applied to resistors.

Example 1

A plastic film capacitor is colour coded, from top to bottom, brown, black, yellow, black, red. Determine the value of the capacitor, its tolerance and working voltage.

From Table 6.5 we obtain the following:

Band (a) – brown has a value 1
Band (b) – black has a value 0
Band (c) – yellow indicates multiply by 10,000
Band (d) – black indicates 20%
Band (e) – red indicates 250V

The capacitor has a value of 1,00,000 pF or 0.1 μF with a tolerance of 20% and a maximum working voltage of 250 V.

Example 2

Determine the value, tolerance and working voltage of a polyester capacitor colour-coded, from top to bottom, yellow, violet, yellow, white, yellow.

From Table 6.5 we obtain the following:

Band (a) – yellow has a value 4
Band (b) – violet has a value 7
Band (c) – yellow indicates multiply by 10,000
Band (d) – white indicates 10%
Band (e) – yellow indicates 400V

The capacitor has a value of 4,70,000 pF or 0.47 μF with a tolerance of 10% and a maximum working voltage of 400 V.

Capacitance value codes

Where the numerical value of the capacitor includes a decimal point, it is standard practice to use the prefix for the multiplication factor in place of the decimal point. This is the same practice as we used earlier for resistors. The abbreviation μ means microfarad, n means nano farad and p means picofarad. Therefore, a 1.8 pF capacitor would be abbreviated to 1 p8, a 10 pF capacitor to 10 p, a 150 pF capacitor to 150 p or n15, a 2200 pF capacitor to 2n2 and a 10,000 pF capacitor to 10 n.

$$1000 \text{ pF} = 1\text{nF} = 0.001 \text{ μF}$$

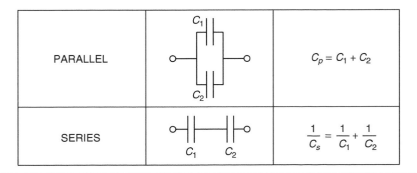

PARALLEL	C_1 C_2	$C_p = C_1 + C_2$
SERIES	C_1 C_2	$\dfrac{1}{C_s} = \dfrac{1}{C_1} + \dfrac{1}{C_2}$

Figure 6.25 Capacitors in parallel are the same as resisters in series: you add up their values. Capacitors in series are equal to resistors in parallel: you add up their reciprocal values.

Semiconductor materials

Modern electronic devices use the semiconductor properties of materials such as silicon or germanium. The atoms of pure silicon or germanium are arranged in a lattice structure, as shown in Fig. 6.26. The outer electron orbits contain four electrons known as *valence* electrons. These electrons are all linked to other valence electrons from adjacent atoms, forming a covalent bond. There are no free electrons in pure silicon or germanium and, therefore, no conduction can take place unless the bonds are broken and the lattice framework is destroyed. To make conduction possible without destroying the crystal it is necessary to replace a four-valent atom with a three- or five-valent atom. This process is known as *doping*.

If a three-valent atom is added to silicon or germanium a hole is left in the lattice framework. Since the material has lost a negative charge, the material becomes positive and is known as a p-type material (p for positive).

If a five-valent atom is added to silicon or germanium, only four of the valence electrons can form a bond and one electron becomes mobile or free to carry charge. Since the material has gained a negative charge it is known as an n-type material (n for negative).

Bringing together a p-type and n-type material allows current to flow in one direction only through the p–n junction. Such a junction is called a diode, since it is the semiconductor equivalent of the vacuum diode valve used by Fleming to rectify radio signals in 1904.

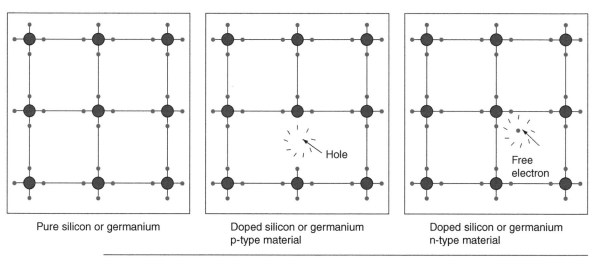

Pure silicon or germanium   Doped silicon or germanium   Doped silicon or germanium
                                    p-type material   n-type material

Figure 6.26 Semiconductor material.

Semiconductor diode

A semiconductor or junction diode consists of a p-type and n-type material formed in the same piece of silicon or germanium. The p-type material forms the anode and the n-type the cathode, as shown in Fig. 6.27. If the anode is made positive with respect to the cathode, the junction will have very little resistance and current will flow. This is referred to as forward bias. However, if reverse bias is applied, that is, the anode is made negative with respect to the cathode, the junction resistance is high and no current can flow, as shown in Fig. 6.28. The characteristics for a forward and reverse bias p–n junction are

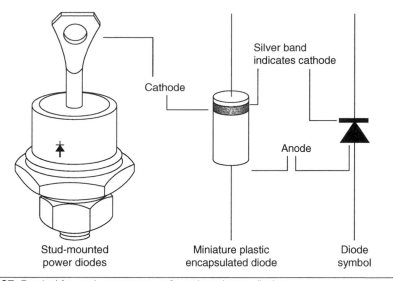

Cathode

Silver band
indicates cathode

Anode

Stud-mounted
power diodes

Miniature plastic
encapsulated diode

Diode
symbol

Figure 6.27 Symbol for and appearance of semiconductor diodes.

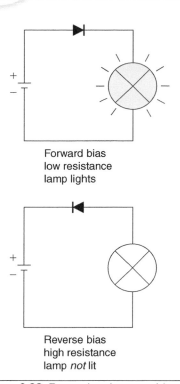

Forward bias
low resistance
lamp lights

Reverse bias
high resistance
lamp *not* lit

Figure 6.28 Forward and reverse bias of a diode.

given in Fig. 6.29. It can be seen that a small voltage is required to forward bias the junction before a current can flow. This is approximately 0.6 V for silicon and 0.2 V for germanium. The reverse bias potential of silicon is about 1200 V and for germanium about 30 V. If the reverse bias voltage is exceeded the diode will break down and current will flow in both directions. Similarly, the diode will break down if the current rating is exceeded, because excessive heat will be generated. Manufacturers' information therefore gives maximum voltage and current ratings for individual diodes which must not be exceeded. However, it is possible to connect a number of standard diodes in series or parallel, thereby sharing current or voltage, as shown in Fig. 6.30, so that the manufacturers' maximum values are not exceeded by the circuit.

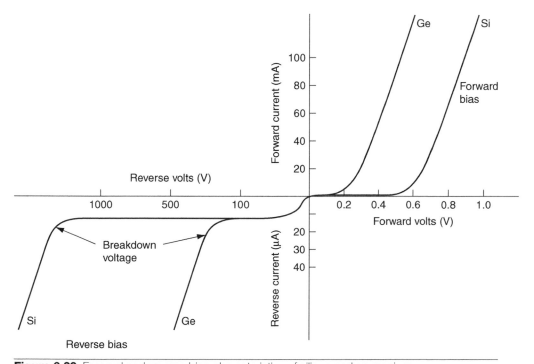

Forward current (mA)

Reverse volts (V)

1000   500   100

Breakdown
voltage

Si

Ge

Reverse bias

Ge   Si

Forward
bias

0.2 0.4 0.6 0.8 1.0

Forward volts (V)

Reverse current (µA)

20
30
40

Figure 6.29 Forward and reverse bias characteristics of silicon and germanium.

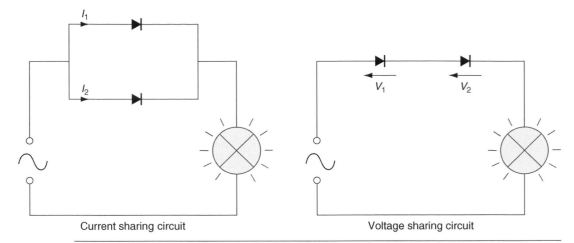

Current sharing circuit   Voltage sharing circuit

Figure 6.30 Using two diodes to reduce the current or voltage applied to a diode.

Diode testing

The p–n junction of the diode has a low resistance in one direction and a very high resistance in the reverse direction. Connecting an ohmmeter, with the red positive lead to the anode of the junction diode and the black negative lead to the cathode, would give a very low reading. Reversing the lead connections would give a high resistance reading in a 'good' component.

Zener diode

Definition

Zener diodes are often used in voltage stability devices in voltage regulators. A Zener will generate a small change in voltage for a given large change in current and therefore makes up for output voltage falling when load is applied.

A Zener diode is a silicon junction diode but with a different characteristic than the semiconductor diode considered previously. It is a special diode with a predetermined reverse breakdown voltage, the mechanism for which was discovered by Carl Zener in 1934. Its symbol and general appearance are shown in Fig. 6.31. In its forward bias mode, that is, when the anode is positive and the cathode negative, the Zener diode will conduct at about 0.6 V, just like an ordinary diode, but it is in the reverse mode that the Zener diode is normally used. When connected with the anode made negative and the cathode positive, the reverse current is zero until the reverse voltage reaches a predetermined

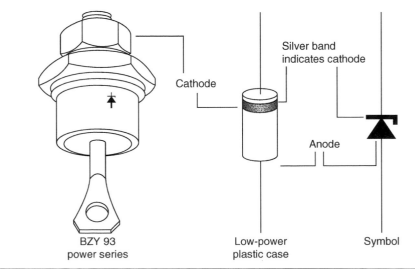

Cathode

Silver band indicates cathode

Anode

BZY 93 power series

Low-power plastic case

Symbol

Figure 6.31 Symbol for and appearance of Zener diodes.

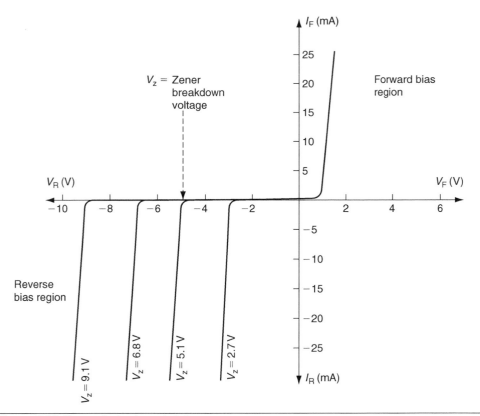

Figure 6.32 Zener diode characteristics.

value, when the diode switches on, as shown by the characteristics given in Fig. 6.32. This is called the Zener voltage or reference voltage. Zener diodes are manufactured in a range of preferred values, for example, 2.7, 4.7, 5.1, 6.2, 6.8, 9.1, 10, 11, 12 V, etc., up to 200 V at various ratings. The diode may be damaged by overheating if the current is not limited by a series resistor, but when this is connected, the voltage across the diode remains constant. It is this property of the Zener diode which makes it useful for stabilizing power supplies and these circuits are considered in Fig. 6.32.

If a test circuit is constructed as shown in Fig. 6.33, the Zener action can be observed. When the supply is less than the Zener voltage (5.1 V in this case) no current will flow and the output voltage will be equal to the input voltage.

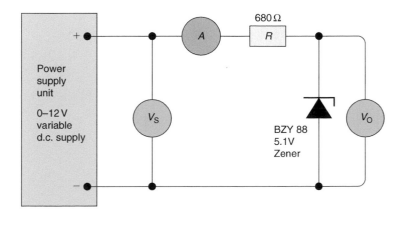

P.S.U. supply volts (V_S)	Current (A)	Output volts (V_O)
1		
2		
3		
4		
5		
6		
7		
8		
9		
10		
11		
12		

Figure 6.33 Experiment to demonstrate the operation of a Zener diode.

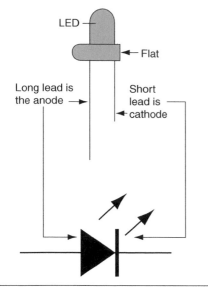

Figure 6.34 Symbol for and general appearance of an LED.

When the supply is equal to or greater than the Zener voltage, the diode will conduct and any excess voltage will appear across the 680 Ω resistor, resulting in a very stable voltage at the output. When connecting this and other electronic circuits you must take care to connect the polarity of the Zener diode as shown in the diagram. Note that current must flow through the diode to enable it to stabilize.

Light-emitting diode

The light-emitting diode (LED) is a p–n junction especially manufactured from a semi conducting material that emits light when a current of about 10 mA flows through the junction.

No light is emitted when the junction is reverse biased and if this exceeds about 5 V the LED may be damaged.

The general appearance and circuit symbol are shown in Fig. 6.34. The LED will emit light if the voltage across it is about 2 V. If a voltage greater than 2 V is to be used then a resistor must be connected in series with the LED.

To calculate the value of the series resistor we must ask ourselves what we know about LEDs. We know that the diode requires a forward voltage of about 2 V and a current of about 10 mA must flow through the junction to give sufficient light. The value of the series resistor R will, therefore, be given by:

$$R = \frac{\text{Supply voltage} - 2V}{10mA} \, \Omega$$

The circuit is, therefore, as shown in Fig. 6.34. LEDs are available in red, yellow and green and, when used with a series resistor, may replace a filament lamp. They use less current than a filament lamp, are smaller, do not become hot and last indefinitely. A filament lamp, however, is brighter and emits white light. LEDs are often used as indicator lamps, to indicate the presence of a voltage. They do not, however, indicate the precise amount of voltage present at that point. Another application of the LED is the seven-segment display used as a numerical indicator in calculators, digital watches and measuring instruments. Seven LEDs are arranged as a figure 8 so that when various segments are illuminated, the numbers 0–9 are displayed as shown in Fig. 6.37.

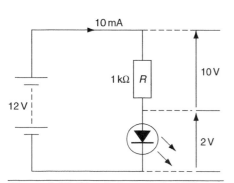

Figure 6.36 Circuit diagrams for LED example.

Figure 6.35 LED signal lamps.

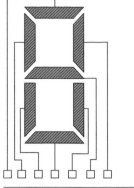

Figure 6.37 LED used in seven-segment display.

Example

Calculate the value of the series resistor required when an LED is to be used to show the presence of a 12 V supply.

$$R = \frac{12V - 2V}{10mA}\ \Omega$$

$$R = \frac{10V}{10mA} = \Omega$$

Try this

LEDs

Make a list in the margin of examples where you have seen LEDs being used.

Recent scientific discoveries have also enabled this humble LED signal lamp to become the super efficient, broad spectrum general lighting service LED lamp we are familiar with today. We will look at these lamps later in this Chapter under the heading Comparison of Light Sources.

Light-dependent resistor

Almost all materials change their resistance with a change in temperature. Light energy falling on a suitable semiconductor material also causes a change in resistance. The semiconductor material of a light-dependent resistor (LDR) is encapsulated as shown in Fig. 6.38 together with the circuit

Definition

To distinguish between an *LED* and a *LDR* look at the way the arrow is pointing. LED points out (emitting), LDR points towards the device (light dependant).

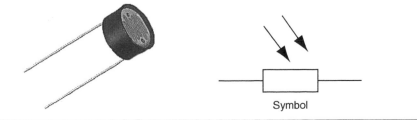

Symbol

Figure 6.38 Symbol and appearance of an LDR.

symbol. The resistance of an LDR in total darkness is about 10 MΩ, in normal room lighting about 5 kΩ and in bright sunlight about 100 Ω. They can carry tens of milliamperes, an amount that is sufficient to operate a relay. The LDR uses this characteristic to switch on automatically street lighting and security alarms.

Photodiode

The photodiode is a normal junction diode with a transparent window through which light can enter. The circuit symbol and general appearance are shown in Fig. 6.39 It is operated in reverse bias mode and the leakage current increases in proportion to the amount of light falling on the junction. This is due to the light energy breaking bonds in the crystal lattice of the semiconductor material to produce holes and electrons.

Photodiodes will only carry microamperes of current but can operate much more quickly than LDRs and are used as 'fast' counters when the light intensity is changing rapidly.

Thermistor

The thermistor is a thermal resistor, a semiconductor device whose resistance varies with temperature. Its circuit symbol and general appearance are shown in Fig. 6.40. They can be supplied in many shapes and are used for the measurement and control of temperature up to their maximum useful temperature limit of about 300°C. They are very sensitive and because the bead of semiconductor material can be made very small, they can measure temperature in the most inaccessible places with very fast response times. Thermistors are embedded in high-voltage underground transmission cables in order to monitor the temperature of the cable. Information about the temperature of a cable allows engineers to load the cables more efficiently. A particular cable can carry a larger load in winter for example, when heat from the cable is being dissipated more efficiently. A thermistor is also used to monitor the water temperature of a motor car.

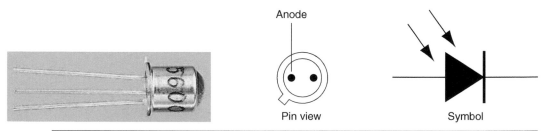

Figure 6.39 Symbol for pin connections and appearance of a photodiode.

Figure 6.40 Symbol for and appearance of a thermistor.

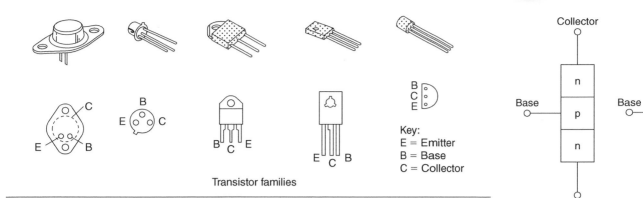

Key:
E = Emitter
B = Base
C = Collector

Transistor families

Figure 6.41 The appearance and pin connections of the transistor family.

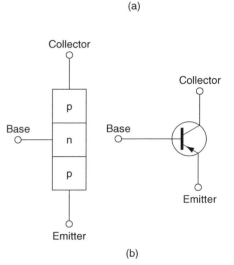

(a)

(b)

Figure 6.42 Structure of and symbol for (a) n-p-n and (b) p-n-p transistors.

Transistors

The transistor has become the most important building block in electronics. It is the modern, miniature, semiconductor equivalent of the thermionic valve and was invented in 1947 by Bardeen, Shockley and Brattain at the Bell Telephone Laboratories in the United States. Transistors are packaged as separate or *discrete* components, as shown in Fig. 6.41.

There are two basic types of transistor, the *bipolar* or junction transistor and the *field-effect* transistor (FET).

The FET has some characteristics that make it a better choice in electronic switches and amplifiers. It uses less power and has a higher resistance and frequency response. It takes up less space than a bipolar transistor and, therefore, more of them can be packed together on a given area of silicon chip.

It is, therefore, the FET that is used when many transistors are integrated on to a small area of silicon chip as in the IC that will be discussed later.

When packaged as a discrete component the FET looks much the same as the bipolar transistor. Its circuit symbol and connections are given in the Appendix. However, it is the bipolar transistor that is much more widely used in electronic circuits as a discrete component.

The bipolar transistor

The bipolar transistor consists of three pieces of semiconductor material sandwiched together as shown in Fig. 6.42. The structure of this transistor makes it a three-terminal device having a base, collector and emitter terminal.

By varying the current flowing into the base connection a much larger current flowing between collector and emitter can be controlled. Apart from the supply connections, the n-p-n and p-n-p types are essentially the same but the n-p-n type is more common.

A transistor is generally considered a current-operated device. There are two possible current paths through the transistor circuit, shown in Fig. 6.43: the base–emitter path when the switch is closed; and the collector–emitter path. Initially, the positive battery supply is connected to the n-type material of the collector, the junction is reverse biased and, therefore, no current will flow. Closing the switch will forward bias the base–emitter junction and current flowing through this junction causes current to flow across the collector–emitter junction and the signal lamp will light.

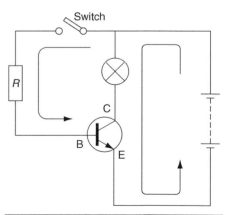

Figure 6.43 Operation of the transistor.

> **Definition**
>
> A *transistor* is used as both an electronic switch and as an amplifying device.

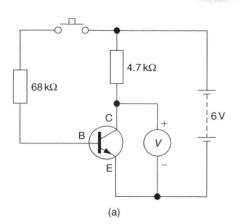

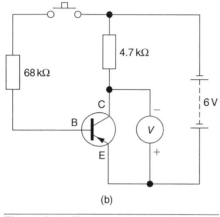

Figure 6.44 Transistor test circuits
(a) n-p-n transistor test; (b) p-n-p transistor test.

A small base current can cause a much larger collector current to flow. This is called the *current gain* of the transistor, and is typically about 100. When I say a much larger collector current, I mean a large current in electronic terms, up to about half an ampere.

We can, therefore, regard the transistor as operating in two ways: as a switch because the base current turns on and controls the collector current; and as a current amplifier because the collector current is greater than the base current.

We could also consider the transistor to be operating in a similar way to a relay. However, transistors have many advantages over electrically operated switches such as relays. They are very small, reliable, have no moving parts and, in particular, they can switch millions of times a second without arcing occurring at the contacts.

Transistor testing

A transistor can be thought of as two diodes connected together and, therefore, a transistor can be tested using an ohmmeter in the same way as was described for the diode.

Assuming that the red lead of the ohmmeter is positive, the transistor can be tested in accordance with Table 6.6.

When many transistors are to be tested, a simple test circuit can be assembled as shown in Fig. 6.44.

With the circuit connected, as shown in Fig. 6.44, a 'good' transistor will give readings on the voltmeter of 6 V with the switch open and about 0.5 V when the switch is made. The voltmeter used for the test should have a high internal resistance, about ten times greater than the value of the resistor being tested – in this case 4.7 kΩ – and this is usually indicated on the back of a multi-range meter or in the manufacturer's information supplied with a new meter.

Integrated circuits

ICs were first developed in the 1960s. They are densely populated miniature electronic circuits made up of hundreds and sometimes thousands of microscopically small transistors, resistors, diodes and capacitors, all connected together on a single chip of silicon no bigger than a baby's fingernail. When assembled in a single package, as shown in Fig. 6.45, we call the device an IC.

There are two broad groups of IC: digital ICs and linear ICs. Digital ICs contain simple switching-type circuits used for logic control and calculators, and linear ICs incorporate amplifier-type circuits which can respond to audio and radio frequency signals. The most versatile linear IC is the operational amplifier, which has applications in electronics, instrumentation and control.

The IC is an electronic revolution. ICs are more reliable, cheaper and smaller than the same circuit made from discrete or separate transistors, and electronically superior. One IC behaves differently than another because of the arrangement of the transistors within the IC. Manufacturers' data sheets describe the characteristics of the different ICs, which have a reference number stamped on the top.

Table 6.6 Transistor testing using an ohmmeter

A 'good' n-p-n transistor will give the following readings:
Red to base and black to collector = low resistance Red to base and black to emitter = low resistance Reversed connections on the above terminals will result in a high resistance reading, as will connections of either polarity between the collector and emitter terminals.
A 'good' p-n-p transistor will give the following readings:
Black to base and red to collector = low resistance Black to base and red to emitter = low resistance Reversed connections on the above terminals will result in a high resistance reading, as will connections of either polarity between the collector and emitter terminals.

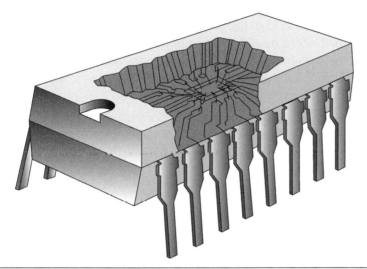

Figure 6.45 Exploded view of an IC.

When building circuits, it is necessary to be able to identify the IC pin connection by number. The number 1 pin of any IC is indicated by a dot pressed into the encapsulation; it is also the pin to the left of the cutout (Fig. 6.46). Since the packaging of ICs has two rows of pins they are called DIL (dual in line) packaged ICs and their appearance is shown in Fig. 6.47.

ICs are sometimes connected into DIL sockets and at other times are soldered directly into the circuit. The testing of ICs is beyond the scope of a practising electrician, and when they are suspected of being faulty an identical or equivalent replacement should be connected into the circuit, ensuring that it is inserted the correct way round, which is indicated by the position of pin number 1 as described above.

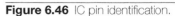

Figure 6.46 IC pin identification.

The thyristor

The *thyristor* was previously known as a 'silicon controlled rectifier' since it is a rectifier that controls the power to a load. It consists of four pieces of semiconductor material sandwiched together and connected to three terminals, as shown in Fig. 6.48.

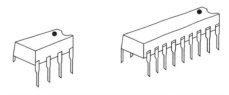

Figure 6.47 DIL packaged ICs.

The word thyristor is derived from the Greek word *thyra* meaning door, because The thyristor behaves like a door. It can be open or shut, allowing or preventing current flow through the device. The door is opened – we say the thyristor is triggered – to a conducting state by applying a pulse voltage to the gate connection. Once the thyristor is in the conducting state, the gate loses all control over the devices. The only way to bring the thyristor back to a non conducting state is to reduce the voltage across the anode and cathode to zero or apply reverse voltage across the anode and cathode.

We can understand the operation of a thyristor by considering the circuit shown in Fig. 6.49. This circuit can also be used to test suspected faulty components. When SWB only is closed the lamp will not light, but when SWA is also closed, the lamp lights to full brilliance. The lamp will remain illuminated even when SWA is opened. This shows that the thyristor is operating correctly. Once a voltage has been applied to the gate the thyristor becomes forward conducting, like a diode, and the gate loses control.

A thyristor may also be tested using an ohmmeter as described in Table 6.7, assuming that the red lead of the ohmmeter is positive. The thyristor has no moving parts and operates without arcing. It can operate at extremely high speeds, and the currents used to operate the gate are very small. The most common application for the thyristor is to control the power supply to a load, for example, lighting dimmers and motor speed control.

The power available to an a.c. load can be controlled by allowing current to be supplied to the load during only a part of each cycle. This can be achieved

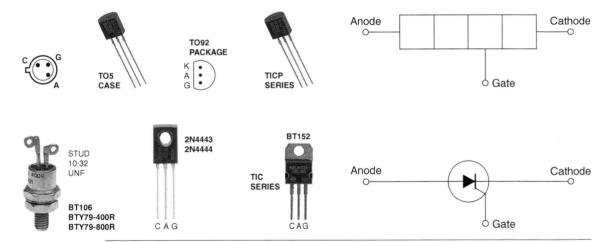

Figure 6.48 Symbol for and structure and appearance of a thyristor.

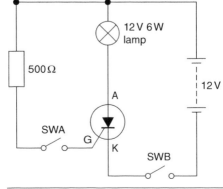

Figure 6.49 Thyristor test circuit.

Table 6.7 Thyristor testing using an ohmmeter

A 'good' thyristor will give the following readings:
Black to cathode and red on gate = low resistance
Red to cathode and black on gate = a higher resistance value
The value of the second reading will depend upon the thyristor, and may vary from only slightly greater to very much greater
Connecting the test instrument leads from cathode to anode will result in a very high resistance reading, whatever polarity is used

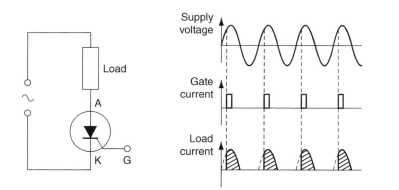

Figure 6.50 Waveforms to show the control effect of a thyristor.

by supplying a gate pulse automatically at a chosen point in each cycle, as shown by Fig. 6.50 Power is reduced by triggering the gate later in the cycle.

The thyristor is only a half-wave device (like a diode) allowing control of only half the available power in an a.c. circuit. This is very uneconomical, and a further development of this device has been the triac, which is considered next.

The triac

The triac was developed following the practical problems experienced in connecting two thyristors in parallel to obtain full-wave control, and in providing two separate gate pulses to trigger the two devices. The triac is a single device containing a back-to-back, two-directional thyristor that is triggered on both halves of each cycle of the a.c. supply by the same gate signal. The power available to the load can, therefore, be varied between zero and full load.

Its symbol and general appearance are shown in Fig. 6.51. Power to the load is reduced by triggering the gate later in the cycle, as shown by the waveforms of Fig. 6.52.

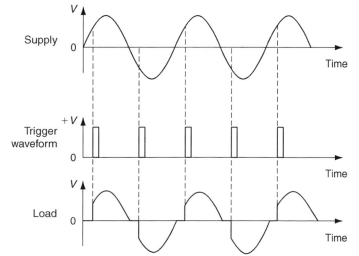

Figure 6.52 Waveforms to show the control effect of a triac.

> **Definition**
>
> A *triac* is used in a.c. motor control since it can be gated by either a positive or negative pulse.

Figure 6.51 Appearance of a triac.

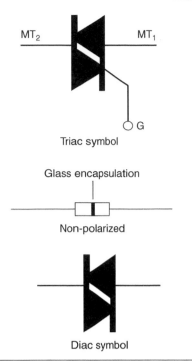

Triac symbol

Glass encapsulation

Non-polarized

Diac symbol

Figure 6.53 Symbol for and appearance of a diac used in triac firing circuits.

The triac is a three-terminal device, just like the thyristor, but the terms anode and cathode have no meaning for a triac. Instead, they are called main terminal one (MT1) and main terminal two (MT2). The device is triggered by applying a small pulse to the gate (G). A gate current of 50 mA is sufficient to trigger a triac switching up to 100 A. They are used for many commercial applications where control of a.c. power is required, for example, motor speed control and lampdimming.

The diac

The diac is a two-terminal device containing a two-directional Zener diode. It is used mainly as a trigger device for the thyristor and triac. The symbol is shown in Fig. 6.53.

The device turns on when some predetermined voltage level is reached, say 30 V, and, therefore, it can be used to trigger the gate of a triac or thyristor each time the input waveform reaches this predetermined value. Since the device contains back-to-back Zener diodes it triggers on both the positive and negative halfcycles.

Voltage divider

Earlier in this chapter we considered the distribution of voltage across resistors connected in series. We found that the supply voltage was divided between the series resistors in proportion to the size of the resistor. If two identical resistors were connected in series across a 12 V supply, as shown in Fig. 6.54(a), both common sense and a simple calculation would confirm that 6 V would be measured across the output. In the circuit shown in Fig. 6.54(b), the 1 and 2 kΩ resistors divide the input voltage into three equal parts. One part, 4 V, will appear across the 1 kΩ resistor and two parts, 8 V, will appear across the 2 kΩ resistor. In Fig. 6.54(c) the situation is reversed and, therefore, the voltmeter will read 4 V.

The division of the voltage is proportional to the ratio of the two resistors and, therefore, we call this simple circuit a *voltage divider* or *potential divider*. The values of the resistors R1 and R2 determine the output voltage as follows:

$$V_{OUT} = V_{IN} \times \frac{R_2}{R_1 + R_2} \text{ (V)}$$

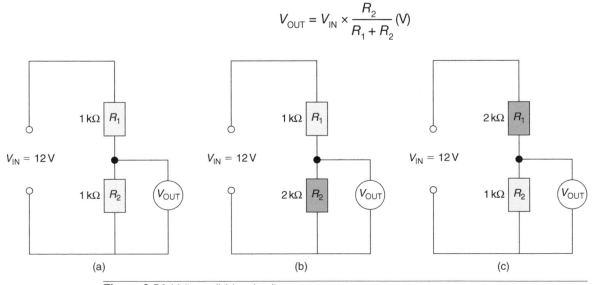

(a)   (b)   (c)

Figure 6.54 Voltage divider circuit.

Example 1

For the circuit shown in Fig. 6.55, calculate the output voltage.

$$V_{OUT} = 6\,V \times \frac{2.2\,k\Omega}{10\,k\Omega + 2.2\,k\Omega} = 1.08\,V$$

Example 2

For the circuit shown in Fig. 6.56(a), calculate the output voltage.

We must first calculate the equivalent resistance of the parallel branch:

$$\frac{1}{R_T} = \frac{1}{R_1} + \frac{1}{R_2}$$

$$\frac{1}{R_T} = \frac{1}{10\,k\Omega} + \frac{1}{10\,k\Omega} = \frac{1+1}{10\,k\Omega} = \frac{2}{10\,k\Omega}$$

$$R_T = \frac{10\,k\Omega}{2} = 5\,k\Omega$$

The circuit may now be considered as shown in Fig. 6.56 (b):

$$V_{OUT} = 6\,V \times \frac{10\,k\Omega}{5\,k\Omega + 10\,k\Omega} - 4\,V$$

Voltage dividers are used in electronic circuits to produce a reference voltage which is suitable for operating transistors and ICs. The volume control in a radio or the brightness control of a cathode-ray oscilloscope requires a continuously variable voltage divider and this can be achieved by connecting a variable resistor or potentiometer, as shown in Fig. 6.57. With the wiper arm making a connection at the bottom of the resistor, the output would be zero. When connection is made at the centre, the voltage would be 6 V, and at the top of the resistor the voltage would be 12 V. The voltage is continuously variable between 0 and 12 V simply by moving the wiper arm of a suitable variable resistor such as those shown in Fig. 6.57.

When a voltmeter is connected to a voltage divider it 'loads' the circuit, causing the output voltage to fall below the calculated value. To avoid this, the resistance of the voltmeter should be at least ten times as great as the value of the resistor across which it is connected. For example, the voltmeter connected across the voltage divider shown in Fig. 6.54(b) must be greater than 20 k, and across 6.54(c) greater than 10 k. This problem of loading the circuit occurs when taking voltage readings in electronic circuits and therefore a high impedance voltmeter should always be used to avoid instrument errors.

Rectification of a.c.

When a d.c. supply is required, batteries or a rectified a.c. supply can be provided. Batteries have the advantage of portability, but a battery supply is more expensive than using the a.c. mains supply suitably rectified. **Rectification**

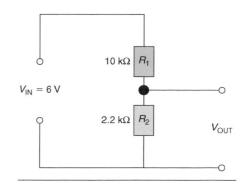

Figure 6.55 Voltage divider circuit for Example 1.

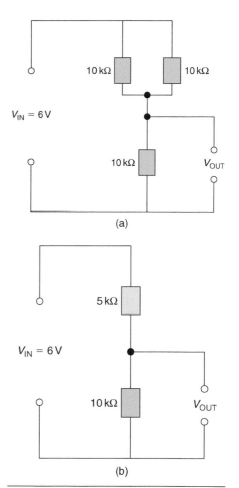
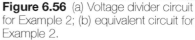

Figure 6.56 (a) Voltage divider circuit for Example 2; (b) equivalent circuit for Example 2.

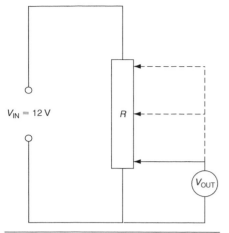

$V_{IN} = 12\,V$ R

V_{OUT}

Figure 6.57 Constantly variable voltage divider circuit.

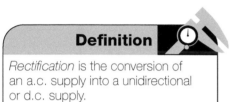

Definition

Rectification is the conversion of an a.c. supply into a unidirectional or d.c. supply.

is the conversion of an a.c. supply into a unidirectional or d.c. supply. This is one of the many applications for a diode which will conduct in one direction only, that is when the anode is positive with respect to the cathode.

Half-wave rectification

The circuit is connected as shown in Fig. 6.58. During the first half-cycle the anode is positive with respect to the cathode and, therefore, the diode will conduct. When the supply goes negative during the second half-cycle, the anode is negative with respect to the cathode and, therefore, the diode will not allow current to flow. Only the positive half of the waveform will be available at the load and the lamp will light at reduced brightness.

Full-wave rectification

Rectification is the conversion of an a.c. supply into a unidirectional or d.c. supply.

Figure 6.59 shows an improved rectifier circuit that makes use of the whole a.c. waveform and is, therefore, known as a full-wave rectifier. When the four diodes are assembled in this diamond-shaped configuration, the circuit is also known as a *bridge rectifier*. During the first half-cycle diodes D_1 and D_3 conduct, and diodes D_2 and D_4 conduct during the second half-cycle. The lamp will light to full brightness.

Full-wave and half-wave rectification can be displayed on the screen of a CRO and will appear as shown in Figs. 6.58 and 6.59.

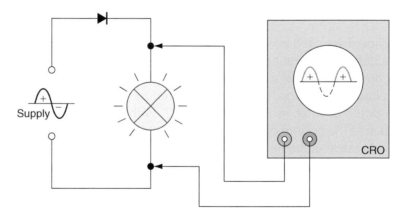

Figure 6.58 Half-wave rectification.

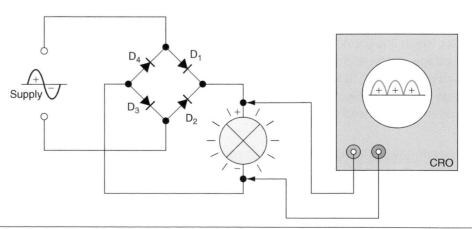

Figure 6.59 Full-wave rectification using a bridge circuit.

Smoothing

The circuits of Figs. 6.58 and 6.59 convert an alternating waveform into a waveform which never goes negative, but they cannot be called continuous d.c. because they contain a large alternating component. Such a waveform is too bumpy to be used to supply electronic equipment but may be used for battery charging. To be useful in electronic circuits the output must be smoothed. The simplest way to smooth an output is to connect a large-value capacitor across the output terminals as shown in Fig. 6.60.

When the output from the rectifier is increasing, as shown by the dotted lines of Fig. 6.61, the capacitor charges up. During the second quarter of the cycle, when the output from the rectifier is falling to zero, the capacitor discharges into the load. The output voltage falls until the output from the rectifier once again charges the capacitor. The capacitor connected to the full-wave rectifier circuit is charged up twice as often as the capacitor connected to the half-wave circuit and, therefore, the output ripple on the full-wave circuit is smaller, giving better smoothing. Increasing the current drawn from the supply increases the size of the ripple. Increasing the size of the capacitor reduces the amount of ripple.

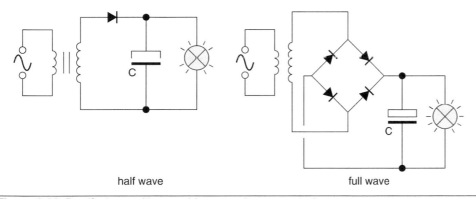

half wave   full wave

Figure 6.60 Rectified a.c. with smoothing capacitor connected.

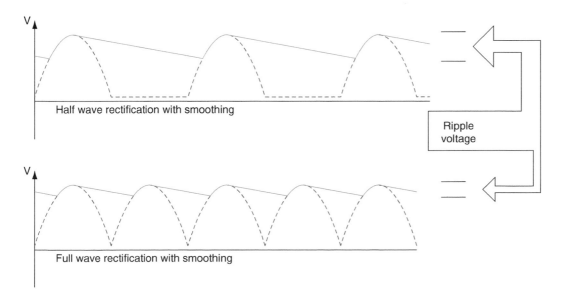

Half wave rectification with smoothing

Ripple voltage

Full wave rectification with smoothing

Figure 6.61 Output waveforms with smoothing, showing reduced ripple with full wave.

Try this

Battery charger

- Do you have a battery charger for your car battery?
- What type of circuit do you think is inside?
- Carefully look inside and identify the components.

Low-pass filter

The ripple voltage of the rectified and smoothed circuit shown in Fig. 6.60 can be further reduced by adding a low-pass filter, as shown in Fig. 6.61. A low-pass filter allows low frequencies to pass while blocking higher frequencies. Direct current has a frequency of zero hertz, while the ripple voltage of a full-wave rectifier has a frequency of 100 Hz. Connecting the low-pass filter will allow the d.c. to pass while blocking the ripple voltage, resulting in a smoother output voltage.

The low-pass filter shown in Fig. 6.62 does, however, increase the output resistance, which encourages the voltage to fall as the load current increases. This can be reduced if the resistor is replaced by a choke, which has a high impedance to the ripple voltage but a low resistance, which reduces the output ripple without increasing the output resistance.

Stabilized power supplies

The power supplies required for electronic circuits must be ripple-free, stabilized and have good regulation; that is, the voltage must not change in value over the whole load range. A number of stabilizing circuits are available which, when connected across the output of the circuit shown in Fig. 6.63, give a constant

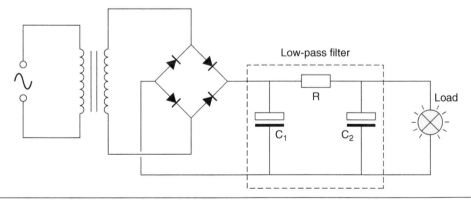

Figure 6.62 Rectified a.c. with low-pass filter connected.

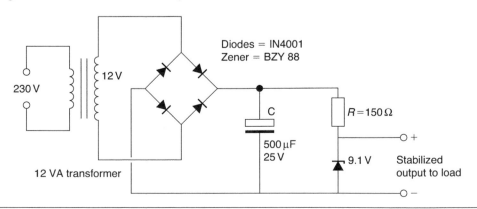

Figure 6.63 Stabilized d.c. supply.

or stabilized voltage output. These circuits use the characteristics of the Zener diode which was described by the experiment in Fig. 6.33.

Figure 6.63 shows an d.c. supply which has been rectified, smoothed and stabilized. You could build and test this circuit at college if your lecturers agree.

Inverter

An inverter is a device that changes direct current (d.c.) to alternating current (a.c). Mechanical versions are in existence, for instance a d.c. motor feeding an a.c. generator will convert the d.c. supplying the motor into an alternating supply as seen below.

Static inverters are much more common and use electronics that are far more reliable since they do not include any moving parts. The static inverter combines an oscillator, amplifier and a transformer in order to produce an alternating output, which is used extensively in solar photovoltaic applications in order to convert the d.c. produced by the solar cells to a mains equivalent alternating supply.

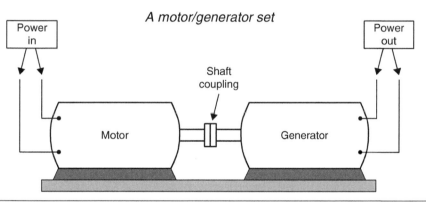

Figure 6.64 A d.c. motor feeding an a.c. generator creates a mechanical invertor: changing direct current to alternating current.

Assessment criteria 2.2

State the applications of components and devices in electrotechnical systems

Assessment criteria 2.3

Describe the function of electronic components and devices in electrotechnical systems

Intruder alarms

The installation of security alarm systems in the United Kingdom is already a multi-million-pound business and yet it is also a relatively new industry. As society becomes increasingly aware of crime prevention, it is evident that the market for security systems will expand.

Most intruders are looking for portable and easily saleable items such as DVD players, television sets, home computers, jewellery, cameras, silverware, money, cheque books or credit cards.

Figure 6.65 Security lighting reduces crime.

Security lighting

Security lighting installed on the outside of the home may be activated by external detectors. These detectors sense the presence of a person outside the protected property and additional lighting is switched on. This will deter most potential intruders while also acting as courtesy lighting for visitors (Fig. 6.65).

Passive infra-red detectors

Passive infra-red (PIR) detector units allow a householder to switch on lighting units automatically whenever the area covered is approached by a moving body whose thermal radiation differs from the background. This type of detector is ideal for driveways or dark areas around the protected property. It also saves energy because the lamps are only switched on when someone approaches the protected area. The major contribution to security lighting comes from the 'unexpected' high-level illumination of an area when an intruder least expects it. This surprise factor often encourages the potential intruder to 'try next door'.

PIR detectors are designed to sense heat changes in the field of view dictated by the lens system. The field of view can be as wide as 180°, as shown by the diagram in Fig. 6.66. Many of the 'better' detectors use a split lens system so that a number of beams have to be broken before the detector switches on the security lighting. This capability overcomes the problem of false alarms, and a typical PIR is shown in Fig.6.67.

PIR detectors are often used to switch LED or tungsten halogen floodlights because, of all available luminaires, LEDs and tungsten halogen lamps offers instant high-level illumination. Light fittings must be installed out of reach of an intruder in order to prevent sabotage of the security lighting system.

Intruder alarm systems

Alarm systems are now increasingly considered to be an essential feature of home security for all types of homes and not just property in high-risk areas.

An **intruder alarm system** serves as a deterrent to a potential thief and often reduces home insurance premiums. In the event of a burglary they alert the

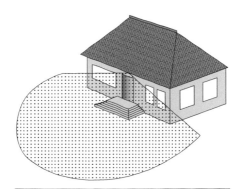

Figure 6.66 PIR detector and field of direction.

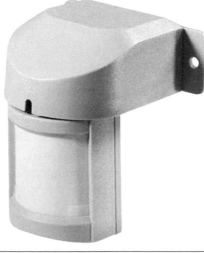

Figure 6.67 A typical PIR detector.

Figure 6.68 Proximity switches for perimeter protection.

occupants, neighbours and officials to a possible criminal act and generate fear and uncertainty in the mind of the intruder which encourages a more rapid departure. Intruder alarm systems can be broadly divided into three categories – those that give perimeter protection, space protection or trap protection. A system can comprise one or a mixture of all three categories.

A **perimeter protection system** places alarm sensors on all external doors and windows so that an intruder can be detected as he or she attempts to gain access to the protected property. This involves fitting proximity switches to all external doors and windows.

A **movement or heat detector** placed in a room will detect the presence of anyone entering or leaving that room. PIR detectors and ultrasonic detectors give space protection. Space protection does have the disadvantage of being triggered by domestic pets but it is simpler and, therefore, cheaper to install. Perimeter protection involves a much more extensive and, therefore, expensive installation, but is easier to live with.

Trap protection places alarm sensors on internal doors and pressure pad switches under carpets on through routes between, for example, the main living area and the master bedroom. If an intruder gains access to one room he cannot move from it without triggering the alarm.

Definition

An *intruder alarm system* serves as a deterrent to a potential thief and often reduces home insurance premiums.

A *perimeter protection system* places alarm sensors on all external doors and windows so that an intruder can be detected as he or she attempts to gain access to the protected property.

A *movement or heat detector* placed in a room will detect the presence of anyone entering or leaving that room.

Trap protection places alarm sensors on internal doors and pressure pad switches under carpets on through routes between, for example, the main living area and the master bedroom.

Proximity switches

These are designed for the discreet protection of doors and windows. They are made from moulded plastic and are about the size of a chewing-gum packet, as shown in Fig. 6.68. One moulding contains a reed switch, the other a magnet, and when they are placed close together the magnet maintains the contacts of the reed switch in either an open or closed position. Opening the door or window separates the two mouldings and the switch is activated, triggering the alarm. The same effect can be achieved using mercury switches which completes the circuit since mercury conducts electricity.

PIR detectors

These are activated by a moving body that is warmer than the surroundings.

The PIR shown in Fig. 6.69 has a range of 12 m and a detection zone of 110° when mounted between 1.8 and 2 m high.

Telephones

A component that converts one form of energy into another is known as a transducer. A telephone uses a microphone and loudspeaker to operate by converting sound into electrical signals when you speak into the phone and the complete opposite when you hear the other person. The loudspeaker uses a diaphragm that vibrates when sound waves strike against it, which in turn causes the diaphragm to create electricity by interacting with a small magnetic field.

The loudspeaker does the complete opposite: when an electrical signal is received it creates a magnetic field, which pushes and pulls a coil back and forth, which creates sound through its vibration.

A telephone also uses a tilting mechanism, which cuts the line to the exchange when the phone is rested in place.

Figure 6.69 PIR intruder alarm detector.

Electrical heating control (thermostats)

Water heating or space heating systems make use of a thermostat, which is a device for maintaining a constant temperature at some predetermined value. The operation of a thermostat is often based upon the principle of differential expansion between dissimilar metals, that is, a bimetal strip, which causes a contact to make or break at a chosen temperature. Figure 6.70 shows the principle of a rod-type thermostat that is often used with water heaters. An invar rod, which has minimal expansion when heated, is housed within a copper tube and the two metals are brazed together at one end. The other end of the copper tube is secured to one end of the switch mechanism. As the copper expands and contracts under the influence of a varying temperature, the switch contacts make and break. In the case of a thermostat such as this, the electrical circuit is broken when the temperature setting is reached. Figure 6.71 shows a room thermostat which works on the same principle.

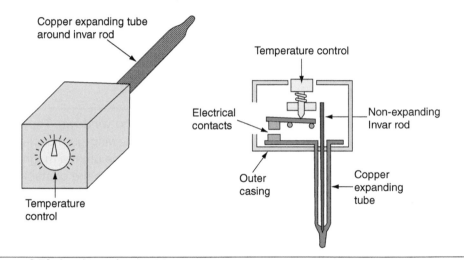

Figure 6.70 A rod-type thermostat.

Figure 6.71 Room thermostat.

Programmers

A heating system may be fuelled by gas, oil, electricity or solar, but whatever the fuel system providing the heat energy, it is very likely that the system will be controlled by an electrical or electronic programmer, thermostats and a circulating pump for a water-based system. Fig 6.72 shows a block diagram for a typical domestic space heating and water heating system. The programmer will incorporate a time clock so that the customer or client can select the number of hours each day that the system will operate. For a working family in a domestic property this will probably be a couple of hours before the working day begins, and then five to seven hours in the evening. In a commercial situation it will probably switch on a couple of hours before the working day begins and remain on until the working day ends. In both situations the programmer unit often incorporates an override or boost facility for changing circumstances and weather conditions.

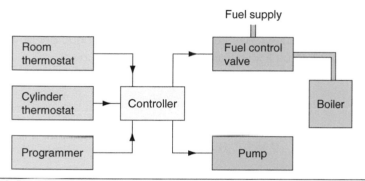

Figure 6.72 Block diagram – space heating control system (Honeywell Y plan).

During the period when the space heating or water heating is in the 'on position', thermostats based upon the bimetal strip principle will maintain the room and water temperature at some predetermined level. Figure 6.72 shows a wiring diagram for a domestic space and water heating system that circulates water through radiators to warm individual rooms. In this type of system, energy conservation now recommends that each radiator be additionally fitted with a TRV (thermostatic radiator valve) to control the temperature in each room.

Assessment criteria 3.1

Determine electrical quantities in alternating current circuits

In this section we will first of all consider the theoretical circuits of pure resistance, inductance and capacitance acting alone in an a.c. circuit before going on to consider the practical circuits of resistance, inductance and capacitance acting together. Let us first define some of our terms of reference.

Resistance

In d.c. circuits the **resistance** is defined as opposition to current flow.

From Ohm's law:

$$R = \frac{V_R}{I_R}(\Omega)$$

Example 1

An electric heater, when connected to a 230 V supply, was found to take a current of 4 A. Calculate the element resistance.

$$R = \frac{V}{I}$$

$$\therefore R = \frac{230 \text{ V}}{4 \text{ A}} = 57.5\,\Omega$$

Example 2

The insulation resistance measured between phase conductors on a 400 V supply was found to be 2 MΩ. Calculate the leakage current.

$$I = \frac{V}{R}$$

$$\therefore I = \frac{400 \text{ V}}{2 \times 10^6 \ \Omega} = 200 \times 10^{-6} \text{ A} = 200 \ \mu\text{A}$$

Example 3

When a 4 Ω resistor was connected across the terminals of an unknown d.c. supply, a current of 3 A flowed. Calculate the supply voltage.

$$V = I \times R$$
$$\therefore V = 3A \times 4\Omega = 12V$$

Resistors in series and in parallel

In an electrical circuit resistors may be connected in series, in parallel, or in various combinations of series and parallel connections.

Series-connected resistors

Top tip

Series resistance

The total value of resistance in a series circuit is always greater than the largest individual value.

In any series circuit a current *I* will flow through all parts of the circuit as a result of the potential difference supplied by a battery V_T. Therefore, we say that in a series circuit the current is common throughout that circuit.

When the current flows through each resistor in the circuit, for example, R_1, R_2 and R_3 in Fig. 6.73, there will be a voltage drop across that resistor whose value will be determined by the values of *I* and *R*, since from Ohm's law $V = I \times R$. The sum of the individual voltage drops, for example, V_1, V_2 and V_3 in Fig. 6.73, will be equal to the total voltage V_T.

We can summarize these statements as follows. For any series circuit, I is common throughout the circuit and

$$V_T = V_1 + V_2 + V_3 \qquad \text{(Equation 1)}$$

Let us call the total circuit resistance R_T. From Ohm's law we know that $V = I \times R$ and therefore

Total voltage $V_T = I \times R_T$
Voltage drop across R_1 is $V_1 = I \times R_1$
Voltage drop across R_2 is $V_2 = I \times R_2$ (Equation 2)
Voltage drop across R_3 is $V_3 = I \times R_3$

We are looking for an expression for the total resistance in any series circuit and, if we substitute equation (2) into equation (1), we have:

$$V_T = V_1 + V_2 + V_3$$
$$\therefore I \times R_T = I \times R_1 + I \times R_2 + I \times R_3$$

Now, since I is common to all terms in the equation, we can divide both sides of the equation by I. This will cancel out I to leave us with an expression for the circuit resistance:

$$R_T = R_1 + R_2 + R_3$$

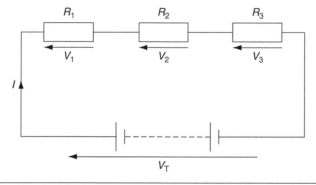

Figure 6.73 A series circuit.

Note that the derivation of this formula is given for information only. Craft students need only state the expression $R_T = R_1 + R_2 + R_3$ for series connections.

Parallel-connected resistors

In any parallel circuit, as shown in Fig. 6.74, the same voltage acts across all branches of the circuit. The total current will divide when it reaches a resistor junction, part of it flowing in each resistor. The sum of the individual currents, for example, I_1, I_2 and I_3 in Fig. 6.74, will be equal to the total current I_T.

We can summarize these statements as follows. For any parallel circuit, V is common to all branches of the circuit and

$$I_T = I_1 + I_2 + I_3 \qquad \text{(Equation 3)}$$

Let us call the total resistance R_T.

From Ohm's law we know, that $I = \dfrac{V}{R}$, and therefore

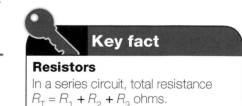

Key fact

Resistors
In a series circuit, total resistance $R_T = R_1 + R_2 + R_3$ ohms.

Top tip

Parallel resistance
The total value of resistance in a parallel circuit is always less than the smallest individual value.

the total current $I_T = \dfrac{V}{R_T}$

the current through R_1 is $I_1 = \dfrac{V}{R_1}$

the current through R_2 is $I_2 = \dfrac{V}{R_2}$

the current through R_3 is $I_3 = \dfrac{V}{R_3}$

(Equation 4)

We are looking for an expression for the equivalent resistance R_T in any *parallel* circuit and, if we substitute equations (4) into equation (3), we have:

$$I_T = I_1 + I_2 + I_3$$

$$\therefore \frac{V}{R_T} = \frac{V}{R_1} + \frac{V}{R_2} + \frac{V}{R_3}$$

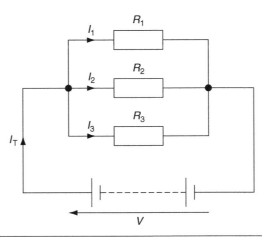

Figure 6.74 A parallel circuit.

Now, since V is common to all terms in the equation, we can divide both sides by V, leaving us with an expression for the circuit resistance:

$$\frac{1}{R_T} = \frac{1}{R_1} + \frac{1}{R_2} + \frac{1}{R_3}$$

Note that the derivation of this formula is given for information only. Craft students need only state the expression $1/R_T = 1/R_1 + 1/R_2 + 1/R_3$ for parallel connections.

Example 1

Three $6\,\Omega$ resistors are connected (a) in series (see Fig. 6.75), and (b) in parallel (see Fig. 6.76), across a $12\,V$ battery. For each method of connection, find the total resistance and the values of all currents and voltages.

For any series connection

$$R_T = R_1 + R_2 + R_3$$
$$\therefore R_T = 6\,\Omega + 6\,\Omega + 6\,\Omega = 18\,\Omega$$

(Continued)

Example 1 (Continued)

Total current $I_T = \dfrac{V_T}{R_T}$

$$\therefore I_T = \dfrac{12V}{18\Omega} = 0.67A$$

The voltage drop across R_1 is

$$V_1 = I_T \times R_1$$
$$\therefore V_1 = 0.67A \times 6\Omega = 4V$$

The voltage drop across R_2 is

$$V_2 = I_T \times R_2$$
$$\therefore V_2 = 0.67A \times 6\Omega = 4V$$

The voltage drop across R_3 is

$$V_3 = I_T \times R_3$$
$$\therefore V_3 = 0.67A \times 6\Omega = 4V$$

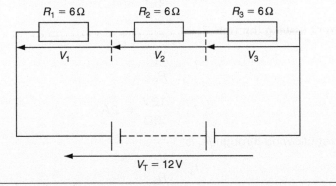

Figure 6.75 Resistors in series.

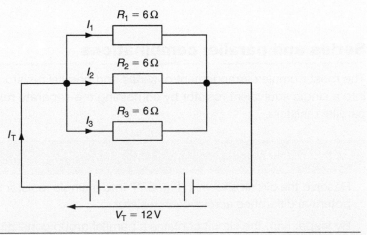

Figure 6.76 Resistors in parallel.

(Continued)

Example 1 (Continued)

For any parallel connection

$$\frac{1}{R_T} = \frac{1}{R_1} + \frac{1}{R_2} + \frac{1}{R_3}$$

$$\therefore \frac{1}{R_T} = \frac{1}{6\Omega} + \frac{1}{6\Omega} + \frac{1}{6\Omega}$$

$$\frac{1}{R_T} = \frac{1+1+1}{6\Omega} = \frac{3}{6\Omega}$$

$$R_T = \frac{6\Omega}{3} = 2\Omega$$

Total current $I_T = \dfrac{V_T}{R_T}$

$$\therefore I_T = \frac{12V}{2\Omega} = 6A$$

The current flowing through R_1 is

$$I_1 = \frac{V_T}{R_1}$$

$$\therefore I_1 = \frac{12V}{6\Omega} = 2A$$

The current flowing through R_2 is

$$I_2 = \frac{V_T}{R_2}$$

$$\therefore I_2 = \frac{12V}{6\Omega} = 2A$$

The current flowing through R_3 is

$$I_3 = \frac{V_T}{R_3}$$

$$\therefore I_3 = \frac{12V}{6\Omega} = 2A$$

Series and parallel combinations

The most complex arrangement of series and parallel resistors can be simplified into a single equivalent resistor by combining the separate rules for series and parallel resistors.

Example 2

Resolve the circuit shown in Fig. 6.77 into a single resistor and calculate the potential difference across each resistor.

By inspection, the circuit contains a parallel group consisting of R_3, R_4 and R_5 and a series group consisting of R_1 and R_2 in series with the equivalent resistor for the parallel branch.

(Continued)

Example 2 (Continued)

Consider the parallel group. We will label this group R_P. Then

$$\frac{1}{R_P} = \frac{1}{R_3} + \frac{1}{R_4} + \frac{1}{R_5}$$

$$\frac{1}{R_P} = \frac{1}{2\Omega} + \frac{1}{3\Omega} + \frac{1}{6\Omega}$$

$$\frac{1}{R_P} = \frac{3+2+1}{6\Omega} = \frac{6}{6\Omega}$$

$$R_P = \frac{6\Omega}{6} = 1\Omega$$

Figure 6.77 may now be represented by the more simple equivalent shown in Fig. 6.78.

Since all resistors are now in series,

$$R_T = R_1 + R_2 + R_P$$
$$\therefore R_T = 3\Omega + 6\Omega + 1\Omega = 10\Omega$$

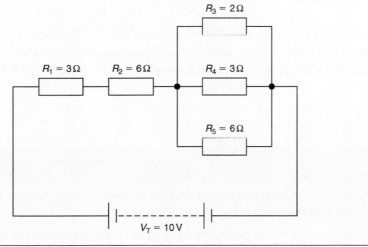

Figure 6.77 A series/parallel circuit.

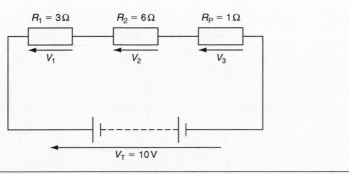

Figure 6.78 Equivalent series circuit.

(Continued)

Example 2 (Continued)

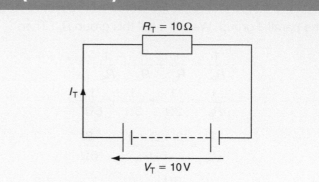

Figure 6.79 Single equivalent resistor for Fig 6.78.

Thus, the circuit may be represented by a single equivalent resistor of value $10\,\Omega$ as shown in Fig. 6.79. The total current flowing in the circuit may be found by using Ohm's law:

$$I_T = \frac{V_T}{R_T} = \frac{10V}{10\Omega} = 1A$$

The potential differences across the individual resistors are

$$V_1 = I_T \times R_1 = 1A \times 3\Omega = 3V$$
$$V_2 = I_T \times R_2 = 1A \times 6\Omega = 6V$$
$$V_P = I_T \times R_P = 1A \times 1\Omega = 1V$$

Since the same voltage acts across all branches of a parallel circuit the same p.d. of 1 V will exist across each resistor in the parallel branch R_3, R_4 and R_5.

Example 3

Determine the total resistance and the current flowing through each resistor for the circuit shown in Fig. 6.80.

By inspection, it can be seen that R_1 and R_2 are connected in series while R_3 is connected in parallel across R_1 and R_2. The circuit may be more easily understood if we redraw it as in Fig. 6.81.

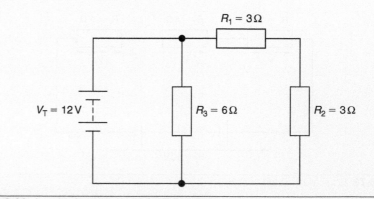

Figure 6.80 A series/parallel circuit for Example 3.

(Continued)

Example 3 (Continued)

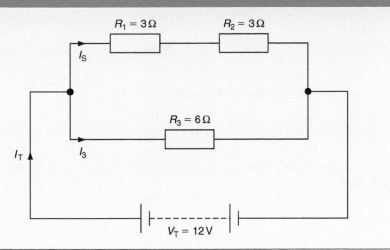

Figure 6.81 Equivalent circuit for Example 3.

For the series branch, the equivalent resistor can be found from

$$R_S = R_1 + R_2$$
$$\therefore R_S = 3\Omega + 3\Omega = 6\Omega$$

Figure 6.81 may now be represented by a more simple equivalent circuit, as in Fig. 6.82.

Since the resistors are now in parallel, the equivalent resistance may be found from

$$\frac{1}{R_T} = \frac{1}{R_S} + \frac{1}{R_3}$$

$$\therefore \frac{1}{R_T} = \frac{1}{6\Omega} + \frac{1}{6\Omega}$$

$$\frac{1}{R_T} = \frac{1+1}{6\Omega} = \frac{2}{6\Omega}$$

$$R_T = \frac{6\Omega}{2} = 3\Omega$$

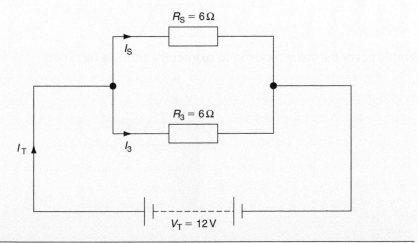

Figure 6.82 Simplified equivalent circuit for Example 2.

(Continued)

Example 3 (Continued)

The total current is

$$I_T = \frac{V_T}{R_T} = \frac{12V}{3\Omega} = 4A$$

Let us call the current flowing through resistor $R_3 I_3$

$$\therefore I_3 = \frac{V_T}{R_3} = \frac{12V}{6\Omega} = 2A$$

Let us call the current flowing through both resistors R_1 and R_2, as shown in Fig. 6.81, I_S

$$\therefore I_S = \frac{V_T}{R_S} = \frac{12V}{6\Omega} = 2A$$

Electrical power

Power and energy are related when consideration is given to the fact that energy is thought of as the amount of work that is being done or being converted and is sometimes classified as horsepower. However, when an electrical current flows in a circuit, heat is generated (energy) and the amount of heat generated depends directly on how much current is flowing (amps), which in turn depends on how much voltage or electrical pressure there is (volts).

This means that we can calculate the rate at which heat is generated through a formula for electrical Power: P = V × I, with the unit for power being represented by the watt, with one horsepower being equivalent to 746 W.

This formula is very important for an electrician since all design calculation start off by calculating how much current an installation requires. Therefore: P = V × I should be thought of as the main formula for Power for an electrician.

An effective method of remembering this formula is to use the power formula triangle

Simply cover the value required to extract the formula required.

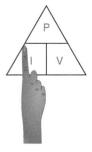

$$I = \frac{P}{V}$$

Another two formulas for Power can be obtained by extracting values from Ohm's law as seen below:

If $V = I \times R$, by transposition we get $I = \dfrac{V}{R}$

Then substitute what I represents $\left(I = \dfrac{V}{R}\right)$ into $P = V \times I$

And you get $P = \dfrac{V \times V}{R}$

Which can be written as $P = \dfrac{V^2}{R}$

Example

Calculate the current demanded by a 60 W lamp when connected to domestic mains supply of 230 V supply.

Use the main formula for power for an electrician

$P = V \times I$

Transpose to find I

$I = \dfrac{P}{V}$

$I = \dfrac{60}{230}$

$I = 0.26\,A$

Resistivity

The resistance or opposition to current flow varies for different materials, each having a particular constant value. If we know the resistance of, say, 1 m of a material, then the resistance of 5 m will be five times the resistance of 1 m. The **resistivity** (symbol ρ – the Greek letter 'rho') of a material is defined as the resistance of a sample of unit length and unit cross-section. Typical values are given in Table 6.8. The resistivity of the conducting material, as well as its length and cross-sectional area are used to determine the resistance of a conductor

Table 6.8 gives the resistivity of silver as $16.4 \times 10^{-9}\,\Omega m$, which means that a sample of silver 1 m long and 1 m in cross-section will have a resistance of $16.4 \times 10^{-9}\,\Omega$.

It is vital that we emphasize the difference between resistance and resistivity.

- Resistance is the opposition to current flow and will include every element within the circuit including the conductor and all the internal connections linking the protective device, busbar, switch and load.
- Resistivity is concerned with the conducting material only and remains constant for a given type of material measured in Ω/m.

The resistivity formula is very important in installation work since it impacts on many aspects. Looking at the formula again

$R = \dfrac{\rho L}{A}$

the length of the conductor is directly proportional to the overall resistance of the circuit.

A = area

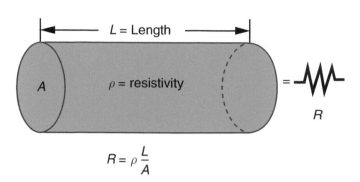

$$R = \rho \frac{L}{A}$$

Figure 6.83 Calculating the resistivity of an object.

Where
    ρ = the resistivity constant for the material (Ωm)
    l = the length of the material (m)
    a = the cross-sectional area of the material (m²)

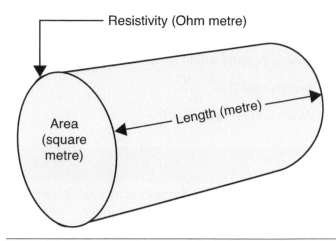

Figure 6.84 As the length of the conductor increases its resistance will increase. Resistivity, however, is based on conductor material therefore remains constant.

Definition

Resistivity is concerned with conductor material.

Directly proportional means that if the length increases so will the resistance.

Therefore, if the length of a circuit is very long, the circuit resistance will also increase. This means that more voltage will be dropped across the cable, depriving the load such as a lamp working voltage. This means that a lamp can appear dim, or motor circuits could run slower than expected. How do we compensate for this?

Again, looking at the formula the cross-sectional area of the conductor is indirectly proportional to the overall resistance of the circuit.

$$R = \frac{\rho L}{A}$$

Indirectly proportional means that if the area increases the resistance will decrease accordingly. Therefore when we encounter volt drop problems we counter this by increasing the cross-sectional area of a cable, in other words: select a bigger size. To summarize the relationship between resistivity and resistance:

Increasing the length *increases* the resistance.
Increasing the area *decreases* the resistance.

Decreasing the length *decreases* the resistance.
Decreasing the area *increases* the resistance.

Doubling the length *doubles* the resistance.
Doubling the area *halves* the resistance.

Calculating resistivity

The general rule that you need to remember when working out any problem with regard to resistivity is to make sure that the units that you use are consistent, i.e. all the units have to be in metres or millimetres. However, whilst cable length and resistivity tend to be represented in metres and Ω/m respectively, cable size is normally represented in mm².

A handy rule to remember when converting millimetres to metres is to attach the magic number × 10⁻⁶. We obtain this number from the fact that 1 metre contains 1000 mm. But cable size is actually cross-sectional area which is measured in m^2 therefore the difference between millimetres and metres can be shown by 6 decimal places or the magic number 10^{-6}. 1.5 mm² would therefore become 1.5×10^{-6} m².

Table 6.8 Resistivity values

Material	Resistivity (Ω m)
Silver	16.4×10^{-9}
Copper	17.5×10^{-9}
Aluminium	28.5×10^{-9}
Brass	75.0×10^{-9}
Iron	100.0×10^{-9}

Example 1

Calculate the resistance of 100 m of copper cable of 1.5 mm² cross-sectional area (csa) if the resistivity of copper is taken as $17.5 \times 10^{-9}\,\Omega$ m.

$$R = \frac{\rho l}{a}\,(\Omega)$$

$$\therefore R = \frac{17.5 \times 10^{-9}\,\Omega\ m \times 100\ m}{1.5 \times 10^{-6}\ m^2} = 1.16\ \Omega$$

Example 2

Calculate the resistance of 100 m of aluminium cable of 1.5 mm² cross-sectional area if the resistivity of aluminium is taken as $28.5 \times 10^{-9}\,\Omega$ m.

$$R = \frac{\rho l}{a}\,(\Omega)$$

$$\therefore R = \frac{28.5 \times 10^{-9}\,\Omega\ m \times 100\ m}{1.5 \times 10^{-6}\ m^2} = 1.9\ \Omega$$

Key fact

Resistance: If the length is doubled, resistance will be doubled. If the csa is doubled, the resistance will be halved.

The above examples show that the resistance of an aluminium cable is some 60% greater than a copper conductor of the same length and cross-section. Therefore, if an aluminium cable is to replace a copper cable, the conductor size must be increased to carry the rated current as given by the tables in Appendix 4 of the IET Regulations and Appendix 6 of the *On-Site Guide*.

The other factor which affects the resistance of a material is the temperature, and we will consider this later.

Try this

Resistance

- Take two 100 m lengths of single cable (2 coils)
- Measure the resistance of 100 m of cable (1 coil)
- Value _____ Ω
- Join the two lengths together (200 m) and again measure the resistance
- Value _____ Ω
- Does this experiment prove resistance is proportional to length?
- If the resistance is doubled, it is proved (QED)! (QED, *quod erat demonstradum*, is the Latin for 'which was to be demonstrated'.)

Assessment criteria 3.2

Explain the relation between resistance, inductance, impedance and capacitance

a.c. circuits

In an a.c. circuit, resistance is only part of the opposition to current flow. The inductance and capacitance of an a.c. circuit also cause an opposition to current flow, which we call *reactance*.

Inductive reactance (X_L) is the opposition to an a.c. current in an inductive circuit. It causes the current in the circuit to lag behind the applied voltage, as shown in Fig. 6.85. It is given by the formula:

$$X_L = 2\pi f L \ (\Omega)$$

Where

$\pi = 3.142$ (a constant)

f = the frequency of the supply

L = the inductance of the circuit, or by:

$$X_L = \frac{V_L}{I_L}$$

Capacitive reactance (X_C) is the opposition to an a.c. current in a capacitive circuit. It causes the current in the circuit to lead ahead of the voltage, as shown in Fig. 6.85. It is given by the formula:

$$X_C = \frac{1}{2\pi f C} \ (\Omega)$$

where π and f are defined as before and C is the capacitance of the circuit. It can also be expressed as:

$$X_C = \frac{V_C}{I_C}$$

Definition

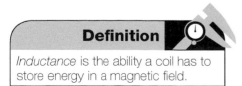

Inductive reactance (X_L) is the opposition to an a.c. current in an inductive circuit. It causes the current in the circuit to lag behind the applied voltage.

Definition

In any circuit, *resistance* is defined as opposition to current flow.

Definition

Capacitive reactance (X_C) is the opposition to an a.c. current in a capacitive circuit. It causes the current in the circuit to lead ahead of the voltage.

Definition

Inductance is the ability a coil has to store energy in a magnetic field.

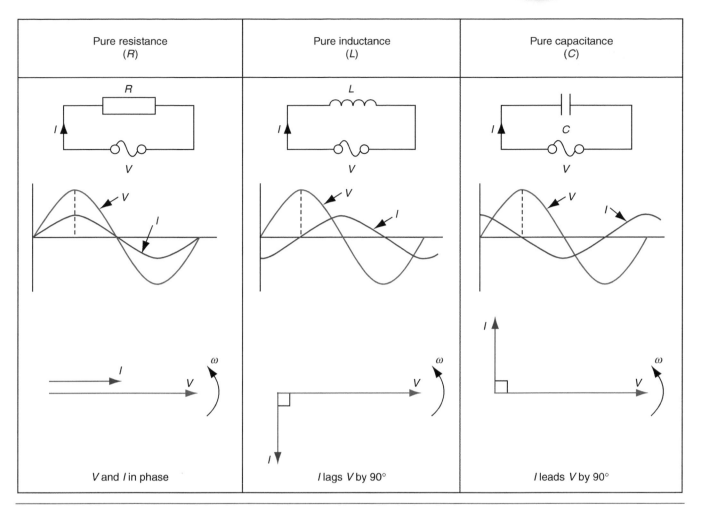

Figure 6.85 Voltage and current relationships in resistive, capacitive and inductive circuits.

Example

Calculate the reactance of a 150 μF capacitor and a 0.05 H inductor if they were separately connected to the 50 Hz mains supply.

For capacitive reactance:

$$X_C = \frac{1}{2\pi fC}$$

where $f = 50$ Hz and $C = 150\,\mu F = 150 \times 10^{-6}$ F

$$\therefore X_C = \frac{1}{2 \times 3.142 \times 50\,\text{Hz} \times 150 \times 10^{-6}\,\text{F}} = 21.2\,\Omega$$

For inductive reactance:

$$X_L = 2\pi fL$$

where $f = 50$ Hz and $L = 0.05$ H

$$\therefore X_L = 2 \times 3.142 \times 50\,\text{Hz} \times 0.05\,\text{H} = 15.7\,\Omega$$

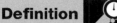

> **Definition**
>
> The total opposition to current flow in an a.c. circuit is called *impedance* and given the symbol Z.

> **Definition**
>
> *Self inductance* is when a coil produces a back e.m.f. which actually opposes the current that produced it.

> **Definition**
>
> *Mutual inductance* is when a rate of change in current in one coil produces an e.m.f in another coil.

Impedance

The total opposition to current flow in an a.c. circuit is called **impedance** and given the symbol Z. Thus impedance is the combined opposition to current flow of the resistance, inductive reactance and capacitive reactance of the circuit and can be calculated from the formula:

$$Z = \sqrt{R^2 + X^2}\,(\Omega)$$

or

$$Z = \frac{V_T}{I_T}$$

Example 1

Calculate the impedance when a $5\,\Omega$ resistor is connected in series with a $12\,\Omega$ inductive reactance.

$$Z = \sqrt{R^2 + X_L^2}\,(\Omega)$$
$$\therefore Z = \sqrt{5^2 + 12^2}$$
$$Z = \sqrt{25 + 144}$$
$$Z = \sqrt{169}$$
$$Z = 13\,\Omega$$

Example 2

Calculate the impedance when a $48\,\Omega$ resistor is connected in series with a $55\,\Omega$ capacitive reactance.

$$Z = \sqrt{R^2 + X_C^2}\,(\Omega)$$
$$\therefore Z = \sqrt{48^2 + 55^2}$$
$$Z = \sqrt{2304 + 3025}$$
$$Z = \sqrt{5329}$$
$$Z = 73\,\Omega$$

Impedance can also be represented when we have a combination of resistance, inductive reactance as well as capacitive reactance. The formula for impedance becomes:

$$Z = \sqrt{R^2 + \left(X_C - X_L\right)^2}$$

Resistance, inductance and capacitance in an a.c. circuit

When a resistor only is connected to an a.c. circuit the current and voltage waveforms remain together, starting and finishing at the same time. We say that the waveforms are *in phase*.

When a pure inductor is connected to an a.c. circuit the current lags behind the voltage waveform by an angle of 90°. We say that the current *lags* the voltage by 90°. When a pure capacitor is connected to an a.c. circuit the current *leads* the voltage by an angle of 90°. These various effects can be observed on an oscilloscope, but the circuit diagram, waveform diagram and phasor diagram for each circuit are shown in Fig. 6.85.

Phasor diagrams

Phasor diagrams and a.c. circuits are an inseparable combination. Phasor diagrams allow us to produce a model or picture of the circuit under consideration, which helps us to understand the circuit. A phasor is a straight line, having definite length and direction, which represents to scale the magnitude and direction of a quantity such as a current, voltage or impedance. To find the combined effect of two quantities we combine their phasors by adding the beginning of the second phasor to the end of the first. The combined effect of the two quantities is shown by the resultant phasor, which is measured from the original zero position to the end of the last phasor.

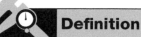

Definition

A *phasor* is a straight line, having definite length and direction, which represents to scale the magnitude and direction of a quantity such as a current, voltage or impedance.

Example

Find by phasor addition the combined effect of currents *A* and *B* acting in a circuit. Current *A* has a value of 4 A, and current *B* a value of 3 A, leading *A* by 90°. We usually assume phasors to rotate anticlockwise and so the complete diagram will be as shown in Fig. 6.86. Choose a scale of, for example, 1 A = 1 cm and draw the phasors to scale; that is, *A* = 4 cm and *B* = 3 cm, leading *A* by 90°.

The magnitude of the resultant phasor can be measured from the phasor diagram and is found to be 5 A acting at a phase angle ϕ of about 37° leading *A*. We therefore say that the combined effect of currents *A* and *B* is a current of 5 A at an angle of 37° leading *A*.

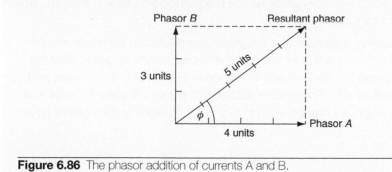

Figure 6.86 The phasor addition of currents A and B.

Phase angle ϕ

In an a.c. circuit containing resistance only, such as a heating circuit, the voltage and current are in phase, which means that they reach their peak and zero values together, as shown in Fig. 6.87a.

In an a.c. circuit containing inductance, such as a motor or discharge lighting circuit, the current often reaches its maximum value after the voltage, which means that the current and voltage are out of phase with each other, as shown in

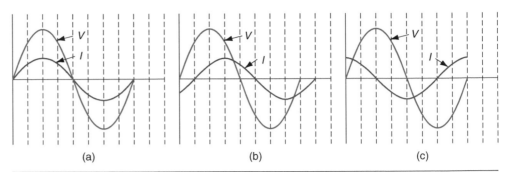

Figure 6.87 Phase relationship of a.c. waveform: (a) *V* and *I* in phase, phase angle $\phi = 0°$ and power factor = cos $\phi = 1$; (b) *V* and *I* displaced by 45°, $\phi = 45°$ and p.f. = 0.707; and (c) *V* and *I* displaced by 90°, $\phi = 90°$ and p.f. = 0.

Fig. 66.87(a), the phase difference, measured in degrees between the current and voltage, is called the phase angle of the circuit, and is denoted by the symbol ϕ, the lower-case Greek letter phi. When circuits contain two or more separate elements, such as RL, RC or RLC, the phase angle between the total voltage and total current will be neither 0° nor 90° but will be determined by the relative values of resistance and reactance in the circuit. This is shown by figures 6.87(b) and 6.87(c) respectively whereby the phase angle between applied voltage and current is represented by the angle ϕ.

Alternating current series circuits

In a circuit containing a resistor and inductor connected in series as shown in Fig. 6.88, the current I will flow through the resistor and the inductor causing the voltage V_R to be dropped across the resistor and V_L to be dropped across the inductor. The sum of these voltages will be equal to the total voltage V_T but because this is an a.c. circuit the voltages must be added by phasor addition. The result is shown in Fig. 6.88, where V_R is drawn to scale and in phase with the current and V_L is drawn to scale and leading the current by 90°. The phasor addition of these two voltages gives us the magnitude and direction of VT, which leads the current by some angle ϕ.

In a circuit containing a resistor and capacitor connected in series as shown in Fig. 6.89, the current *I* will flow through the resistor and capacitor causing voltage drops V_R and V_C. The voltage V_R will be in phase with the current and V_C will lag the current by 90°. The phasor addition of these voltages is equal to the total voltage V_T which, as can be seen in Fig. 6.89, is lagging the current by an angle ϕ.

Definition

An *inductor* causes the circuit current to lag the circuit voltage.

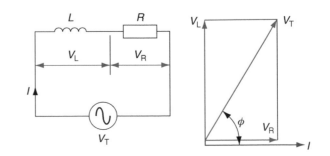

Figure 6.88 A series RL circuit and phasor diagram.

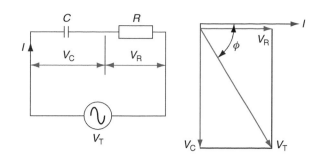

Figure 6.89 A series RC circuit and phasor diagram.

The impedance triangle

We have now established the general shape of the phasor diagram for a series a.c. circuit. Figures 6.88 and 6.89 show the voltage phasors, but we know that $V_R = IR$, $V_L = IX_L$, $V_C = IX_C$ and $V_T = IZ$, and therefore the phasor diagrams (a) and (b) of Fig. 6.90 must be equal. From Figure 6.90(b), by the theorem of Pythagoras, we have:

$$(IZ)^2 = (IR)^2 + (IX)^2$$
$$I^2Z^2 = I^2R^2 + I^2X^2$$

If we now divide throughout by I^2 we have:

$$Z^2 = R^2 + X^2$$
$$\text{or } Z = \sqrt{R^2 + X^2} \ \Omega$$

The phasor diagram can be simplified to the impedance triangle given in Fig. 6.90(c).

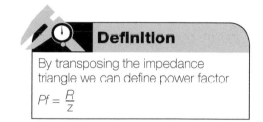

Definition

By transposing the impedance triangle we can define power factor

$$Pf = \frac{R}{Z}$$

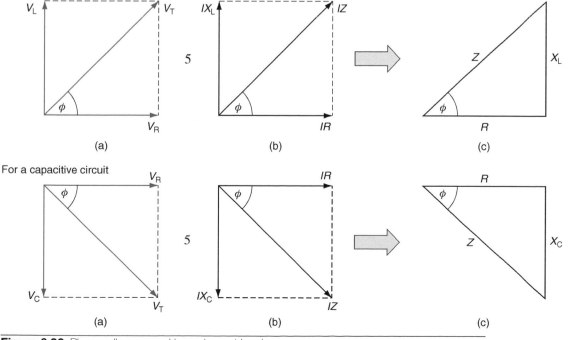

Figure 6.90 Phasor diagram and impedance triangle.

Example 1

A coil of 0.15 H is connected in series with a 50 Ω resistor across a 100 V 50 Hz supply. Calculate (a) the reactance of the coil, (b) the impedance of the circuit, and (c) the current.

For (a)

$$X_L = 2\pi fL \ (\Omega)$$
$$\therefore X_L = 2 \times 3.142 \times 50\text{Hz} \times 0.15\text{H} = 47.1\Omega$$

For (b)

$$Z = \sqrt{R^2 + X^2} \ (\Omega)$$
$$\therefore Z = \sqrt{(50\,\Omega)^2 + (47.1\,\Omega)^2} = 68.69\,\Omega$$

For (c)

$$I = \frac{V}{Z}\,(\text{A})$$
$$\therefore I = \frac{100\,V}{68.69\,\Omega} = 1.46\,\text{A}$$

Example 2

A 60 μF capacitor is connected in series with a 100 Ω resistor across a 230 V 50 Hz supply. Calculate (a) the reactance of the capacitor, (b) the impedance of the circuit and (c) the current.

For (a)

$$X_c = \frac{1}{2\pi fC}\ (\Omega)$$
$$\therefore X_c \frac{1}{2\pi \times 50\text{Hz} \times 60 \times 10^{-6}\,F} = 53.05\,\Omega$$

For (b)

$$Z = \sqrt{Z^2 + X^2}\ (\Omega)$$
$$\therefore Z = \sqrt{(100\,\Omega)^2 + (53.05\,\Omega)^2} = 113.2\,\Omega$$

For (c)

$$I = \frac{V}{Z}\,(\text{A})$$
$$\therefore I = \frac{230\,V}{113.2\,\Omega} = 2.03\,\text{A}$$

Assessment criteria 3.3

Determine power quantities in alternating current circuits

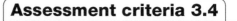

Assessment criteria 3.4

Explain the relation between power quantities in alternating current circuits

Assessment criteria 3.5

Calculate power factor

Power and power factor

Power factor (p.f.) is defined as the cosine of the phase angle between the current and voltage:

$$\text{p.f.} = \cos \phi$$

If the current lags the voltage as shown in Fig. 6.88, we say that the p.f. is lagging, and if the current leads the voltage as shown in Fig. 6.89, the p.f. is said to be leading. From the trigonometry of the impedance triangle shown in Fig. 6.90, p.f. is also equal to:

$$\text{p.f.} = \cos \phi = \frac{R}{Z} = \frac{V_R}{V_T}$$

The electrical power in a circuit is the product of the instantaneous values of the voltage and current. Figure 6.91 shows the voltage and current waveform for a pure inductor and pure capacitor. The power waveform is obtained from the product of *V* and *I* at every instant in the cycle. It can be seen that the power waveform reverses every quarter cycle, indicating that energy is alternately being fed into and taken out of the inductor and capacitor. When considered over one complete cycle, the positive and negative portions are equal, showing that the average power consumed by a pure inductor or capacitor is zero. This shows that inductors and capacitors store energy during one part of the voltage cycle and feed it back into the supply later in the cycle. Inductors store energy as a magnetic field and capacitors as an electric field.

In an electric circuit more power is taken from the supply than is fed back into it, since some power is dissipated by the resistance of the circuit, and therefore:

$$P = I^2 R \text{ (W)}$$

In any d.c. circuit the power consumed is given by the product of the voltage and current, because in a d.c. circuit voltage and current are in phase. In an a.c. circuit the power consumed is given by the product of the current and that part of the voltage which is in phase with the current. The in-phase component of the voltage is given by $V \cos \varphi$, and so power can also be given by the equation:

$$P = VI \cos \phi \text{ (W)}$$

The power factor of most industrial loads is lagging because the machines and discharge lighting used in industry are mostly inductive. This causes an

Definition

Power factor (p.f.) is defined as the cosine of the phase angle between the current and voltage.

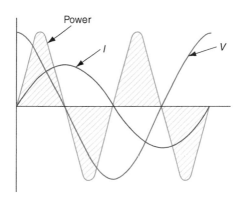

Pure inductor

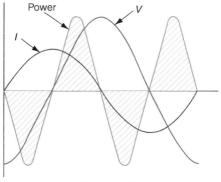

Pure capacitor

Figure 6.91 Waveform for the a.c. power in purely inductive and purely capacitive circuits.

Key fact

If power factor is not corrected then electrical machines draw more current, cables have to be made bigger, the system is very inefficient and in some cases companies can be fined.

Example 1

A coil has a resistance of $30\,\Omega$ and a reactance of $40\,\Omega$ when connected to a $250\,V$ supply. Calculate (a) the impedance, (b) the current, (c) the p.f., and (d) the power.

For (a)

$$Z = \sqrt{R^2 + X^2}\,(\Omega)$$
$$\therefore Z = \sqrt{(30\,\Omega)^2 + (40\,\Omega)^2} = 50\,\Omega$$

For (b)

$$I = \frac{V}{Z}\,(A)$$
$$\therefore I = \frac{250\,V}{50\,\Omega} = 5\,A$$

For (c)

$$\text{p.f.} = \cos\phi = \frac{R}{Z}$$
$$\therefore \text{p.f.} = \frac{30\,\Omega}{50\,\Omega} = 0.6 \text{ lagging}$$

For (d)

$$P = VI\cos\phi\,(W)$$
$$\therefore P = 250\,V \times 5\,A \times 0.6 = 750\,W$$

Example 2

A capacitor of reactance $12\,\Omega$ is connected in series with a $9\,\Omega$ resistor across a $150\,V$ supply. Calculate (a) the impedance of the circuit, (b) the current, (c) the p.f., and (d) the power.

For (a)

$$Z = \sqrt{R^2 + X^2}\,(\Omega)$$
$$\therefore Z = \sqrt{(9\,\Omega)^2 + (12\,\Omega)^2} = 15\,\Omega$$

For (b)

$$I = \frac{V}{Z}\,(A)$$
$$\therefore I = \frac{150\,V}{15\,\Omega} = 10\,A$$

For (c)

$$\text{p.f.} = \cos\phi = \frac{R}{Z}$$
$$\therefore \text{p.f.} = \frac{9\,\Omega}{15\,\Omega} = 0.6 \text{ leading}$$

For (d)

$$P = VI\cos\phi\,(W)$$
$$\therefore P = 150\,V \times 10\,A \times 0.6 = 900\,W$$

additional magnetizing current to be drawn from the supply, which does not produce power, but does need to be supplied, making supply cables larger.

Power factor can also be represented through the power triangle as seen in Fig. 6.92.

Example 3

A 230 V supply feeds three 1.84 kW loads with power factors of 1, 0.8 and 0.4. Calculate the current at each power factor.

The current is given by:

$$I = \frac{P}{V \cos \phi}$$

where $P = 1.84$ kW $= 1840$ W and $V = 230$ V. If the p.f. is 1, then:

$$I = \frac{1840\,\text{W}}{230\,\text{V} \times 1} = 8\,\text{A}$$

For a p.f. of 0.8:

$$I = \frac{1840\,\text{W}}{230\,\text{V} \times 0.8} = 10\,\text{A}$$

For a p.f. of 0.4:

$$I = \frac{1840\,\text{W}}{230\,\text{V} \times 0.4} = 20\,\text{A}$$

It can be seen from these calculations that a 1.84 kW load supplied at a power factor of 0.4 would require a 20 A cable, while the same load at unity power factor could be supplied with an 8 A cable. There may also be the problem of higher voltage drops in the supply cables. As a result, the supply companies encourage installation engineers to improve their power factor to a value close to 1 and sometimes charge penalties if the power factor falls below 0.8.

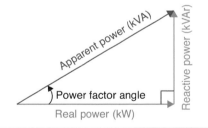

Figure 6.92 From the power triangle we can extract the formula for power factor (pF) = Apparent power / Real power.

As explained previously, due to the inherent nature of inductive loads and their reactive component, a back e.m.f. is created that then acts against the circuit current and causes it to lag the voltage. This produces reactive power and is represented by KVAr, but because the current and voltage are not acting together, the power produced is known as wattles. In other words, a form of useless power. When the power factor operates at unity (1) then these issues are not evident but this is simply not possible since all loads tend to include an inductive component and can be seen in such items as: motors, generators, fluorescent lights and relays etc.

In comparison, true power is also known by many aliases such as real power and active power but the one I prefer is working power measured in Watts since it's what actually powers our equipment and unlike reactive power is very useful.

However, what we actually pay for is KVA apparent power, in other words the combination of working and reactive power, which is shown through the pint of beer analogy since although you pay for a whole pint, the froth is useless.

Figure 6.93 The beer analogy – you pay for the whole pint (kVA), the froth is useless power (kVAr) and the real power is the useful or worthwhile element (kW).

Given that power factor is always represented by cosine, this means we can also write this as an equation.

Whereby the impedance triangle gave us $pf = \dfrac{R}{Z}$

The power triangle gives us $pf = \dfrac{\text{Working (Real) power}}{\text{Apparent power}}$

An example involving the power triangle can be shown below:

If the circuit working power (real power is) 9 kW and the apparent power is 10 kW, calculate the power factor.

$$pf = \frac{\text{Working (Real) power}}{\text{Apparent power}}$$

$$pf = \frac{9000}{10000}$$

$$pf = 0.9$$

Definition

By transposing the power triangle we can define power factor

$$Pf = \frac{\text{Working power}}{\text{Apparent power}}$$

Assessment criteria 3.6

Explain the term power factor correction

Assessment criteria 3.7

Specify the methods of power factor correction

Power factor correction

Most installations have a low or bad power factor because of the inductive nature of the load. A capacitor has the opposite effect of an inductor, and so it seems reasonable to add a capacitor to a load that is known to have a lower or bad power factor, for example, a motor. Referring back to the power triangle, power factor correction is also endeavouring to make apparent power equal to working power by eliminating or reducing the effects of reactive power.

Figure 6.94(a) shows an industrial load with a low power factor. If a capacitor is connected in parallel with the load, the capacitor current I_C leads the applied voltage by 90°. When this capacitor current is added to the load current as shown in Fig.6.94(b) the resultant load current has a much improved power factor. However, using a slightly bigger capacitor, the load current can be pushed up until it is 'in phase' with the voltage as can be seen in Fig. 6.94(c).

Capacitors may be connected across the main bus-bars of industrial loads in order to provide power factor improvement; this is called bulk correction. Alternatively, smaller capacitors may also be connected across an individual piece of equipment, as is the case for fluorescent light fittings.

A further method of correcting power factor is to use synchronous motors, a type of motor used in precision and timing applications. The field and excitation can be adjusted so that the motor effectively behaves likes a capacitor and causes the current to lead and therefore counters the effect of other inductive loads. A disadvantage of using synchronous motors to correct power factor issues is the relatively high cost involved compared with other methods.

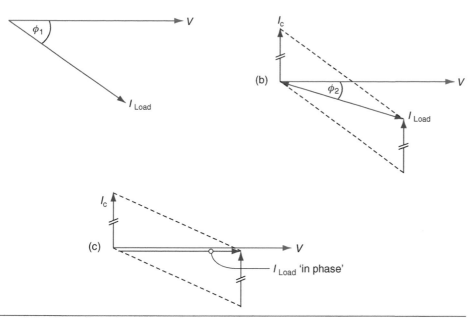

(a)

(b)

(c)

Figure 6.94 Power factor improvement using capacitors.

Assessment criteria 3.8

Determine the neutral current in a three-phase four wire system

Star and delta connections

The three-phase windings of an a.c. generator may be star connected or delta connected, as shown in Fig. 6.95. The important relationship between phase and line currents and voltages is also shown. The square root of 3 ($\sqrt{3}$) is simply a constant for three-phase circuits, and has a value of 1.732. The delta connection is used for electrical power transmission because only three conductors are required. Delta connection is also used to connect the windings of most three-phase motors because the phase windings are perfectly balanced and, therefore, do not require a neutral connection.

Making a star connection at the local substation has the advantage that two voltages become available – a line voltage of 400 V between any two phases, and a phase voltage of 230 V between line and neutral which is connected to the star point.

In any star-connected system currents flow along the lines (I_L), through the load and return by the neutral conductor connected to the star point. In a *balanced* three-phase system all currents have the same value and, when they are added up by phasor addition, we find the resultant current is zero. Therefore, no current flows in the neutral and the star point is at zero volts. The star point of the distribution transformer is earthed because earth is also at zero potential. A star connected system is also called a three-phase four-wire system and allows us to also derive single-phase loads from a three-phase system as shown in Fig 6.95.

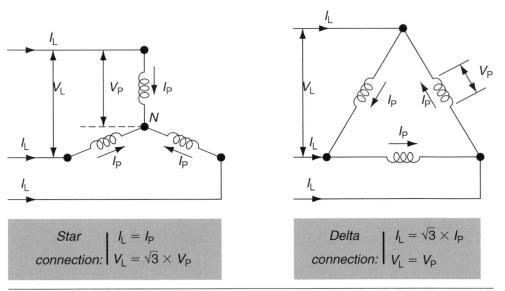

| Star connection: | $I_L = I_P$ |
| | $V_L = \sqrt{3} \times V_P$ |

| Delta connection: | $I_L = \sqrt{3} \times I_P$ |
| | $V_L = V_P$ |

Figure 6.95 Star and delta connections.

Every effort should be made to balance three-phase loads since this ensures that all line conductors can be designed to the same size.

Assessment criteria 3.9

Calculate voltage and current in star and delta connected systems

Three-phase power

We know from our single-phase alternating current theory earlier in this chapter that power can be found from the following formula:

$$\text{Power} = VI \cos\phi \text{ (W)}$$

In any balanced three-phase system, the total power is equal to three times the power in any one phase.

$$\therefore \text{Total three-phase power} = 3V_P I_P \cos\phi \text{ (W)} \qquad \text{(Equation 1)}$$

Now for a star connection:

$$V_P = \frac{V_L}{\sqrt{3}} \quad \text{and} \quad I_L = I_P \qquad \text{(Equation 2)}$$

Substituting equation (2) into equation (1), we have:

$$\text{Total three-phase power} = \sqrt{3}\, V_L I_L \cos\phi \text{ (W)}$$

Now consider a delta connection:

$$V_P = V_L \quad \text{and} \quad I_P = \frac{I_L}{\sqrt{3}} \qquad \text{(Equation 3)}$$

Substituting equation (3) into equation (1) we have, for any balanced three-phase load,

$$\text{Total three-phase power} = \sqrt{3}\ V_L L_L\ \cos\phi\ \text{(W)}$$

So, a general equation for three-phase power is:

$$\text{Power} = \sqrt{3}\ V_L I_L\ \cos\phi$$

Example 1

A balanced star-connected three-phase load of $10\,\Omega$ per phase is supplied from a $400\,\text{V}$, $50\,\text{Hz}$ mains supply at unity power factor. Calculate (a) the phase voltage, (b) the line current and (c) the total power consumed.

For a star connection

$$V_L = \sqrt{3}\ V_P \text{ and } I_L = I_P$$

For (a)

$$V_P = \frac{V_L}{\sqrt{3}}\ \text{(V)}$$

$$V_P = \frac{400\,\text{V}}{1.732} = 230.9\,\text{V}$$

For (b)

$$I_L = I_P = \frac{V_P}{R_P}\ \text{(A)}$$

$$I_L = I_P = \frac{230.9\,\text{V}}{10\,\Omega} = 23.09\,\text{A}$$

For (c)

$$\text{Power} = \sqrt{3}\ V_L\ I_L\ \cos\phi\ \text{(W)}$$

$$\therefore \text{Power} = 1.732 \times 400\,\text{V} \times 23.09\,\text{A} \times 1 = 16\ \text{kW}$$

Example 2

A $20\,\text{kW}$, $400\,\text{V}$ balanced delta-connected load has a power factor of 0.8. Calculate (a) the line current and (b) the phase current.

We have that:

$$\text{Three-phase power} = \sqrt{3}\ V_L\ I_L\ \cos\phi\ \text{(W)}$$

For (a)

$$I_L = \frac{\text{Power}}{\sqrt{3}\ V_L\ \cos\phi}\ \text{(A)}$$

$$\therefore I_L = \frac{20{,}000\,\text{W}}{1.732 \times 400\,\text{V} \times 0.8}$$

$$I_L = 36.08\ \text{(A)}$$

(Continued)

Example 2 (Continued)

For delta connection

$$I_L = \sqrt{3}\, I_P \text{ (A)}$$

Thus, for (b)

$$I_P = I_L / \sqrt{3} \text{ (A)}$$

$$\therefore I_P = \frac{36.08\,\text{A}}{1.732} = 20.83\,\text{A}$$

Example 3

Three identical loads each having a resistance of $30\,\Omega$ and inductive reactance of $40\,\Omega$ are connected first in star and then in delta to a 400 V three-phase supply. Calculate the phase currents and line currents for each connection.

For each load

$$Z = \sqrt{R^2 + X_L^2} \text{ (}\Omega\text{)}$$

$$\therefore Z = \sqrt{30^2 + 40^2}$$

$$Z = \sqrt{2500} = 50\,\Omega$$

For star connection

$$V_L = \sqrt{3}\, V_P \quad \text{and} \quad I_L = I_P$$

$$V_P = \frac{V_L}{\sqrt{3}} \text{ (V)}$$

$$\therefore V_P = \frac{400\,\text{V}}{1.732} = 230.9\,\text{V}$$

$$I_P = \frac{V_P}{Z_P} \text{ (A)}$$

$$\therefore I_P = \frac{230.9\,\text{V}}{50\,\Omega} = 4.62\,\text{A}$$

$$I_P = I_L$$

Therefore phase and line currents are both equal to 4.62 A.

For delta connection

$$V_L = V_P \quad \text{and} \quad I_L = \sqrt{3}\, I_P$$

$$V_L = V_P = 400\,\text{V}$$

$$I_P = V_P / Z_P \text{ (A)}$$

$$\therefore I = \frac{400\,\text{V}}{50\,\Omega} = 8\,\text{A}$$

$$I_L = \sqrt{3}\, I_P \text{ (A)}$$

$$\therefore I_L = 1.732 \times 8\,\text{A} = 13.86\,\text{A}$$

Summary

Single-phase circuits

To calculate power factor in a single phase circuit the formula becomes:

$$P = V \times I \times \cos \phi$$

Remembering, of course, that the voltage represents phase voltage V_p (line to neutral), its value will be 230 V and $\cos \phi$ represents power factor.

Three-phase circuits

To calculate power factor in a three-phase circuit the formula becomes:

$$P = \sqrt{3} \times V \times I \times \cos \phi$$

whereby the $\sqrt{3}$ is a constant value used because in any balanced three-phase system, the total power is equal to three times the power in any one phase.

Remembering, of course, that the voltage represents line voltage (V_L), its value will be 400V and $\cos \phi$ again represents power factor.

Perfect power factor is known as unity: 1

Star connected systems gives us two sets of voltages: 230 V (line to neutral) and 400 V (line to line).

This can be seen through the formula $V_L = \sqrt{3} \times V_p$

The introduction of a neutral is why it's also referred to as a four-wire system. When completely balanced the line and phase currents through the loads are equal ($I_L = I_p$), and the neutral current will be zero.

Delta tends to be used in circumstances where all three loads are manufactured to high tolerance and are matched exactly in value as in, for example, a motor. Voltage does not change in delta ($V_L = V_p$) but the line current $I_L = \sqrt{3} \times I_p$.

Definition

Star configuration produces two sets of voltages: line to neutral = 230 V, line to line = 400 V. Line and phase current, however, does not change.

Definition

In *delta configuration* line and phase voltage does not change but line current is $\sqrt{3} \times$ (times) the phase current.

Assessment criteria 4.1

Understand the principles of the laws of illumination

Lighting laws, lamps and luminaires

In ancient times, much of the indoor work done by humans depended upon daylight being available to light the interior and this is still the case today in many developing nations where communities live beyond the reach of the national grid. Today, life in the wealthy nations carries on after dark where almost all buildings have electric lighting installed. We automatically assume that we can work indoors or out of doors at any time of the day or night, and that light will always be available.

Good lighting is important in all building interiors, helping work to be done efficiently and safely and also playing an important part in creating pleasant and comfortable surroundings.

Figure 6.96 Modern classrooms have better lighting systems than ever before.

Lighting schemes are designed using many different types of light fitting or luminaire. 'Luminaire' is the modern term given to the equipment that supports and surrounds the lamp and may control the distribution of the light. The 18th Edition of the IET Regulations has introduced a new Regulation at 411.3.4. This now requires additional protection within domestic (household) premises for **all circuits supplying luminaires** to be protected by an RCD rated at 30 mA. Modern lamps use the very latest technology to provide illumination cheaply and efficiently. To begin to understand the lamps and lighting technology used today, we must first define some of the terms we will be using.

Luminous intensity – symbol *I*

This is the illuminating power of the light source to radiate luminous flux in a particular direction. The earliest term used for the unit of luminous intensity was the candle power because the early standard was the wax candle. The SI unit is the candela (abbreviated as cd).

Luminous flux – symbol *F*

This is the flow of light that is radiated from a source. The SI unit is the lumen, one lumen being the light flux that is emitted within a unit solid angle (volume of a cone) from a point source of 1 candela.

Illuminance – symbol *E*

This is a measure of the light falling on a surface, which is also called the incident radiation. The SI unit is the lux (lx) and is the illumination produced by 1 lumen over an area of 1 m^2.

Luminance – symbol *L*

Since this is a measure of the brightness of a surface it is also a measure of the light that is reflected from a surface. The objects we see vary in appearance according to the light that they emit or reflect towards the eye. The SI units of luminance vary with the type of surface being considered. For a diffusing surface such as blotting paper or a matt white painted surface the unit of luminance is the lumen per square metre. With polished surfaces such as a silvered glass reflector, the brightness is specified in terms of the light intensity and the unit is the candela per square metre.

Illumination laws

Assessment criteria 4.2

Determine illumination quantities

Rays of light falling upon a surface from some distance d will illuminate that surface with an illuminance of say 1 lx. If the distance d is doubled as shown in Fig. 6.97, the illuminance of 1 lx will fall over four square units of area. Thus the illumination of a surface follows the **inverse square law**, where

$$E = \frac{I}{d^2} \text{(lx)}$$

The illumination of surface A in Fig. 6.98 will follow the inverse square law described above. If this surface were removed, the same luminous flux would then fall on surface B. Since the parallel rays of light falling on the inclined surface B are spread over a larger surface area, the illuminance will be reduced by a factor θ, and therefore:

$$E = \frac{I\cos\theta}{d^2} \text{(lx)}$$

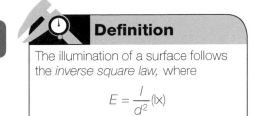

Definition

The illumination of a surface follows the *inverse square law,* where

$$E = \frac{I}{d^2} \text{(lx)}$$

Definition

According to the *inverse square law,* if a lamp's distance to the surface is doubled, then its light intensity will reduce by a factor of 4

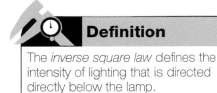

Definition

The *inverse square law* defines the intensity of lighting that is directed directly below the lamp.

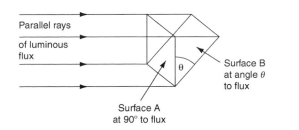

Figure 6.97 The inverse square law.

Figure 6.98 The cosine law.

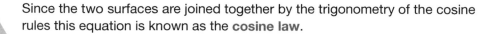

Definition

The *cosine law* is similar to the inverse square law but represents the amount of light that is directed at an angle.

Since the two surfaces are joined together by the trigonometry of the cosine rules this equation is known as the **cosine law**.

Example 1

A lamp of luminous intensity 1000 cd is suspended 2 m above a laboratory bench. Calculate the illuminance directly below the lamp:

$$E = \frac{I}{d^2} \,(\text{lx})$$

$$\therefore E = \frac{1000 \text{ cd}}{(2 \text{ m})^2} = 250 \text{ lx}$$

Example 2

A street lantern suspends a 2000 cd light source 4 m above the ground. Determine the illuminance directly below the lamp and 3 m to one side of the lamp base.

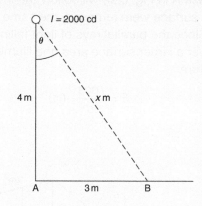

The illuminance below the lamp, E_A, is:

$$E_A = \frac{I}{d^2} \,(\text{lx})$$

$$\therefore E_A = \frac{2000 \text{ cd}}{(4 \text{ m})^2} = 125 \text{ lx}$$

To work out the illuminance at 3 m to one side of the lantern, E_B, we need the distance between the light source and the position on the ground at B; this can be found by Pythagoras' theorem:

$$x \,(\text{m}) = \sqrt{(4\text{m})^2 + (3\text{m})^2} = \sqrt{25\text{m}}$$

$$x = 5\text{m}$$

$$\therefore E_B = \frac{I \cos \theta}{d^2} \,(\text{lx}) \text{ and } \cos \theta = \frac{4}{5}$$

$$\therefore E_B = \frac{2000 \text{ cd} \times 4}{(5 \text{ m})^2 \times 5} = 64 \text{ lx}$$

Example 3

A discharge lamp is suspended from a ceiling 4 m above a bench. The illuminance on the bench below the lamp was 300 lx. Find:

(a) the luminous intensity of the lamp
(b) the distance along the bench where the illuminance falls to 153.6 lx.

For (a)

$$E_A = \frac{I}{d^2} \,(\text{lx})$$

$$\therefore I = E_A d^2 \,(\text{cd})$$

$$I = 300\,\text{lx} \times 16\,\text{m} = 4800\,\text{cd}$$

For (b)

$$E_B = \frac{I}{d^2} \cos\theta \,(\text{lx})$$

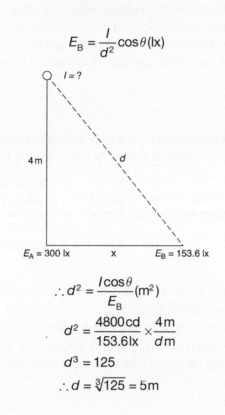

$I = ?$

$4\,\text{m}$

d

$E_A = 300\,\text{lx}$ x $E_B = 153.6\,\text{lx}$

$$\therefore d^2 = \frac{I\cos\theta}{E_B} \,(\text{m}^2)$$

$$d^2 = \frac{4800\,\text{cd}}{153.6\,\text{lx}} \times \frac{4\,\text{m}}{d\,\text{m}}$$

$$d^3 = 125$$

$$\therefore d = \sqrt[3]{125} = 5\,\text{m}$$

By Pythagoras

$$x = \sqrt{5^2 - 4^2} = 3\,\text{m}$$

The recommended levels of illuminance for various types of installation are given by the IES (Illumination Engineers Society). Some examples are given in Table 6.9.

Table 6.9 Illuminance values

Task	Working situation	Illuminance (Lx)
Casual vision	Storage rooms – Stairs – Washrooms	100
Rough Assembly	Workshops – Garages	300
Reading/writing/drawing	Classrooms – Offices	500
Fine assembly	Electronic component assembly	1000
Minute assembly	Watch making	3000

The activities being carried out in a room will determine the levels of illuminance required, since different levels of illumination are required for the successful operation or completion of different tasks. The assembly of electronic components in a factory will require a higher level of illumination than, say, the assembly of engine components in a garage because the electronic components are much smaller and finer detail is required for their successful assembly.

The inverse square law calculations considered earlier are only suitable for designing lighting schemes where there are no reflecting surfaces producing secondary additional illumination. This method could be used to design an outdoor lighting scheme for a cathedral, bridge or public building.

Interior luminaires produce light directly on to the working surface but additionally there is a secondary source of illumination from light reflected from the walls and ceilings. When designing interior lighting schemes the method most frequently used depends upon a determination of the total flux required to provide a given value of illuminance at the working place. This method is generally known as the lumen method.

Lumen method

To determine the total number of luminires required to produce a given illuminance by the lumen method we apply the following formula:

$$\text{Total number of Luminaires required to provide a chosen level of illumination at a surface} = \frac{\text{illuminance level (lx)} \times \text{Area (m}^2\text{)}}{\text{Lumen output of each luminair} \times \text{UF} \times \text{LLF}}$$

Where:

- the level of illuminace at the surface is chosen after consideration of the IES code or the architects specifation;
- the area is the total working area to be illuminated;
- the lumen output of each luminaire is that given in the manufacturer's specification and may be found by reference to manufacturers tables or table 6.10.
- UF is the utilization factor
- LLF is the light loss factor

Utilization factor

The light flux reaching the working plane is always less than the lumen output of the lamp, since some of the light is absorbed by the various surface textures. The method of calculating the utilization factor (UF) is detailed in Chartered Institution of Building Services Engineers (CIBSE) Technical Memorandum No 5, although lighting manufacturers' catalogues give factors for standard conditions. The UF is expressed as a number which is always less than unity; a typical value might be 0.9 for a modern office building.

Light loss factor

The light output of a luminaire is reduced during its life because of an accumulation of dust and dirt on the lamp and fitting. Decorations also deteriorate with time, and this results in more light flux being absorbed by the walls and ceiling.

You can see from Table 6.9 that the output lumens of the lamp decrease with time – for example, a warm white tube gives out 4950 lumens after the first 100 hours of its life but this falls to 4600 lumens after 2000 hours.

The total light loss can be considered under four headings:

1 light loss due to luminaire dirt depreciation (LDD),
2 light loss due to room dirt depreciation (RDD),
3 light loss due to lamp failure factor (LFF),
4 light loss due to lamp lumen depreciation (LLD).

The LLF is the total loss due to these four separate factors and typically has a value between 0.8 and 0.9.

When using the LLF in lumen method calculations we always use the manufacturer's initial lamp lumens for the particular lamp because the LLF takes account of the depreciation in lumen output with time. Let us now consider a calculation using the lumen method.

Table 6.10 Characteristics of a Thorn Lighting 1500 mm 65 W bi-pin tube

Tube colour	Initial lamp lumens*	Lighting design lumens†	Colour rendering quality	Colour appearance
Artificial daylight	2600	2100	Excellent	Cool
De luxe natural	2900	2500	Very Good	Intermediate
De luxe warm white	3500	3200	Good	Warm
Natural	3700	3400	Good	Intermediate
Daylight	4800	4450	Fair	Cool
Warm white	4950	4600	Fair	Warm
White	5100	4750	Fair	Warm
Red	250*	250	Poor	Deep red

Coloured tubes are intended for decorative purposes only
*The initial lumens are the measured lumens after 100 hours of life
†The lighting design lumens are the output lumens after 2000 hours of life

Burning position	Lamp may be operated in any position
Rated life	7500 hours
Efficacy	30–70 l m/W depending upon the tube colour

Example

It is proposed to illuminate an electronic workshop of dimensions $9 \times 8 \times 3\,m$ to an illuminance of 550 lx at the bench level. The specification calls for luminaires having one 1500 mm 65 W natural tube with an initial output of 3700 lumens (see Table 6.8). Determine the number of luminaires required for this installation when the UF and LLF are 0.9 and 0.8, respectively.

$$\text{The number of luminaires required} = \frac{E\ (\text{lx}) \times \text{area}\ (\text{m}^2)}{\text{lumens from each luminaire} \times \text{UF} \times \text{LLF}}$$

$$\text{The number of luminaires} = \frac{550\ (\text{lx}) \times 9\,m \times 8\,m}{3700 \times 0.9 \times 0.8} = 14.86$$

Therefore 15 luminaires will be required to illuminate this workshop to a level of 550 lx. The electrical designer will set out the positions of the 15 luminaires as is most appropriate for the room size and layout of the work areas.

Assessment criteria 4.3

Explain the operating principles of luminaires

Assessment criteria 4.4

Distinguish applications for luminaires

Comparison of light sources

When comparing one light source with another we are interested in the colour-reproducing qualities of the lamp and the efficiency with which the lamp converts electricity into illumination. These qualities are expressed by the lamp's efficacy and colour rendering qualities.

Lamp efficacy

Definition

The performance of a lamp is quoted as a ratio of the number of lumens of light flux that it emits to the electrical energy input that it consumes. Thus *efficacy* is measured in lumens per watt; the greater the efficacy the better the lamp's performance in converting electrical energy into light energy.

The performance of a lamp is quoted as a ratio of the number of lumens of light flux that it emits to the electrical energy input that it consumes. Thus **efficacy** is measured in lumens per watt; the greater the efficacy the better is the lamp's performance in converting electrical energy into light energy.

A general lighting service (GLS) lamp, for example, has an efficacy of 14 lumens per watt, while a fluorescent tube, which is much more efficient at converting electricity into light, has an efficacy of about 50 lumens per watt.

Colour rendering

We recognize various materials and surfaces as having a particular colour because luminous flux of a frequency corresponding to that colour is reflected from the surface to our eye, which is then processed by our brain. White light is made up of the combined frequencies of the colours red, orange, yellow, green, blue, indigo and violet. Colours can only be seen if the lamp supplying the illuminance is emitting light of that particular frequency. The ability to show colours faithfully as they would appear in daylight is a measure of the colour-rendering property of the light source.

GLS lamps

Definition

A *GLS lamp* works by simply heating up an element that becomes white hot. This is why it operates as an incandescent lamp.

GLS lamps produce light as a result of the heating effect of an electrical current. Most of the electricity goes to producing heat and a little to producing light. A fine tungsten wire is first coiled and coiled again to form the incandescent filament of the GLS lamp. The coiled coil arrangement reduces filament cooling and increases the light output by allowing the filament to operate at a higher temperature. The light output covers the visible spectrum, giving a warm white to yellow light with a colour rendering quality classified as fairly good. The efficacy of the GLS lamp is 14 lumens per watt over its intended lifespan of 1000 h. The filament lamp in its simplest form is a purely functional light source that is unchallenged on the domestic market despite the manufacture of more efficient lamps. One factor that may have contributed to its popularity is that lamp designers have been able to modify the glass envelope of the lamp to give a very pleasing decorative appearance, as shown by Fig. 6.99.

Figure 6.99 Some decorative GLS lamp shapes.

Tungsten halogen lamps and the halogen cycle

The high operating temperature of the tungsten filament in the GLS lamp causes some evaporation of the tungsten which is carried by convection currents onto the bulb wall. When the lamp has been in service for some time, evaporated tungsten darkens the bulb wall, the light output is reduced, and the filament becomes thinner and eventually fails.

To overcome these problems the envelope of the tungsten halogen lamp contains a trace of one of the halogen gases, iodine, chlorine, bromine or fluorine. This allows a reversible chemical action to occur between the tungsten filament and the halogen gas.

When tungsten is evaporated from the incandescent filament, some part of it spreads out towards the bulb wall but at a point close to the wall where the temperature conditions are favourable, the tungsten combines with the halogen. This tungsten halide molecule then drifts back towards the filament where it once more separates, depositing the tungsten back onto the filament, thereby leaving the halogen available for a further reaction cycle. Since all the evaporated tungsten is returned to the filament, the bulb blackening normally associated with tungsten lamps is completely eliminated and a high efficacy is maintained throughout the life of the lamp.

A minimum bulb wall temperature of 250°C is required to maintain the halogen cycle and consequently a small glass envelope is required. This also permits a much higher gas pressure to be used, which increases the lamp life to 2,000 hours and allows the filament to be operated at a higher temperature, giving more light. The lamp is very small and produces a very white intense light giving it a colour rendering classification of good and an efficacy of 20 lumens per watt.

The tungsten halogen lamp shown in Fig. 6.100 was a major development in lamp design and resulted in Thorn Lighting gaining the Queens Award for Technical Innovation in Industry in 1972.

Definition

Tungsten halogen lamps use the re-generation effect to prolong life.

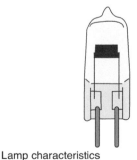

Lamp characteristics

Watts	Lighting design lumens
300	5000 at 230 V
500	9500 at 230 V
Burning position	Linear lamps must be operated horizontally or within 4° of the horizontal
Rated life	2000 hours
Efficacy	20 lm/W
Colour rendering	Good

Figure 6.100 A tungsten halogen lamp.

When installing the lamp, grease contamination of the glass envelope by touching must be avoided. Any grease present on the outer surface will cause cracking and premature failure of the lamp because of the high operating temperatures. The lamp should be installed using the paper sleeve and if accidentally touched with bare hands, the lamp should be cleaned with methylated spirit to remove the grease.

Tungsten halogen dichroic reflector miniature spot lamps

Figure 6.101 Tungsten halogen dichroic lamp.

Tungsten halogen dichroic reflector miniature spot lamps such as the one shown in Fig. 6.101 are extremely popular in the lighting schemes of the new millennium.

Their small size and bright white illumination makes them very popular in both commercial and domestic installations. They are available as a 12 V bi-pin package in 20, 35 and 50 W and as a 230 V bayonet type cap (called a GU10 or GZ10 cap) in 20, 35 and 50 W. At 20 lumens of light output over its intended lifespan of 2,000 h they are more energy efficient than GLS lamps. However, only lamps offering more than 40 lumens of light output are considered energy efficient by the government's new criteria.

Discharge lamps

Definition

Discharge lamps do not produce light by means of an incandescent filament but by the excitation of a gas or metallic vapour contained within a glass envelope.

Discharge lamps do not produce light by means of an incandescent filament but by the excitation of a gas or metallic vapour contained within a glass envelope. A voltage applied to two terminals or electrodes sealed into the end of a glass tube containing a gas or metallic vapour will excite the contents and produce light directly. The colour of the light produced depends upon the type of gas or metallic vapour contained within the tube. Some examples are:

Gases	neon	red
	argon	green/blue
	hydrogen	pink
	helium	ivory
	mercury	blue
Metallic vapours	sodium	yellow
	magnesium	grass green

Fluorescent tubes and CFLs operate on this principle.

Fluorescent luminaires and lamps

Definition

A *fluorescent lamp* produces light by passing a current through a gas.

A **luminaire** is equipment that supports an electric lamp and distributes or filters the light created by the lamp. It is essentially the 'light fitting'. A lamp is a device for converting electrical energy into light energy. There are many types of lamps. General lighting service (GLS) lamps and tungsten halogen lamps use a very hot wire filament to create the light and so they also become very hot in use. Fluorescent tubes operate on the 'discharge' principle; that is, the excitation of a gas within a glass tube. They are cooler in operation and very efficient in converting electricity into light. They form the basic principle of most energy-efficient lamps.

Fluorescent lamps are linear arc tubes, internally coated with a fluorescent powder, containing a little low-pressure mercury vapour and argon gas. The lamp construction is shown in Fig. 6.102.

Passing a current through the electrodes of the tube produces a cloud of electrons that ionize the mercury vapour and the argon in the tube, producing invisible ultraviolet light and some blue light. The fluorescent powder on the inside of the glass tube is very sensitive to ultraviolet rays and converts this radiation into visible light.

Fluorescent luminaires require a simple electrical circuit to initiate the ionization of the gas in the tube and a device to control the current once the arc is struck and the lamp is illuminated. Such a circuit is shown in Fig. 6.103.

A typical application for a fluorescent luminaire is in suspended ceiling lighting modules used in many commercial buildings. Energy efficient lamps use electricity much more efficiently.

Compact fluorescent lamps

CFLs are miniature fluorescent lamps designed to replace ordinary GLS lamps. They are available in a variety of shapes and sizes so that they can be fitted into existing light fittings. Figure 6.104 shows three typical shapes. The 'stick' type

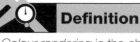
Definition

Colour rendering is the ability of a light source to reproduce natural colours.

Definition

CFLs are miniature fluorescent lamps designed to replace ordinary GLS lamps.

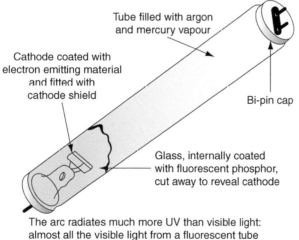

Tube filled with argon and mercury vapour

Cathode coated with electron emitting material and fitted with cathode shield

Bi-pin cap

Glass, internally coated with fluorescent phosphor, cut away to reveal cathode

The arc radiates much more UV than visible light: almost all the visible light from a fluorescent tube comes from the phosphors

Figure 6.102 Fluorescent lamp construction.

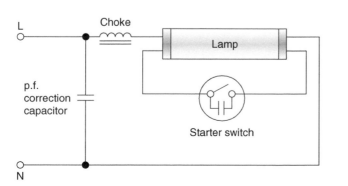

L

Choke

Lamp

p.f. correction capacitor

Starter switch

N

Figure 6.103 Fluorescent lamp circuit arrangement.

Figure 6.104 LED lamp.

give most of their light output radially while the flat 'double D' type give most of their light output above and below. See Figures 6.105 and 6.106.

LED lamps

Light emitting diode lamps have been around for more than fifty years as small signal and indicator lamps, but recent scientific research and new technology have enabled the super efficient LED lamps that we now know, to be created. LED lamps have come such a long way in recent years and are now probably the first choice for energy efficient lamp and luminaire replacement.

A light emitting diode (LED) is a very small, semi-conductor source of illumination. It is a P-N junction diode that emits photons of light when activated, called electroluminence.

LED lamps have many advantages over GLS and compact fluorescent lamps (CFLs) including very low energy consumption, greater robustness, smaller size and longer life. They are available in a range of shapes and sizes with Edison Screw, Bayonet Cap, Bi-Pin, GU 10 and GX 53 connections so that they may easily replace existing lamps.

When installing LED Lamps and luminaires, the installer must take account of the colour of the light emitted by the LED lamp, white or warm white, for a particular

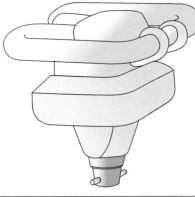
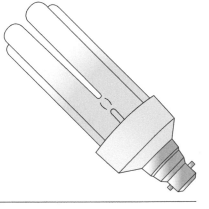

Figure 6.105 Energy-efficient lamps.

Figure 6.106 Energy-efficient lamps use less electricity.

application. The colour of the light emitted by an LED lamp is measured in Kelvin, a temperature measurement. The higher the number, the whiter is the light output. A light output of 3,000K and above gives a bright white light suitable for installation in contemporary kitchens and dentist surgeries. A 2,700K lamp gives a much warmer light, suitable for table lamps with shades in a domestic situation. A 2,700K LED lamp almost replicates the soft yellow/white light output of a GLS filament lamp but consumes much less energy.

The electrical contractor, in discussion with a customer, must balance the advantages and disadvantages of energy-efficient lamps compared to other sources of illumination for each individual installation.

High-pressure mercury vapour lamp

The high-pressure mercury discharge takes place in a quartz glass arc tube contained within an outer bulb which, in the case of the lamp classified as MBF, is internally coated with fluorescent powder. The lamp's construction and characteristics are shown in Fig. 6.107.

The inner discharge tube contains the mercury vapour and a small amount of argon gas to assist starting. The main electrodes are positioned at either end of the tube and a starting electrode is positioned close to one main electrode.

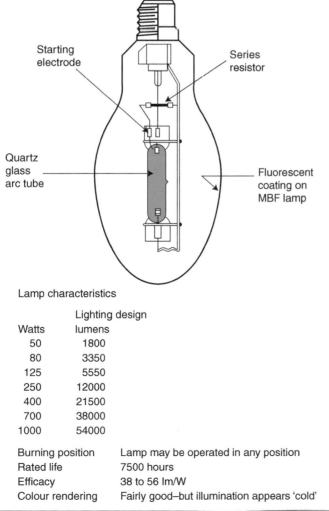

Lamp characteristics

Watts	Lighting design lumens
50	1800
80	3350
125	5550
250	12000
400	21500
700	38000
1000	54000

Burning position	Lamp may be operated in any position
Rated life	7500 hours
Efficacy	38 to 56 lm/W
Colour rendering	Fairly good–but illumination appears 'cold'

Figure 6.107 High-pressure mercury vapour lamp.

When the supply is switched on the current is insufficient to initiate a discharge between the main electrodes, but ionization does occur between the starting electrode and one main electrode in the argon gas.

This spreads through the arc tube to the other main electrode. As the lamp warms the mercury is vaporized, the pressure builds up and the lamp achieves full brilliance after about five to seven minutes.

If the supply is switched off the lamp cannot be relit until the pressure in the arc tube has reduced. It may take a further five minutes to restrike the lamp.

The lamp is used for commercial and industrial installations, street lighting, shopping centre illumination and area floodlighting.

Metal halide lamps

Metal halide lamps are high-pressure mercury vapour lamps in which metal halide chemical compounds have been added to the arc tube. This improves the colour-rendering properties of the lamp making it a better artificial light source for photography.

Low-pressure sodium lamps

Definition

A *low-pressure sodium* is not good at reflecting natural colour (poor colour rendering) due to operating at one predominant frequency which is shown through its colour (yellow). It is very efficient, however, and is used in street and motorway lighting.

The low-pressure sodium discharge takes place in a U-shaped arc tube made of special glass which is resistant to sodium attack. This U-tube is encased in a tubular outer bulb of clear glass as shown in Fig. 6.108. Lamps classified as type SOX have a BC lampholder while the SL1/H lamp has a bi-pin lampholder at each end.

Since at room temperature the pressure of sodium is very low, a discharge cannot be initiated in sodium vapour alone. Therefore, the arc tube also contains neon gas to start the lamp. The arc path of the low-pressure sodium lamp is much longer than that of mercury lamps and starting is achieved by imposing a high voltage equal to about twice the main voltage across the electrodes by means of a leakage transformer. This voltage initiates a red discharge in the neon gas which heats up the sodium. The sodium vaporizes and over a period of six to 11 minutes the lamp reaches full brilliance, changing colour from red to bright yellow.

The lamp must be operated horizontally so that when the lamp is switched off the condensing sodium is evenly distributed around the U-tube.

The light output is yellow and has poor colour-rendering properties but this is compensated by the fact that the wavelength of the light is close to that at which the human eye has its maximum sensitivity, giving the lamp a high efficacy. The main application for this lamp is street lighting where the light output meets the requirements of the Ministry of Transport.

High-pressure sodium lamp

The high-pressure sodium discharge takes place in a sintered aluminium oxide arc tube contained within a hard glass outer bulb. Until recently no suitable material was available which would withstand the extreme chemical activity of sodium at high pressure. The construction and characteristics of the high-pressure sodium lamp classified as type SON are given in Fig. 6.109.

The arc tube contains sodium and a small amount of argon or xenon to assist starting. When the lamp is switched on an electronic pulse igniter of 2 kV or more initiates a discharge in the starter gas. This heats up the sodium and in about

SOX lamp　　　　SLI/H lamp

Lamp characteristics

Watts	Lighting design lumens
Type SOX	
35	4300
55	7500
90	12500
135	21500
Type SLI/H	
140	20000
200	25000
200 HO	27500
Burning position	Horizontal or within 20° of the horizontal
Rated life	6000 hours
Guaranteed life	4000 hours
Efficacy	61 to 160 lm/W
Colour rendering	Very poor – illumination very yellow

Figure 6.108 Low-pressure sodium lamp.

five to seven minutes the sodium vaporizes and the lamp achieves full brilliance. Both colour and efficacy improve as the pressure of the sodium rises, giving a pleasant golden white colour to the light which is classified as having a fair colour-rendering quality.

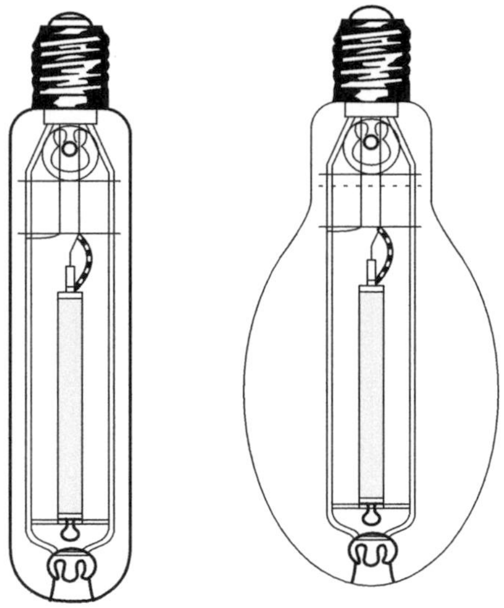

| | SON/T | SON |

Lamp characteristics

	Lighting design lumens
Watts	
Tubular clear (SON/T)	
250	21000
400	38000
Elliptical coated (SON)	
250	19500
400	36000
Burning position	Universal
Rated life	6000 hours
Guaranteed life	4000 hours
Efficacy	100 to 120 lm/W
Colour rendering	Fair – illumination appears 'warm' and golden

Figure 6.109 High-pressure sodium lamp.

The SON lamp is suitable for many applications. Because of the warming glow of the illuminance it is used in food halls and hotel reception areas. Also, because of the high efficacy and long lamp life it is used for high bay lighting in factories and warehouses and for area floodlighting at airports, car parks and dockyards.

Control gear for lamps

Luminaires are wired using the standard lighting circuits, but discharge lamps require additional control gear and circuitry for their efficient and safe operation. The circuit diagrams for high-pressure mercury vapour and high- and low-pressure sodium lamps are given in Fig. 6.110. Each of these circuits requires the inclusion of a choke or transformer, creating a lagging power factor which must be corrected. This is usually achieved by connecting a capacitor across the supply to the luminaire as shown.

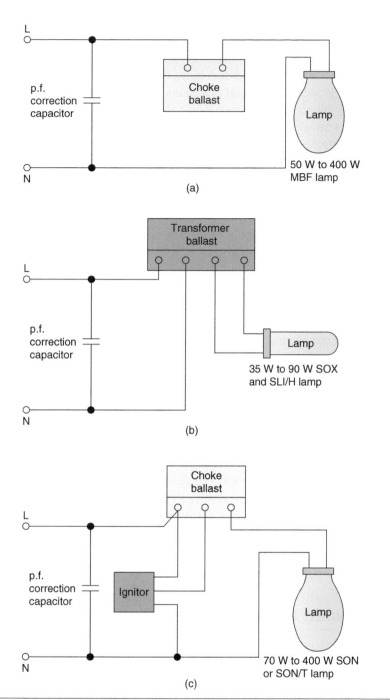

Figure 6.110 Discharge lamp control gear circuits: (a) high-pressure mercury vapour lamp; (b) low-pressure sodium lamp; (c) high-pressure sodium-lamp.

Fluorescent lamp control circuits

A fluorescent lamp requires some means of initiating the discharge in the tube, and a device to control the current once the arc is struck. Since the lamps are usually operated on a.c. supplies, these functions are usually achieved by means of a choke ballast. Three basic circuits are commonly used to achieve starting – switch-start, quick-start and semi-resonant.

Switch-start fluorescent lamp circuit

Figure 6.111 shows a switch-start fluorescent lamp circuit in which a glow-type starter switch is now standard. A glow-type starter switch consists of two bimetallic strip electrodes encased in a glass bulb containing an inert gas. The starter switch is initially open-circuit. When the supply is switched on the full mains voltage is developed across these contacts and a glow discharge takes place between them. This warms the switch electrodes and they bend towards each other until the switch makes contact. This allows current to flow through the lamp electrodes, which become heated so that a cloud of electrons is formed at each end of the tube, which in turn glows.

When the contacts in the starter switch are made the glow discharge between the contacts is extinguished since no voltage is developed across the switch. The starter switch contacts cool and after a few seconds spring apart. Since there is a choke in series with the lamp, the breaking of this inductive circuit causes a voltage surge across the lamp electrodes which is sufficient to strike the main arc in the tube. If the lamp does not strike first time the process is repeated. When the main arc has been struck in the low-pressure mercury vapour, the current is limited by the choke. The capacitor across the mains supply provides power-factor correction and the capacitor across the starter switch contact is for radio interference suppression.

Quick-start fluorescent lamp circuit

When the circuit is switched on the tube cathodes are heated by a small auto-transformer. After a short pre-heating period the mercury vapour is ionized and spreads rapidly through the tube to strike the arc. The luminaire or some earthed metal must be in close proximity to the lamp to assist in the striking of the main arc.

When the main arc has been struck the current flowing in the circuit is limited by the choke. A capacitor connected across the supply provides power-factor correction. The circuit is shown in Fig. 6.112.

Semi-resonant start fluorescent lamp circuit

In this circuit a specially wound transformer takes the place of the choke. When the circuit is switched on, a current flows through the primary winding to one cathode of the lamp, through the secondary winding and a large capacitor to the other cathode. The secondary winding is wound in opposition to the primary

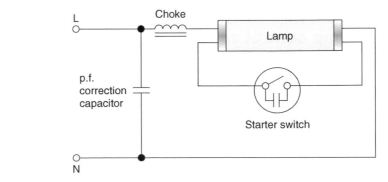

Figure 6.111 Switch-start fluorescent lamp circuit.

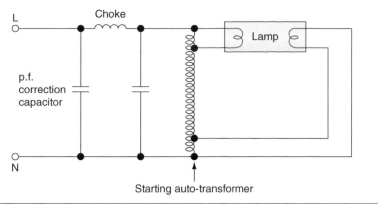

Figure 6.112 Quick-start fluorescent lamp circuit.

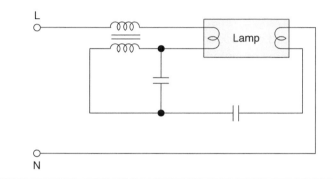

Figure 6.113 Semi-resonant start fluorescent lamp circuit.

winding. Therefore, the voltage developed across the transformer windings is 180° out of phase. The current flowing through the electrodes causes an electron cloud to form around each cathode. This cloud spreads rapidly through the tube due to the voltage across the tube being increased by winding the transformer windings in opposition. When the main arc has been struck the current is limited by the primary winding of the transformer, which behaves as a choke. A power factor correction capacitor is not necessary since the circuit is predominantly capacitive and has a high power factor. With the luminaire earthed to assist starting this circuit will start very easily at temperatures as low as –5°C. The circuit is shown in Fig. 6.113.

Installation of discharge luminaires

Operating position

Some lamps, particularly discharge lamps, have limitations placed upon their operating position. Since the luminaire is designed to support the lamp, any restrictions upon the operating position of the lamp will affect the position of the luminaire. Some indications of the operating position of lamps were given earlier under the individual lamp characteristics. The luminaire must be suitable for the environment in which it is to operate. This may be a corrosive atmosphere, an outdoor situation or a low-temperature zone. On the other hand, the luminaire may be required to look attractive in a commercial environment. It must satisfy all these requirements while at the same time providing adequate illumination without glare.

Many lamps contain a wire filament and delicate supports enclosed in a glass envelope. The luminaire is designed to give adequate support to the lamp under normal conditions, but a luminaire subjected to excessive vibration will encourage the lamp to fail prematurely by either breaking the filament, cracking the glass envelope or breaking the lamp holder seal.

Control gear for discharge luminaires

Chokes and ballasts for discharge lamps have laminated sheet steel cores in which a constant reversal of the magnetic field due to the a.c. supply sets up vibrations. In most standard chokes the noise level is extremely low, and those manufactured to BS EN 60598 have a maximum permitted noise level of 30 dB. This noise level is about equal to the sound produced by a Swatch watch at a distance of one metre in a very quiet room.

Chokes must be rigidly fixed, otherwise metal fittings can amplify choke noise. Plasterboard, hardboard or wooden panels can also act as a sounding board for control gear or luminaires mounted upon them, thereby amplifying choke noise. The background noise will obviously affect people's ability to detect choke noise, and so control gear and luminaires that would be considered noisy in a library or church may be unnoticeable in a busy shop or office. The cable, accessories and fixing box must be suitable for the mass suspended and the ambient temperature in accordance with IET Regulations Section 559. Self-contained luminaires must have an adjacent means of isolation provided in addition to the functional switch to facilitate safe maintenance and repair (Regulation 537.2). Control gear should be mounted as close as possible to the lamp. Where it is liable to cause overheating it must be:

- enclosed in a suitably designed non-combustible enclosure or;
- mounted so as to allow heat to dissipate or;
- placed at a sufficient distance from adjacent materials to prevent the risk of fire (Regulations 559.4.1).

Discharge lighting may also cause a stroboscopic effect where rotating or reciprocating machinery is being used. This effect causes rotating machinery to appear stationary, and we will consider the elimination of this dangerous effect next.

Stroboscopic effect (IET Regulation 559.9)

The flicker effect of any light source can lead to the risk of a stroboscopic effect. This causes rotating or reciprocating machinery to appear to be running at speeds other than their actual speed, and in extreme cases a circular saw or lathe chuck may appear stationary when rotating. A stroboscopic light is used to good effect when electronically 'timing' a car, by making the crank shaft appear stationary when the engine is running so that the top dead centre position (TDC) may be found.

All discharge lamps used on a.c. circuits flicker, often unobtrusively, due to the arc being extinguished every half-cycle as the lamp current passes through zero. This variation in current and light output is illustrated in Fig. 6.114. The elimination of this flicker is desirable in all commercial installations and particularly those that use rotating machinery. Fluorescent phosphors with a long afterglow help to eliminate the brightness peaks. Where a three-phase supply is available, discharge lamps may be connected to a different phase, so that they reach their brightness peaks at different times. The combined effect is to produce a much more uniform overall level of illumination, which eliminates the flicker effect.

When only a single-phase supply is available, adjacent fluorescent tubes in twin fittings may be connected to a lead-lag circuit, as illustrated in Fig. 6.115. This is one lamp connected in series with a choke and the other in series with a choke and capacitor, which causes the currents in the two lamps to differ by between 120 and 180 degrees. The lamps flicker out of phase, producing a more uniform level of illumination which eliminates the stroboscopic effect, as shown in Fig. 6.116.

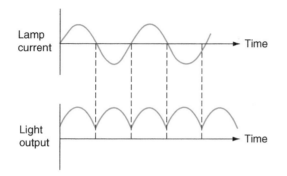

Figure 6.114 Variation of current and light output for a discharge lamp.

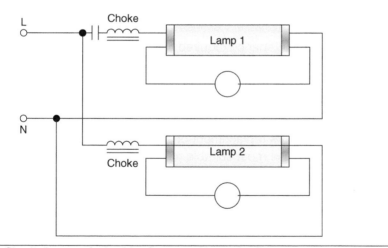

Figure 6.115 Circuit diagrams for lead-lag fluorescent lamp circuits.

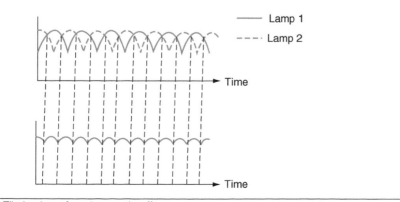

Figure 6.116 Elimination of stroboscopic effect.

GLS lamps do not flicker because the incandescent filament produces a carrying-over of the light output as the current reverses. Thus the light remains almost constant and, therefore, the stroboscopic flicker effect is not apparent.

Loading and switching of discharge circuits

Discharge circuits must be capable of carrying the total steady current (the current required by the lamp plus the current required by any control gear). Appendix 1 of the *On-Site Guide* states that where more exact information is not available, the rating of the final circuits for discharge lamps may be taken as the rated lamp wattage multiplied by 1.8. Therefore, an 80 W fluorescent lamp luminaire will have an assumed demand of 80 × 1.8 = 144 W.

All discharge lighting circuits are inductive and will cause excessive wear of the functional switch contacts. Where discharge lighting circuits are to be switched, the rating of the functional switch must be suitable for breaking an inductive load. To comply with this requirement we usually assume that the rating of the functional switch should be *twice* the total steady current of the inductive circuit from information previously given in the 15th edition of the IET Regulations. The 18th edition of the Regulations at 537.3.1.2 says it must be suitable for the most onerous duty it is intended to perform.

Maintenance of discharge lighting installations

The extent of the maintenance required will depend upon the size and type of lighting installation and the installed conditions – for example, whether the environment is dusty or clean. However, lamps and luminaires will need cleaning, and lamps will need to be replaced. They can be replaced either by 'spot replacement' or 'group replacement'.

Spot replacement is the replacement of individual lamps as and when they fail. This is probably the most suitable method of maintaining small lighting installations in shops, offices and nursing homes, and is certainly the preferred method in domestic property. But each time a lamp is replaced or cleaned, a small disturbance occurs to the normal environment. Access equipment must be set up, furniture must be moved and the electrician chats to the people around him. In some environments this small disturbance and potential hazard from someone working on steps or a mobile tower is not acceptable and, therefore, group replacement of lamps must be considered.

Group replacement is the replacement of all the lamps at the same time. The time interval for lamp replacement is determined by taking into account the manufacturer's rated life for the particular lamp and the number of hours each week that the lamp is illuminated.

Group replacement and cleaning can reduce labour costs and inconvenience to customers and is the preferred method for big stores and major retail outlets. Group replacement in these commercial installations is often carried out at night or at the weekend to avoid a disturbance of the normal working environment.

Assessment criteria 5.1

State the basic types of direct current machines

Electrical machines

Electrical machines are energy converters. If the machine input is mechanical energy and the output electrical energy then that machine is a generator, as shown in Fig. 6.118a. Alternatively, if the machine input is electrical energy and the output mechanical energy then the machine is a motor, as shown in Fig.6.118b. An electrical machine may be used as a motor or a generator, although in practice the machine will operate more efficiently when operated in the mode for which it was designed.

Definition

A *motor* accepts electrical energy in and produces mechanical energy or torque which is defined as a turning force.

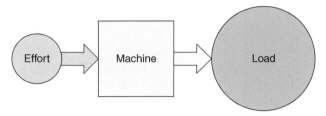

Figure 6.117 Simple machine concept.

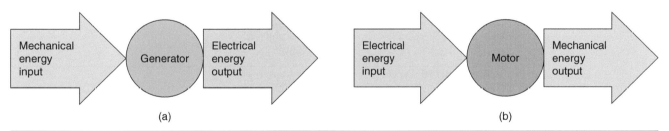

(a)   (b)

Figure 6.118 Electrical machines as energy converters.

Simple a.c. generator or alternator

If a simple loop of wire is rotated between the poles of a permanent magnet, as shown in Fig. 6.119, the loop of wire will cut the lines of magnetic flux between the north and south poles. This flux cutting will induce an electromotive force (e.m.f.) in the wire by **Faraday's law,** which states that *when a conductor cuts or is cut by a magnetic field, an e.m.f. is induced in that conductor*. If the generated e.m.f. is collected by carbon brushes at the slip rings and displayed on the screen of a cathode ray oscilloscope, the waveform will be seen to be approximately sinusoidal. Alternately changing, first positive and then negative, then positive again, gives an alternating output.

Definition

Faraday's law states that *when a conductor cuts or is cut by a magnetic field, an e.m.f. is induced in that conductor.*

Simple d.c. generator or dynamo

If the slip rings of Fig. 6.119 are replaced by a single split ring, called a commutator, the generated e.m.f. will be seen to be in one direction, as shown in Fig. 6.120. The action of the commutator is to reverse the generated e.m.f. every half-cycle, rather like an automatic changeover switch. However, this simple arrangement produces a very bumpy d.c. output. In a practical machine, the commutator would contain many segments and many windings to produce a smoother d.c. output similar to the unidirectional battery supply.

The creation of electricity when a conductor is moved across a magnetic field is known as the generator effect and is shown through Fleming's right hand rule. The formula in use with this rule is e = B L V, whereby the generation of an

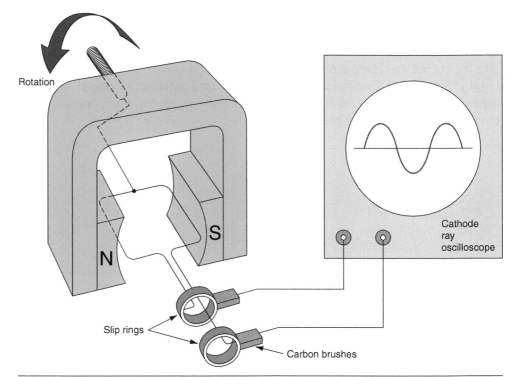

Figure 6.119 Simple a.c. generator or alternator.

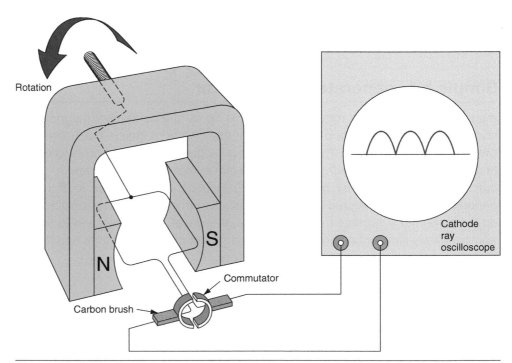

Figure 6.120 Simple d.c. generator or dynamo.

e.m.f. is directly proportional to B (density of the magnetic field), L (length of the conductor) and the velocity that the conductor is cutting the magnetic field.

Fleming's Right Hand Rule (Generator Rule)

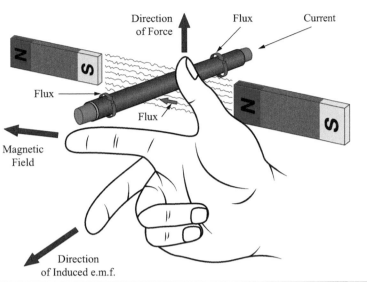

Figure 6.121 Fleming's right hand rule.

Assessment criteria 5.2

Describe the operating principles of direct current machines

Assessment criteria 5.3

State the applications of direct current machines

Direct current motors

All electric motors work on the principle that when a current-carrying conductor is placed in a magnetic field it will experience a force. An electric motor uses this magnetic force to turn the shaft of the electric motor. Let us try to understand this action. If a current-carrying conductor is placed into the field of a permanent magnet as shown in Fig. 6.124(c) a force F will be exerted on the conductor to push it out of the magnetic field.

To understand the force, let us consider each magnetic field acting alone. Figure 6.124(a) shows the magnetic field due to the current carrying conductor only. Figure 6.124 (b) shows the magnetic field due to the permanent magnet in which is placed the conductor carrying no current. Figure 6.124 (c) shows the effect of the combined magnetic fields which are distorted and, because lines of magnetic flux never cross, but behave like stretched elastic bands always trying to find the shorter distance between a north and south pole, the force F is exerted on the conductor, pushing it out of the permanent magnetic field. This is the basic motor principle, and the force F is dependent upon the strength of the magnetic field B, the magnitude of the current flowing in the conductor I and the length of conductor within the magnetic field I. The following equation expresses this relationship:

Fleming's Left Hand Rule (Motor Rule)

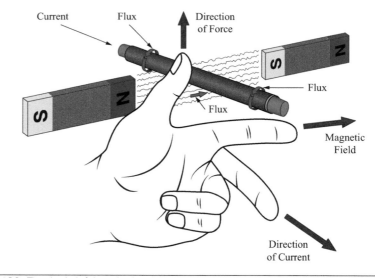

Figure 6.122 Fleming's left hand rule.

$$F = BLI \text{ (N)}$$

where B is in tesla, l is in metres, I is in amperes and F is in newtons.

Example

A coil which is made up of a conductor some 15 m in length lies at right angles to a magnetic field of strength 5 T. Calculate the force on the conductor when 15 A flows in the coil.

$$F = BLI \text{ (N)}$$

$$F = 5T \times 15m \times 15A = 1125N$$

Practical d.c. motors

Practical motors are constructed as shown in Fig. 6.123. All d.c. motors contain a field winding wound on pole pieces attached to a steel yoke. The armature winding rotates between the poles and is connected to the commutator. The commutator is made from copper pieces but interspersed by an insulator such as mica.

Contact with the external circuit is made through carbon brushes rubbing on the commutator segments. Direct current motors are classified by the way in which the field and armature windings are connected, which may be in series or in parallel.

Series motor

The field and armature windings are connected in series and consequently share the same current. The series motor has the characteristics of a high starting torque but a speed that varies with load. Figure 6.125 shows series motor

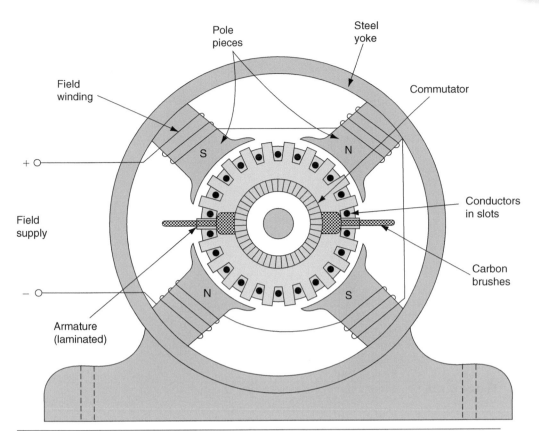

Figure 6.123 Showing d.c. machine construction.

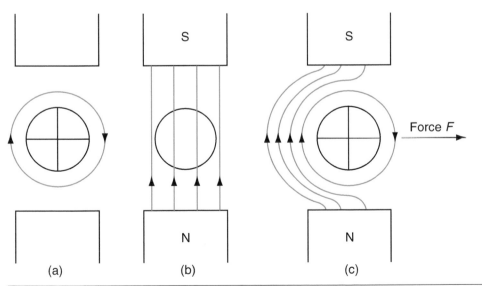

Figure 6.124 Force on a conductor in a magnetic field.

connections and characteristics. For this reason the motor is only suitable for direct coupling to a load, except in very small motors, such as vacuum cleaners and hand drills, and is ideally suited for applications where the machine must start on load, such as electric trains, cranes and hoists.

Reversal of rotation may be achieved by reversing the connections of either the field or armature windings but not both. This characteristic means that the machine will run on both a.c. or d.c. and is, therefore, sometimes referred to as a 'universal' motor.

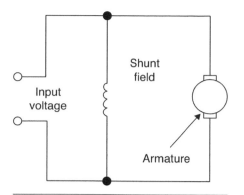

Figure 6.126 The output of a shunt motor will remain constant irrespective of its loading.

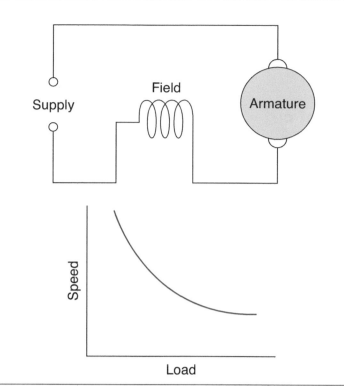

Figure 6.125 Series motor connections and characteristics.

A shunt type d.c. motor is used to maintain machinery that requires constant speed irrespective of the loading such as an assembly line.

Another type is one that combines the properties of both d.c. series and shunt and is known as a d.c. compound motor and is used extensively with elevators. The efficiency of machines such as motors is calculated using the formula seen below, whereby the ratio of the output power to the input power indicates how much energy is lost and is expressed as a %. The symbol for efficiency is the Greek letter 'eta' (η).

$$\eta = \frac{\text{Power output}}{\text{Power input}} \times 100$$

Example

A transformer feeds the 9.81 kW motor driving the mechanical hoist of the previous example. The input power to the transformer was found to be 10.9 kW. Find the efficiency of the transformer.

$$\eta = \frac{\text{Power output}}{\text{Power input}} \times 100$$

$$\eta = \frac{9.81 \text{ kw}}{10.9 \text{ kw}} \times 100 = 90 \text{ \%}$$

Thus the transformer is 90% efficient. Note that efficiency has no units, but is simply expressed as a percentage.

Assessment criteria 5.6

Describe the operating principles of three phase alternating current motors

Assessment criteria 5.7

State the applications of alternating current motors

Three-phase a.c. motors

If a three-phase supply is connected to three separate windings equally distributed around the stationary part or stator of an electrical machine, an alternating current circulates in the coils and establishes a magnetic flux. The magnetic field established by the three-phase currents travels around the stator, establishing a rotating magnetic flux, creating magnetic forces on the rotor, which turns the shaft on the motor.

A good analogy for this would be to think of the rotor as a simple magnetic compass, whereby the compass needle will rotate in unison with the rotating magnetic field, which is set up by the supply current. Any machine that therefore synchronized a magnetic field with the frequency of the supply current is known as a synchronous motor.

An advantage of synchronous motors is that they can be operated with leading power factor, which means that they can be used to counter the lag induced by inductors. Other applications include being used as servo mechanisms and clocks.

Calculating synchronous speed

To calculate synchronous speed we use the formula below and it states that the synchronous speed is directly proportional to the frequency but inversely proportional to the number of pole pairs.

$$\text{Synchronous speed } (n_s) \text{ in revs/second} = \frac{\text{Frequency (f) in Hz}}{\text{Number of pole pairs } (p)}$$

Example 1

If a 100 Hz supply is connected to a four-pole synchronous motor calculate its synchronous speed.

$$n_s = \frac{f}{p}$$

The machine has four poles, which equates to: two pole pairs.

$$n_s = \frac{100}{2}$$

n_s therefore = 50 revolutions per second (rps).

To convert this answer to revolutions per minute multiply the answer by 60.

Remember to divide by 60 when carrying out the reverse calculation: rpm to rps.

Three-phase induction motor

When a three-phase supply is connected to insulated coils set into slots in the inner surface of the stator or stationary part of an induction motor as shown in Fig. 6.128 a rotating magnetic flux is produced. The rotating magnetic flux cuts the conductors of the rotor and induces an e.m.f. in the rotor conductors by Faraday's law, which states that when a conductor cuts or is cut by a magnetic field, an e.m.f. is induced in that conductor, the magnitude of which is proportional to the rate at which the conductor cuts or is cut by the magnetic flux. This induced e.m.f. causes rotor currents to flow and establish a magnetic flux, which reacts with the stator flux and causes a force to be exerted on the rotor conductors, turning the rotor as shown in Fig. 6.127 The turning force or torque experienced by the rotor is produced by inducing an e.m.f. into the rotor conductors due to the *relative* motion between the conductors and the rotating field. The torque produces rotation in the same direction as the rotating magnetic field.

Rotor construction

There are two types of induction motor rotor – the wound rotor and the cage rotor. The cage rotor consists of a laminated cylinder of silicon steel with copper or aluminium bars slotted in holes around the circumference and short circuited at each end of the cylinder as shown in Fig. 6.128. In small motors the rotor is cast in aluminium. Better starting and quieter running are achieved if the bars are slightly skewed. This type of rotor is extremely robust and since there are no external connections there is no need for slip rings or brushes. A machine fitted with a cage rotor does suffer from a low starting torque and a machine must be chosen that has a higher starting torque than the load, as shown by curve (b) in Fig. 6.130. A machine with the characteristic shown by curve (a) in Fig. 6.130 would not start since the load torque is greater than the machine starting torque. Alternatively, the load may be connected after the motor has been run up to full speed.

The wound rotor consists of a laminated cylinder of silicon steel with copper coils embedded in slots around the circumference. The windings may be connected

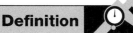

Definition

Cage rotors are three-phase constant speed machines. *Wound rotors* however, are used for variable speed applications whereby resistors are used to start them and then control their speed.

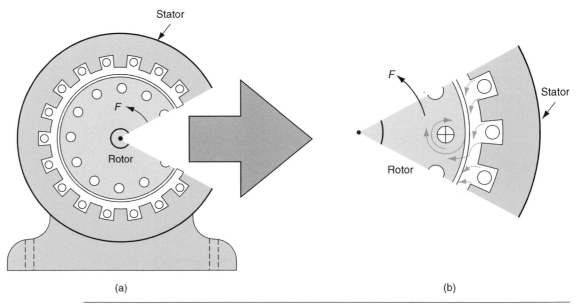

(a)   (b)

Figure 6.127 Segment taken out of an induction motor to show turning force: (a) construction of an induction motor; and (b) production of torque by magnetic fields.

in star or delta and the end connections brought out to slip rings mounted on the shaft. Connection by carbon brushes can then be made to an external resistance to improve starting.

The cage induction motor has a small starting torque and should be used with light loads or started with the load disconnected. The speed is almost constant.

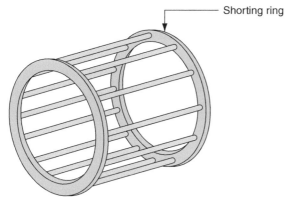
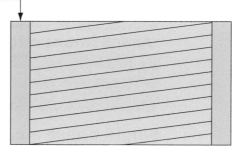

Shorting ring

Arrangement of conductor bars in a cage rotor

Skewed rotor conductors

Figure 6.128 Construction of a cage rotor.

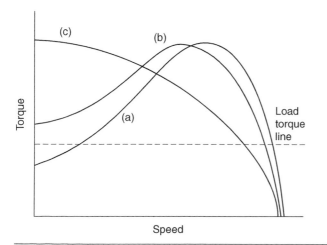

Figure 6.129 (a) The internal workings of an induction motor; (b) the external shell of an induction motor.

Figure 6.130 Various speed-torque characteristics for an induction motor.

Its applications are for constant speed machines such as fans and pumps. Reversal of rotation is achieved by reversing any two of the stator winding connections.

> ## Assessment criteria 5.4
> ### State the basic types of alternating current motors
>
> ## Assessment criteria 5.5
> ### Describe the operating principles of single phase alternating current motors

Single-phase a.c. motors

A single-phase a.c. supply produces a pulsating magnetic field, not the rotating magnetic field produced by a three-phase supply. All a.c. motors require a rotating field to start. Therefore, single-phase a.c. motors have two windings which are electrically separated by about 90°. The two windings are known as the start and run windings. The magnetic fields produced by currents flowing through these out-of-phase windings create the rotating field and turning force required to start the motor. Once rotation is established, the pulsating field in the run winding is sufficient to maintain rotation and the start winding is disconnected by a centrifugal switch which operates when the motor has reached about 80% of the full load speed.

A cage rotor is used on single-phase a.c. motors, the turning force being produced in the way described previously for three-phase induction motors and shown in Fig. 6.127. Because both windings carry currents that are out of phase with each other, the motor is known as a 'split-phase' motor. The phase displacement between the currents in the windings is achieved in one of two ways:

- by connecting a capacitor in series with the start winding, as shown in Fig. 6.131(a), which gives a 90° phase difference between the currents in the start and run windings;
- by designing the start winding to have a high resistance and the run winding a high inductance, once again creating a 90° phase shift between the currents in each winding, as shown in Fig. 6.131(b).

When the motor is first switched on, the centrifugal switch is closed and the magnetic fields from the two coils produce the turning force required to run the rotor up to full speed. When the motor reaches about 80% of full speed, the centrifugal switch clicks open and the machine continues to run on the magnetic flux created by the run winding only.

Split-phase motors are constant speed machines with a low starting torque and are used on light loads such as fans, pumps, refrigerators and washing machines. Reversal of rotation may be achieved by reversing the connections to the start or run windings, but not both.

Other types of single phase motors in existence include the capacitor start and run. This type is similar to the capacitor start but produces higher starting torque and is more efficient due to improved power factor that the capacitors bring. Its uses includes applications such as: air compressors and woodworking machines.

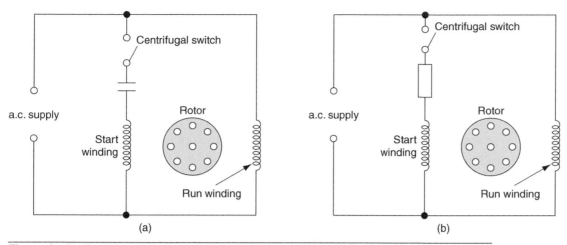

Figure 6.131 Circuit diagram of: (a) capacitor split-phase motors and (b) resistance split-phase motors.

Permanent split capacitor (PSC) motors do not require a centrifugal switch to disable the staring mechanism but use a capacitor that is hard wired in place. This type of motor is used with high-frequency or cyclic use such as mechanisms that control the opening and closing of gates and garage doors.

Shaded pole motors

The shaded pole motor is a simple, robust single-phase motor, which is suitable for very small machines with a rating of less than about 50 W. Figure 6.133 shows a shaded pole motor. It has a cage rotor and the moving field is produced by enclosing one side of each stator pole in a solid copper or brass ring, called a shading ring, which displaces the magnetic field and creates an artificial phase shift.

Shaded pole motors are constant speed machines with a very low starting torque and are used on very light loads such as oven fans, record turntable motors and electric fan heaters. Reversal of rotation is theoretically possible by moving the shading rings to the opposite side of the stator pole face. However, in practice this is often not a simple process, but the motors are symmetrical and it is sometimes easier to reverse the rotor by removing the fixing bolts and reversing the whole motor. There are more motors operating from single-phase supplies than all other types of motor added together. Most of them operate as very small motors in domestic and business machines where single-phase supplies are most common.

TYPICAL PSC MOTOR

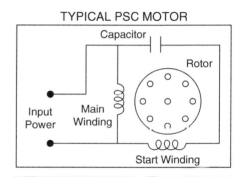

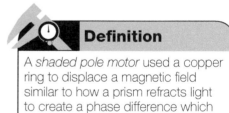

Figure 6.132 PSC motors are used on high cyclic or repetitive use applications such as garage doors.

Definition

A *shaded pole motor* used a copper ring to displace a magnetic field similar to how a prism refracts light to create a phase difference which leads to motion.

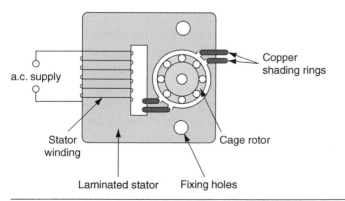

Figure 6.133 Shaded pole motor.

Motor starters

When the armature conductors of an electric motor cut the magnetic flux of the main stator field, an e.m.f. is induced in the armature winding as described by Faraday's Law 'when a conductor cuts, or is cut by a magnetic field, an e.m.f. is induced in that conductor'. The induced e.m.f. is known as a back e.m.f., because it acts in opposition to the supply voltage. During normal running the back e.m.f. is always a little smaller than the supply voltage and acts as a limit to the motor current. However, at the instant that the motor is first switched on, the back e.m.f. does not exist because the conductors are stationary and so a very large current flows for just a few milliseconds.

The magnetic flux generated in the stator of an induction motor such as that shown in Fig. 6.127 earlier in this chapter rotates immediately the supply is switched on, and therefore the machine is self-starting. The purpose of the motor starter is not to start the machine as the name implies, but to reduce heavy starting currents and to provide overload and no-volt protection as required by IET Regulations 552.

Thermal overload protection is usually provided by means of a bimetal strip bending under overload conditions and breaking the starter contactor coil circuit. This de-energizes the coil and switches off the motor under fault conditions such as overloading or single phasing. Once the motor has automatically switched off under overload conditions or because a remote stop/start has been operated, it is an important safety feature that the motor cannot restart without the operator going through the normal start up procedure. Therefore, no-volt protection is provided by incorporating the safety devices into the motor starter control circuit which energizes the contactor coil.

Electronic thermistors (thermal transistors) provide an alternative method of sensing if a motor is overheating. These tiny heat-sensing transistors, about the size of a matchstick head, are embedded in the motor windings to sense the internal temperature. The thermistor then either trips out the contactor coil as described above or operates an alarm. All electric motors with a rating above 0.37 kW must be supplied from a suitable motor starter and we will now consider the more common types.

Direct on line (d.o.l.) starters

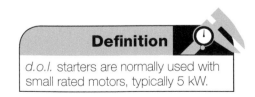

Definition

d.o.l. starters are normally used with small rated motors, typically 5 kW.

The d.o.l. starter switches the main supply directly onto the motor. Since motor starting currents can be seven or eight times greater than the running currents, the d.o.l. starter is only used for small motors of less than about 5 kW rating. When the start button is pressed, current will flow from the brown phase through the control circuit and contactor coil to the grey phase which energizes the contactor coil and the contacts close, connecting the three-phase supply to the motor as can be seen in Fig. 6.134. If the start button is released the control circuit is maintained by the hold on contact. If the stop button is pressed or the overload coils operate, the control circuit is broken and the contactor drops out, breaking the supply to the load. Once the supply is interrupted, the supply to the motor can only be reconnected by pressing the start button. Therefore, this type of arrangement also provides no-volt protection. When large industrial motors

Key fact

Motor starters do not actually start but control the starting sequence of motors.

have to be started, a way of reducing the excessive starting currents must be found. One method is to connect the motor to a star delta motor starter.

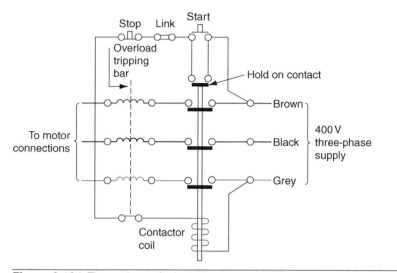

Figure 6.134 Three-phase d.o.l. starter.

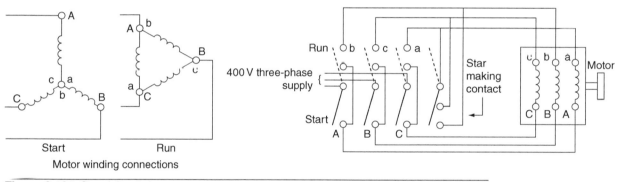

Figure 6.135 Star delta starter.

Star delta motor starters

A star delta motor starter is commonly used on the larger types of motor such as controlling motors rated between 5 kW and 20 kW but it is normally used where the requirements is to start up very light or no-load situations. When three loads, such as the three windings of a motor, are connected in star, the line current has only one-third of the value it has when the same load is connected in delta. A starter that can connect the windings in star during the initial starting period and then switch to a delta connection for normal running will reduce the problem of an excessive starting current. This arrangement is shown in Fig. 6.135 where the six connections to the three stator phase windings are brought out to the starter. For starting, the motor windings are star connected at the a-b-c end of the winding by the star making contacts. This reduces the phase voltage to about 58% of the running voltage, which reduces the current and the motor's torque. Once the motor is running, a double–throw switch makes the changeover from star starting to delta running, thereby achieving a minimum starting current and maximum running torque. The starter will incorporate overcurrent and overload as well as no-volt protection, but these have not been shown in Fig. 6.135 in the interests of showing more clearly the principle of operation.

Auto-transformer motor starter

An auto-transformer motor starter provides another method of reducing the starting current by reducing the voltage during the initial starting period. Since this also reduces the starting torque, the voltage is only reduced by a sufficient amount to reduce the starting current, being permanently connected to the tapping found to be most appropriate by the installing electrician. Switching the changeover switch to start position connects the auto-transformer windings in series with the delta connected motor starter winding. When sufficient speed has been achieved by the motor, the changeover switch is moved to the run connections which connect the three-phase supply directly onto the motor as shown in Fig. 6.136.

This starting method has the advantage of only requiring three connecting conductors between the motor starter and the motor. The starter will incorporate overload and no-volt protection in addition to a method of preventing the motor being switched to the run position while the motor is stationary. These protective devices are not shown in Fig. 6.136 in order to show more clearly the principle of operation. An application for auto-transformer motor starters would be to start large squirrel cage motors.

Rotor resistance motor starter

Any motor containing a cage rotor such as that shown in Fig. 6.128 suffers from a low starting torque. Therefore an induction motor must be matched to the load in a way that the starting torque of the motor is greater than the load torque as shown by line (b) in Fig. 6.130. An induction motor with the characteristic shown by curve (a) in Fig. 6.130 would not self start because the load torque is greater than the motor starting torque.

There are three ways to overcome this problem: first by matching the motor to the load torque as shown by curve (b) in Fig. 6.130; second, by connecting the load after the load has run up to full speed; or third, extra resistance can be added to a wound rotor induction motor through slip rings and brushes. The extra resistance of the rotor increases the starting torque as shown by curve (c) in Fig. 6.130, which is why this type of motor starter is used to start heavy loads. When the motor is first switched on the external rotor resistance is at a maximum. As the motor speed increases the resistance is reduced until at full speed the external resistance is cut out and the machine continues to run

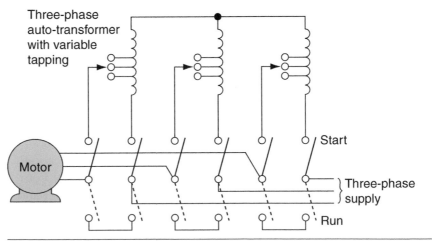

Figure 6.136 Auto-transformer starting.

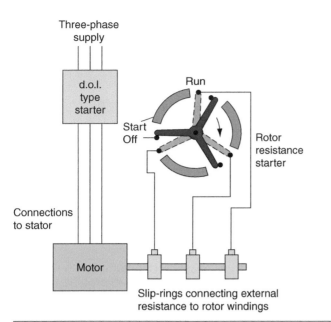

Figure 6.137 Rotor resistance starter for a wound rotor machine.

as a cage induction motor. The starter is provided with overload and no-volt protection and an interlock to prevent the machine being switched on with no rotor resistance connected. These are not shown in Fig. 6.130 because the purpose of the diagram is to show the principle of operation.

Soft start

Another device that reduced the loading effect suffered when starting up electrical motors is known as a soft start device, whereby in essence mechanical devices such as clutches and couplings control the amount of torque, which in turn reduces the amount of mechanical stress. An equivalent electrical system would look to control the amount of torque by limiting voltage and current.

Variable frequency drive

A variable frequency drive (VFD) is a very effective system used to drive and control the rotational speed of a motor when engaged in running loads. In essence, the VFD varies the current in relation to the amount of torque and loading experienced. The system consists of: a rectifier unit, a direct current link, and an inverter.

Three-phase motors are also controlled using inverters, which regulate both the staring sequence through adjustment of voltage and current and operational influence on its speed by altering its voltage and frequency. Circuit sensors are also used to detect any dangerous situations arising from under voltage or excess current.

Definition

Torque is defined as a turning force.

Assessment criteria 6.1

Specify the operating principles of electrical components

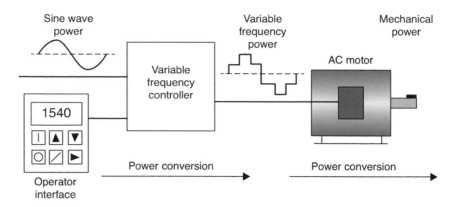

Figure 6.138 A variable frequency drive is used to control and adjust the operational speed of a motor.

Operating principle of electrical devices

Electricity and magnetism have been inseparably connected since the experiments by Oersted and Faraday in the early nineteenth century. An electric current flowing in a conductor produces a magnetic field 'around' the conductor that is proportional to the current. Thus a small current produces a weak magnetic field, while a large current will produce a strong magnetic field.

The electrical solenoid

Definition

A *relay* can be defined as either a device that uses a small rated signal to control a bigger circuit, or when one input can switch a multiple number of outputs.

A current flowing in a *coil* of wire or solenoid establishes a magnetic field that is very similar to that of a bar magnet. Winding the coil around a soft iron core increases the flux density because the lines of magnetic flux concentrate on the magnetic material. The advantage of the electromagnet when compared with the permanent magnet is that the magnetism of the electromagnet can be switched on and off by a functional switch controlling the coil current. This effect is put to practical use in the contactor coil as used in a motor starter or alarm circuit. Figure 6.139 shows the structure and one application of the solenoid, a simple relay.

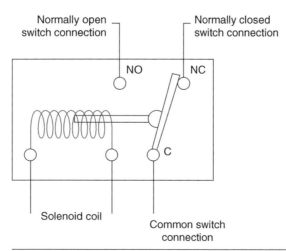

Figure 6.139 A simple relay.

Assessment criteria 6.2

Distinguish the application of electrical components in electrical systems

The electrical relay

A **relay** is an electromagnetic switch operated by a solenoid. The solenoid in a relay operates a number of switch contacts as it moves under the electromagnetic forces. Relays can be used to switch circuits on or off at a distance remotely. The energizing circuit, the solenoid, is completely separate to the switch contacts and, therefore, the relay can switch high-voltage, high power circuits, from a low-voltage switching circuit. This gives the relay many applications in motor control circuits, electronics and instrumentation systems. Figure 6.139 shows a simple relay.

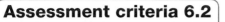

Lines of magnetic flux

Lines of magnetic flux around a solenoid with no core

Lines of magnetic flux around a solenoid with a soft iron core

Spring

Pivot

Solenoid

Movable armature

Air gap

Movement of arm

Simple relay

Figure 6.140 The solenoid and one practical application: the relay.

Assessment criteria 7.2

Define the basic principles of electric water heating

Water heating circuits

A small, single-point over-sink type water heater may be considered as a permanently connected appliance and so may be connected to a ring circuit through a fused connection unit. A water heater of the immersion type is usually rated at a maximum of 3 kW, and could be considered as a permanently connected appliance, fed from a fused connection unit. However, many immersion heating systems are connected into storage vessels of about 150 litres in domestic installations, and the *On-Site Guide* states that immersion heaters fitted to vessels in excess of 15 litres should be supplied by their own circuit.

Therefore, immersion heaters must be wired on a separate radial circuit when they are connected to water vessels that hold more than 15 litres. Figure 6.141 shows the wiring arrangements for an immersion heater. Every switch must be a double-pole (DP) switch and out of reach of anyone using a fixed bath or shower when the immersion heater is fitted to a vessel in a bathroom.

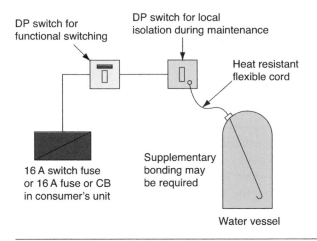

DP switch for functional switching

DP switch for local isolation during maintenance

Heat resistant flexible cord

16 A switch fuse or 16 A fuse or CB in consumer's unit

Supplementary bonding may be required

Water vessel

Figure 6.141 Immersion heater wiring.

Supplementary equipotential bonding to pipework *will only be required* as an addition to fault protection (IET Regulation 415.2) if the immersion heater vessel is in a bathroom that *does not have*:

- all circuits protected by a 30 mA RCD; and
- protective equipotential bonding (IET Regulation 701.415.2) as shown in Chapter 3.

Assessment criteria 7.1

Define the basic principles of electric space heating

Electric space heating systems

Electrical heating systems can be broadly divided into two categories: unrestricted local heating and off-peak heating.

Unrestricted local heating may be provided by portable electric heaters that plug into the socket outlets of the installation. Fixed heaters that are wall mounted or inset must be connected through a fused connection and incorporate a local switch, either on the heater itself or as a part of the fuse connecting unit. Heating appliances where the heating element can be touched must have a DP switch that disconnects all conductors. This requirement includes radiators that have an element inside a silica-glass sheath.

Off-peak heating systems may provide central heating from storage heaters, ducted warm air or underfloor heating elements. All three systems use the thermal storage principle, whereby a large mass of heat-retaining material is heated during the off-peak period and allowed to emit the stored heat throughout the day. The final circuits of all off-peak heating installations must be fed from a separate supply controlled by the electricity supplier's time clock. When calculating the size of cable required to supply a single-storage radiator, it is good practice to assume a current demand equal to 3.4 kW at each point. This will allow the radiator to be changed at a future time with the minimum disturbance to the installation. Each storage heater must have a 20 A DP means of isolation adjacent to the heater and the final connection should be via a flex outlet. See Fig. 6.142 for wiring arrangements.

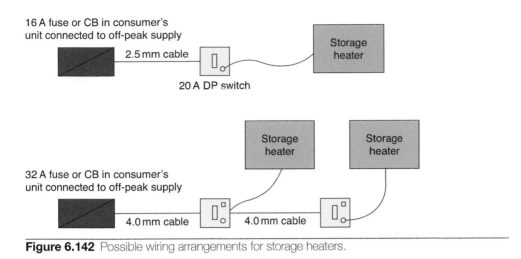

16 A fuse or CB in consumer's
unit connected to off-peak supply

2.5 mm cable

20 A DP switch

Storage heater

32 A fuse or CB in consumer's
unit connected to off-peak supply

4.0 mm cable   4.0 mm cable

Storage heater   Storage heater

Figure 6.142 Possible wiring arrangements for storage heaters.

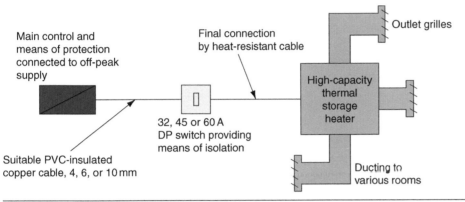

Main control and
means of protection
connected to off-peak
supply

Final connection
by heat-resistant cable

Outlet grilles

High-capacity
thermal
storage
heater

32, 45 or 60 A
DP switch providing
means of isolation

Suitable PVC-insulated
copper cable, 4, 6, or 10 mm

Ducting to
various rooms

Figure 6.143 Ducted warm air heating system.

Ducted warm air systems have a centrally sited thermal storage heater with a high storage capacity. The unit is charged during the off-peak period, and a fan drives the stored heat in the form of warm air through large air ducts to outlet grilles in the various rooms. The wiring arrangements for this type of heating are shown in Fig. 6.143.

The single-storage heater is heated by an electric element embedded in bricks and rated between 6 and 15 kW depending upon its thermal capacity. A radiator of this capacity must be supplied on its own circuit, in cable capable of carrying the maximum current demand and protected by a fuse or circuit breaker (CB) of 30, 45 or 60 A as appropriate. At the heater position, a DP switch must be installed to terminate the fixed heater wiring. The flexible cables used for the final connection to the heaters must be of the heat-resistant type.

Floor-warming installations use the thermal storage properties of concrete. Special cables are embedded in the concrete floor screed during construction. When current is passed through the cables they become heated, the concrete absorbs this heat and radiates it into the room. The wiring arrangements are shown in Fig. 6.144. Once heated, the concrete will give off heat for a long time after the supply is switched off and is, therefore, suitable for connection to an off-peak supply.

Underfloor heating cables installed in bathrooms or shower rooms must incorporate an earthed metallic sheath or be covered by an earthed metallic grid connected to the protective conductor of the supply circuit (IET Regulation 701.753).

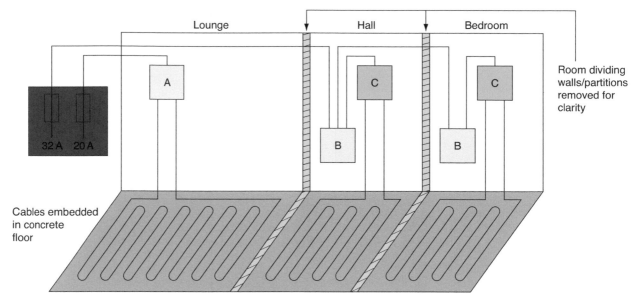

A = Thermostat incorporating DP switch fed by 2.5 mm PVC/copper
B = DP switch fuse fed by 4.0 mm PVC/copper
C = Thermostat fed by 2.5 mm PVC/copper

Figure 6.144 Floor-warming installations.

Assessment criteria 7.3

Explain the operating principles of heating systems

Electrical heating controls

Water heating or space heating systems can be controlled by thermostats, timer switches, programmers, or a combination of the three. Let us first look at the science of temperature and heat.

Temperature and heat transfer

Heat has the capacity to do work and is a form of energy. Temperature is not an energy unit but describes the hotness or coldness of a substance or material. Heat energy is transferred by three separate processes, which can occur individually or in combination. The processes are convection, radiation and conduction.

Convection

Air which passes over a heated surface expands, becomes lighter and warmer and rises, being replaced by descending cooler air. These circulating currents of air are called *convection currents*. In circulating, the warm air gives up some of its heat to the surfaces over which it passes and so warms a room and its contents. The same process takes place in water. The immersion heater in a water vessel such as that shown in Fig. 6.143 heats up the water in contact with the heater element which rises and is replaced by cold water, which then heats up and rises and so on. This process sets up circulating 'convection' currents in the water, which heats up the contents of the whole vessel.

Radiation

Molecules on a metal surface vibrating with thermal energy generate electromagnetic waves. The waves travel away from the surface at the speed of light, taking energy with them and leaving the surface cooler. If the electromagnetic waves meet another body or material they produce a disturbance of the surface molecules; this then raises the temperature of that body or material. Radiated heat requires no intervening medium between the transmitter and receiver, obeys the same laws as light energy and is the method by which the energy from the sun reaches the earth.

Conduction

Heat transferred through a material by conduction occurs because there is direct contact between the vibrating molecules of the material. The application of a heat source to the end of a metal bar causes the atoms to vibrate rapidly within the lattice framework of the material. This violent shaking causes adjacent atoms to vibrate and liberates any loosely bound electrons, which also pass on the heat energy. Thus the heat energy travels through the material by *conduction*.

Temperature scales

When planning any scale of measurement a set of reference points must first be established. Two obvious reference points on any temperature scale are the temperature at which ice melts and water boils. Between these two points the scale is divided into a convenient number of divisions. In the case of the Celsius or Centigrade scale the lower fixed point is zero degrees and the upper fixed point 100 degrees Celsius. The Kelvin takes as its lower fixed point the lowest possible temperature, which has a value of 273°C, called absolute zero. A temperature change of one Kelvin is exactly the same as one degree Celsius and so we can say that

$$0°C = 273 \text{ K} \qquad \text{or} \qquad 0 \text{ K} = -273°C$$

Thermostats

A thermostat is a device for maintaining a constant temperature at some predetermined value. The operation of a thermostat is often based upon the principle of differential expansion between dissimilar metals, that is, a bimetal strip, which causes a contact to make or break at a chosen temperature. Figure 6.145 shows the principle of a rod-type thermostat, which is often used with water heaters. An invar rod, which has minimal expansion when heated, is housed within a copper tube and the two metals are brazed together at one end. The other end of the copper tube is secured to one end of the switch mechanism. As the copper expands and contracts under the influence of a varying temperature, the switch contacts make and break. In the case of a thermostat such as this, the electrical circuit is broken when the temperature setting is reached. Figure 6.146 shows a room thermostat that works on the same principle.

Programmers

A heating system may be fuelled by gas, oil, electricity or solar, but whatever the fuel system providing the heat energy, it is very likely that the system will be controlled by an electrical or electronic programmer, thermostats and a

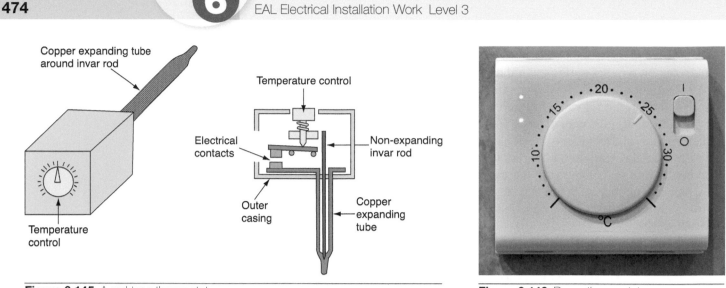

Figure 6.145 A rod-type thermostat.

Figure 6.146 Room thermostat.

circulating pump for a water-based system. Fig 6.147 shows a block diagram for a typical domestic space heating and water heating system called a Y plan, which comprises a controller, separate thermostats for hot water and central heating and one three-port mid-position motorized valve.

The programmer will incorporate a time clock so that the customer or client can select the number of hours each day that the system will operate. For a

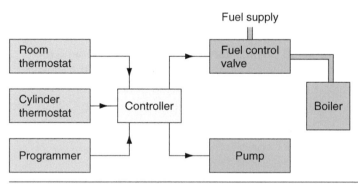

Figure 6.147 Block diagram: space heating control system (Honeywell Y plan).

working family in a domestic property this will probably be a couple of hours before the working day begins, and then five to seven hours in the evening. In a commercial situation it will probably switch on a couple of hours before the working day begins and remain on until the working day ends. In both situations the programmer unit often incorporates an override or boost facility for changing circumstances and weather conditions.

During the period when the space heating or water heating is in the 'on position', thermostats based upon the bimetal strip principle will maintain the room and water temperature at some predetermined level. Figure 6.148 and figure 6.149 show a wiring diagram for a domestic space and water heating system that circulates water through radiators to warm individual rooms. In this type of system, energy conservation now recommends that each radiator be additionally fitted with a TRV (thermostatic radiator valve) to control the temperature in each room.

In order to determine how something works a circuit diagram is ideal so that it helps to gauge how the main components are linked in operation.

A further system is known as an S plan but differs in that it comprises separate water control valves in the regulation of hot water and central heating.

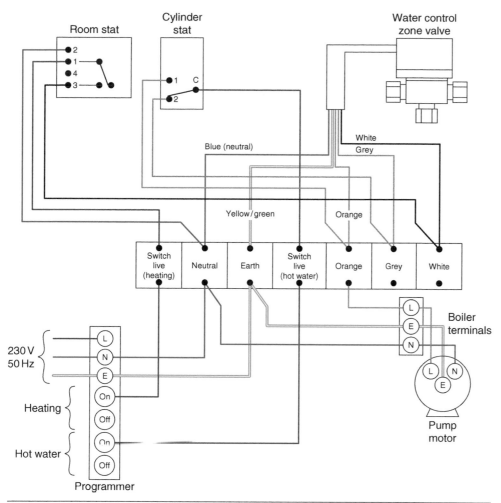

Figure 6.148 Wiring diagram – space heating control system (Honeywell Y Plan).

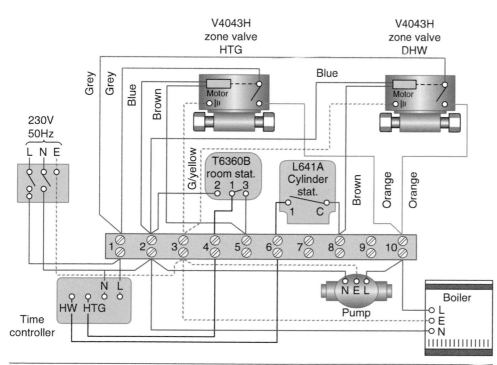

Figure 6.149 Wiring diagram – space heating control system (Honeywell S Plan).

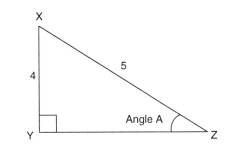
When you have completed the questions, check out the answers at the back of the book.

Note: more than one multiple-choice answer may be correct.

Learning outcome 1

1 Solve the following calculation $-20 \div -4$
    a. -5
    b. 5
    c. 8
    d. 6.

2 Which of the following is correct?
    a. $22\ kW = 22 \times 10^3 = 22{,}000\ W$
    b. $22\ kW = 22 \times 10^3 = 2200\ W$
    c. $22\ kW = 22 \times 10^1 = 22{,}000\ W$
    d. $22\ kW = 22 \times 10^2 = 220{,}000\ W$

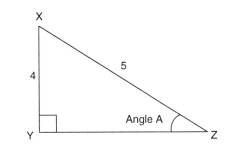

3 Work out angle A.
    a. 53.1°
    b. 36.87°
    c. 38.65°
    d. 90°.

4 What is the result of the following calculation $20 + 4 \times 3 = ?$
    a. 72
    b. 32
    c. 83
    d. 64.

5 What is the result of the following calculation $(42 \times 4) + (6 \div 3) - 2 = ?$
    a. 42
    b. 56
    c. 420
    d. 168.

6   What is the result of the following calculation 10000 × 100?
   a.   100,000
   b.   1000
   c.   10,000
   d.   1,000,000.

7   Work out the fraction $\frac{1}{4} \times \frac{3}{7}$
   a. $\frac{5}{3}$

   b. $\frac{3}{28}$

   c. $\frac{3}{5}$

   d. $\frac{21}{3}$

8   What is the result of the following calculation: 8^3?
   a.   8
   b.   24
   c.   512
   d.   16.

9   Transpose the following equation P = √3 V × I x Cos Θ in order to find I
   a. $I = \dfrac{P}{\sqrt{3} \times V \times Cos\ \Theta}$

   b. $I = \dfrac{V}{\sqrt{3} \times P \times I \times Cos\ \Theta}$

   c. $I = \dfrac{V}{\sqrt{3}\ \times P + I + Cos\ \Theta}$

   d. $I = \sqrt{3}\ xV \times P \times Cos\ \Theta$

10   What kind of angle is shown here?
   a.   right angle
   b.   straight line
   c.   obtuse
   d.   acute.

11   If a = 4, c = 8, b = ?
   a.   8.94
   b.   48
   c.   80
   d.   6.92.

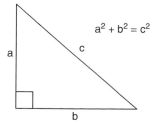

$a^2 + b^2 = c^2$

12 Calculate the mean, median and mode of the following numbers: 4, 8, 7, 2, 6, 8

   a.   mean = 5.83; median= 4.5; mode= 8

   b.   mean = 5; median= 9; mode= 8

   c.   mean = 5.83; median= 6.5; mode= 8

   d.   mean = 5.83; median= 9.5; mode= 7.

13 What kind of triangle is shown?

   a.   scalene

   b.   right angle

   c.   equilateral

   d.   isosceles.

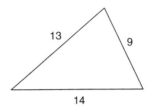

14 Lighting circuits are permitted 3% allowance regarding the amount of voltage that can be dropped. What is 3% of 230 V?

   a.   11.5 V

   b.   12 V

   c.   1.304 V

   d.   6.9 V.

15 Solve the following $3^3 \times 3^6$.

   a. 3^3

   b. 3^{18}

   c. 3^{12}

   d. 3^9.

Learning outcome 2

16 What kind of electronic device only conducts when it is gated and is used in d.c. motor control applications?

   a.   diode

   b.   transistor

   c.   diac

   d.   thyristor.

17 What kind of device changes its resistance with a corresponding temperature change?

   a.   thyristor

   b.   triac

   c.   thermistor

   d.   transistor.

18   A reed switch uses which effect of electricity?
   a.   heat
   b.   chemical
   c.   magnetism
   d.   fusion.

19   Which component is appropriate to be used to stop heat damage to a relay?
   a.   diode
   b.   capacitor
   c.   transistor
   d.   resistor.

20   Which of the following is correct?
   a.   diode = rectifier
   b.   capacitor = rectifier
   c.   transistor = rectifier
   d.   resistor = rectifier.

21   Which of the following is correct?
   a.   diodes are used as transformers
   b.   capacitors are used in smoothing circuits
   c.   transistors are used as electronic switches
   d.   resistors are used as amplifiers.

22   If a diode is forward biased then it will pass current from:
   a.   cathode to anode
   b.   anode to cathode
   c.   base to collector
   d.   gate to anode.

23   An inverter:
   a.   changes a sinusoidal waveform to direct current
   b.   changes d.c. to a.c.
   c.   changes a.c. to d.c.
   d.   changes light to electricity.

24   What kind of electronic device combines two zener diodes and will allow current flow in either direction once it reaches a certain value?
   a.   diode
   b.   transistor
   c.   diac
   d.   thyristor.

25.   NPN and PNP are types of?
   a.   diode
   b.   transistor
   c.   diac
   d.   thyristor.

26 What kind of device reduces its resistance when light falls on it, allowing current to flow?

    a. photo cell

    b. zener diode

    c. light emitting diode

    d. light dependent resistor.

27 What kind of device converts light energy into electrical energy and is used in burglar alarms?

    a. photo cell

    b. zener diode

    c. light emitting diode

    d. light dependent resistor.

28 What kind of semi-conductor device acts like a thyristor but allows current flow in either direction when it is gated?

    a. diac

    b. zener diode

    c. thyristor

    d. triac.

29. If three capacitors are connected in parallel and are rated at 34 μF, 24 ΩF and 17 ΩF, calculate the total capacitance of the circuit.

    a. 75 μF

    b. 17.5 μF

    c. 175 μF

    d. 0.75 μF.

30 If two capacitors are connected in series and are both rated at 10 μF calculate the total capacitance of the circuit.

    a. 7 μF

    b. 5 μF

    c. 15 μF

    d. 20 μF.

31 What kind of rectification circuit would produce the output below?

    a. partial rectification

    b. single diode

    c. half-wave rectification

    d. full wave rectification.

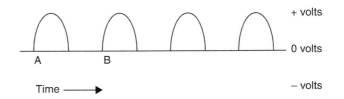

32 What kind of rectification circuit would produce the output below?

    a. partial rectification

    b. single diode

    c. half-wave rectification

    d. full wave rectification.

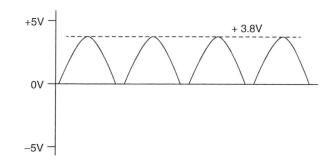

33 An electronic device that emits red, green, or yellow light when a current of
    about 10 mA flows through it is a:
    a. light dependent resistor (LDR)
    b. light-emitting diode (LED)
    c. semiconductor diode
    d. thermistor.

34 What kind of capacitor is polarized?
    a. flux capacitor
    b. electric capacitor
    c. electrolytic capacitor
    d. polaroid capacitor.

Learning outcome 3

35 The opposition to current flow in an a.c. resistive circuit is called:
    a. resistance
    b. inductance
    c. reactance
    d. impedance.

36 The opposition to current flow in an a.c. capacitive or inductive current is
    called:
    a. resistance
    b. inductance
    c. reactance
    d. impedance.

37 The total opposition to current flow in any a.c. circuit is called:
    a. resistance
    b. inductance
    c. reactance
    d. impedance.

38 An a.c. series circuit has an inductive reactance of 4 Ω and a resistance of
    3 Ω. The impedance of this circuit will be:
    a. 5 Ω
    b. 7 Ω
    c. 12 Ω
    d. 25 Ω.

39 An a.c. series circuit has a capacitive reactance of 12 Ω and a resistance of 9 Ω. The impedance of this current will be:
   a. 3 Ω
   b. 15 Ω
   c. 20 Ω
   d. 108 Ω.

40 The inductive reactance of a 100 mH coil when connected to a 50 Hz supply will be:
   a. 5 Ω
   b. 20 Ω
   c. 31.42 Ω
   d. 31.42 kΩ.

41 The capacitive reactance of a 100 μF capacitor when connected to a 50 Hz supply will be:
   a. 5 Ω
   b. 20 Ω
   c. 31.8 Ω
   d. 31.8 kΩ.

42 A circuit with *poor* power factor causes:
   a. the supply voltage to fall
   b. the supply voltage to increase
   c. more current consumption from the supply
   d. less current consumption from the supply.

43 If uncorrected, *poor* power factor means the circuit conductors have to be:
   a. made smaller
   b. made bigger
   c. made longer
   d. made shorter.

44 Calculate the design current in a domestic single phase supply that has a power rating of 5 kW and operates with a power factor of 0.8.
   a. 17.39 A
   b. 27.17 A
   c. 46 A
   d. 36.8 A.

45 Calculate the circuit current in a three-phase supply (400 V) that has a power rating of 5 kW and operates with a power factor of 0.8.
   a. 9.02 A
   b. 12.5 A
   c. 15.6 A
   d. 7.21 A.

46 The effect of an inductor is to cause the current in a circuit to:
   a. be in phase with the voltage
   b. lead the voltage
   c. lag the voltage
   d. double the voltage.

47   Which of the following is used to improve power factor?

   a.   resistor

   b.   inductor

   c.   capacitor

   d.   transformer.

48   In order to carry out bulk correction of power factor, where is the correction item fitted?

   a.   across the line conductor

   b.   across the supply

   c.   across the neutral

   d.   across each individual load.

49   A disadvantage of using synchronous motors to correct for power factor is:

   a.   system uses more voltage

   b.   system uses more current

   c.   the cost of synchronous motors

   d.   the cost of capacitors.

50   Impedance is represented by:

   a.   R

   b.   Z

   c.   XL

   d.   XC.

51   Power factor is shown through which relationship?

   a.   sine (s = O/H)

   b.   cosine (A/H)

   c.   tangent (T = O/A)

   d.   cosine^{-1}.

52   Which of the following represents perfect power factor?

   a.   1

   b.   0.99

   c.   0.5

   d.   0.

53   Given that $\mathrm{Cos} = \dfrac{R}{Z}$ if the resistance is 10 Ω and the impedance is 10.2 Ω,

work out the power factor.

   a.   0.98 Ω

   b.   0.98 V

   c.   0.98

   d.   0.98 A.

54   A star connected system has a line current of 3 A. Work out the phase current.

   a.   3.5 A

   b.   6 A

   c.   3 A

   d.   16 A.

55   A star connected system has a line voltage of 600 V. Work out the phase voltage.
    a.   3.5 kV
    b.   346 V
    c.   250 V
    d.   346 kV.

56   A delta connected system has a line voltage of 600 V. Work out the phase voltage.
    a.   600 V
    b.   346 V
    c.   600 mV
    d.   346 kV.

57   A delta connected system has a phase current of 30 A. Work out the line current.
    a.   30 A
    b.   51.96 A
    c.   45 A
    d.   16 A.

Learning outcome 4

58   A GLS lamp is associated with:
    a.   regeneration effect
    b.   incandescence
    c.   poor colour rendering
    d.   being very inefficient.

59   A tungsten halogen lamp is associated with:
    a.   regeneration effect
    b.   incandescence
    c.   poor colour rendering
    d.   goes from blue to nearly white.

60   A fluorescent lamp is associated with:
    a.   power factor correction capacitor
    b.   incandescence
    c.   choke
    d.   starter.

61   The amount of light reflected on to a surface at any angle from the light source is calculated by using the:
    a.   inverse-square law
    b.   maintenance factor
    c.   cosine law
    d.   lumen method.

62   The intensity of a light is measured in:
    a.   candela
    b.   flux
    c.   candarel
    d.   lux.

63   What advantage does the inclusion of the halogen gas have inside a tungsten lamps?
   a.   increases life expectancy
   b.   increases ceramic
   c.   enhances power factor
   d.   produces a regeneration effect.

64   In comparison to GLS, an advantage of LEDs is that they are:
   a.   long lasting
   b.   not prone to glare
   c.   more expensive
   d.   very efficient.

65   What kind of lamps are often used in street lighting?
   a.   low-pressure mercury
   b.   cfl
   c.   low-pressure sodium
   d.   high-pressure sodium.

66   When a GLS lamp is at the end of its life, what often happens?
   a.   high circuit current, protective device operates
   b.   low circuit current, protective device operates
   c.   high circuit current, protective device remains set
   d.   low circuit current, protective device remains set.

67   What is the common name for a low-pressure mercury vapour lamp?
   a.   spot lamp
   b.   fluorescent lamp
   c.   high frequency
   d.   low tungsten.

68   Why is there a capacitor connected across the supply terminals of a fluorescent lamp?
   a.   it improves efficiency
   b.   it assists with electromagnetic interference
   c.   it reduces power factor issues
   d.   it reduces starting currents.

69   The light output of a high-pressure mercury vapour lamp is a:
   a.   blue white but cold light with fairly good colour rendering
   b.   very yellow light with poor colour rendering
   c.   warm yellow light with good colour rendering
   d.   cold white light with very good colour rendering properties, making it suitable for colour photography.

70   The light output of a high-pressure sodium lamp is a:
   a.   blue white but cold light with fairly good colour rendering
   b.   very yellow light with poor colour rendering
   c.   warm yellow light with good colour rendering but can also be white
   d.   cold white light with very good colour rendering properties, making it suitable for colour photography.

71 The light output of a metal halide lamp is a:
   a. blue white but cold light with fairly good colour rendering
   b. very yellow light with poor colour rendering
   c. warm yellow light with good colour rendering
   d. cold white light with very good colour rendering properties, making it suitable for colour photography.

72 The light output of a low-pressure sodium lamp is a:
   a. blue white but cold light with fairly good colour rendering
   b. very yellow light with poor colour rendering
   c. warm yellow light with good colour rendering
   d. cold white light with very good colour rendering properties, making it suitable for colour photography.

73 What changes UV light to visible white light inside a fluorescent lamp?
   a. the gas
   b. the starter
   c. the phosphorous inside
   d. the power factor capacitor.

74 Which of the following is a measure of how well a light source produces visible light?
   a. luminous efficacy
   b. glass efficiency
   c. phosphorous efficiency
   d. luminous efficiency.

75 With reference to the Inverse Square Law, if you move an object from 3 m to 6 m what would happen to the amount of light that is reflected on the surface?
   a. decease
   b. decrease by a factor of 2
   c. decrease by a factor of 4
   d. decrease by a factor of 8.

76 A street lamp has a luminous intensity of 2000 cd and is suspended 5 m above the ground. The illuminance on the pavement below the lamp will be:
   a. 40 lx
   b. 80 lx
   c. 400 lx
   d. 800 lx.

Learning outcome 5

77 Which of the following motors can be used with either a.c. or d.c. supply?
   a. d.c. series
   b. shunt
   c. universal
   d. compound.

78 Applications for series d.c. motor include:
   a. an electric train
   b. a microwave oven
   c. a central heating pump
   d. an electric drill.

79 Which of the following applies to a d.c. shunt motor?
   a. provides high starting torque
   b. constant speed irrespective of the load
   c. used where speed is not important
   d. used in electric drills.

80 Which of the following applies to a d.c. compound motor?
   a. provides low starting torque
   b. can provide constant speed
   c. can provide high starting torque
   d. used with elevators.

81 One application for an a.c. induction motor is:
   a. an electric train
   b. a microwave oven
   c. a central heating pump
   d. an electric drill.

82 One application for a shaded pole a.c. motor is:
   a. an electric train
   b. a microwave oven
   c. a central heating pump
   d. an electric drill.

83 Which of the following motors is the cheapest?
   a. capacitor start
   b. capacitor start/run
   c. permanent split capacitor (PSC) motors
   d. split phase.

84 Which of the following motors has low starting currents, low starting torques
   and no starting mechanism?
   a. capacitor start
   b. d.c. series
   c. permanent split capacitor (PSC) motors
   d. split phase.

85 Which of the following motors is used on fans, gate operators, garage-door
   openers?
   a. capacitor start
   b. d.c. series
   c. permanent split capacitor (PSC) motors
   d. split phase.

86   Which of the following motors is used to start loads that are hard to start such as lathes, compressors and small conveyor systems?
    a.   capacitor start
    b.   capacitor start/run
    c.   permanent split capacitor (PSC) motors
    d.   split phase.

87   Which of the following would be dangerous to start up without load?
    a.   capacitor start
    b.   d.c. series
    c.   permanent split capacitor (PSC) motors
    d.   split phase.

88   Capacitor-start capacitor-run induction motors are usually:
    a.   three-phase machines
    b.   single phase machines
    c.   synchronous machines
    d.   shaded pole machines.

89   Squirrel cage motors are:
    a.   three-phase machines
    b.   single phase machines
    c.   assynchronous machines
    d.   constant speed.

90   Which of the following would be used to start up motors <5 kW?
    a.   direct-on-line starter
    b.   star delta starter
    c.   rotor-resistance starter
    d.   autotransformer starter.

91   Which of the following would be used to start up motors with a rating of 5 kW to 20 kW but against either light or no load?
    a.   direct-on-line starter
    b.   star delta starter
    c.   rotor-resistance starter
    d.   autotransformer starter.

92   Which of the following would be used on motors involved with heavy loads?
    a.   direct-on-line starter
    b.   star delta starter
    c.   rotor-resistance starter
    d.   autotransformer starter.

93   Which of the following is used to start up squirrel cage motors?
    a.   direct-on-line starter
    b.   star delta starter
    c.   rotor-resistance starter
    d.   autotransformer starter.

94 An electric motor is connected to a motor starter to:
    a.   improve the efficiency of the motor
    b.   provide the initial starting torque for the motor
    c.   reduce heavy starting currents
    d.   provide overload and no-volt protection.

Learning outcome 6

95 An electromagnetic device that can operate one or several contacts is known as:
    a.   a transformer
    b.   a diode
    c.   a relay
    d.   an inductor.

96 An electromagnetic device that uses a small electrical signal to switch a much larger power circuit is known as:
    a.   a transformer
    b.   a diode
    c.   a relay
    d.   an inductor.

97 An electromagnetic device used in switching heavy duty current applications is known as:
    a.   a transformer
    b.   a diode
    c.   a contactor
    d.   an inductor.

98 A relay is made up of:
    a.   switch and contacts
    b.   solenoid and contacts
    c.   spring and contactor
    d.   field and contactor.

99 A component that is magnetized only when an electric current flows through it is called:
    a.   a switch
    b.   a permanent magnet
    c.   an electromagnet
    d.   a motor.

100 Which effect of electricity is used by an inductor?
    a.   magnetic
    b.   heat
    c.   chemical
    d.   light.

101 In heavy duty applications, contacts are typically made from:
    a.   magnesium
    b.   carbon
    c.   mica
    d.   tungsten.

102 A relay uses which of the following?
   a.   heating effect
   b.   chemical effect
   c.   magnetic effect
   d.   power effect.

103 Which of the following does not operate through magnetism?
   a.   contactor
   b.   relay
   c.   electric motor
   d.   battery.

104 The three effects of an electric current are magnetic, chemical and heat. Which sequence is correct for electro-plating, cooker and relay?
   a.   heating, magnetism and chemical
   b.   magnetism, chemical and heating
   c.   chemical, heating and magnetism
   d.   heating, current and chemical.

Learning outcome 7

105 Which of the following uses cheap electricity to heat up bricks?
   a.   instantaneous heaters
   b.   Economy 7
   c.   thermal storage devices
   d.   combi heaters.

106 Which of the following only heats the water that is needed?
   a.   instantaneous heaters
   b.   Economy 7
   c.   thermal storage devices
   d.   combi heaters.

107 Immersion heaters in domestic situation must be:
   a.   fed from a ring circuit
   b.   fed from a dedicated circuit
   c.   switched by a single pole device
   d.   switched by a double pole device.

108 Where large volumes of hot water are needed in a guest house, for instance, the following would be used to supply enough hot water to several outlets at the same time:
   a.   cistern-type (storage)
   b.   non-pressure (storage)
   c.   combi
   d.   instantaneous.

109 Which of the following devices would use a simmerstat?
   a.   cooker oven
   b.   fridge
   c.   washing machine
   d.   cooker hotplate.

110 Which of the following has a short element to heat water up during the day, and a long one for use at night?

a. water heating storage

b. cistern-type (storage)

c. non-pressure (storage)

d. instantaneous water heater.

111 Which of the following would be used in a hairdresser's situated directly over a sink?

a. cistern-type (storage)

b. direct immersion heater

c. non-pressure water heater

d. instantaneous water heater.

112 Which of the following shows the two common types of heating system?

a. X and Y plan

b. S and X plan

c. S and Y plan

d. A and B plan.

113 Sketch the construction of a simple alternator and label all the parts.

114 State how an e.m.f. is induced in an alternator. Sketch and name the shape of the generated e.m.f.

115 Calculate or state the average r.m.s. and maximum value of the domestic a.c. mains supply and show these values on a sketch of the mains supply.

116 Use a sketch with notes of explanation to describe 'good' and 'bad' power factor.

117 State how power factor correction is achieved on:

a. a fluorescent light fitting

b. an electric motor.

118 Use a sketch to help you describe the meaning of the words:

a. inductance

b. mutual inductance.

119 Use a sketch with notes of explanation to describe how a force is applied to a conductor in a magnetic circuit and how this principle is applied to an electric motor.

120 Use a sketch with notes of explanation to show how a turning force is applied to the rotor and, therefore, the drive shaft of an electric motor.

121 Give three applications for each of the following types of motor:

a. a d.c. series motors

b. an a.c. induction motor

c. an a.c. split-phase motor

d. an a.c. shaded pole motor.

122 Sketch an electronic circuit that will give full wave rectification of the a.c. mains supply. Assume a 12 V output from a 230 V supply.

123 Briefly describe how a fuse operates or blows when a fault occurs in a circuit protected by a fuse.

124 Briefly describe how an CB gives protection to a circuit under overload conditions.

125 Briefly describe how an CB gives short circuit protection to a circuit.

Classroom activities

Motor control

Match one of the numbered boxes to each lettered box on the left.

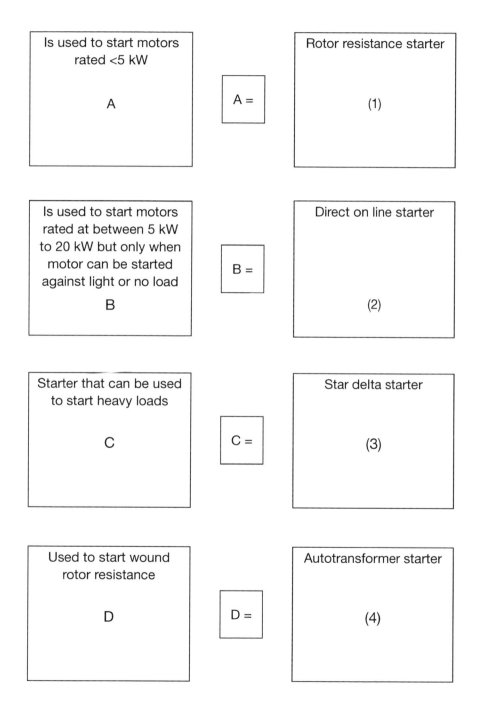

Is used to start motors rated <5 kW A	A =	Rotor resistance starter (1)
Is used to start motors rated at between 5 kW to 20 kW but only when motor can be started against light or no load B	B =	Direct on line starter (2)
Starter that can be used to start heavy loads C	C =	Star delta starter (3)
Used to start wound rotor resistance D	D =	Autotransformer starter (4)

Electrical components

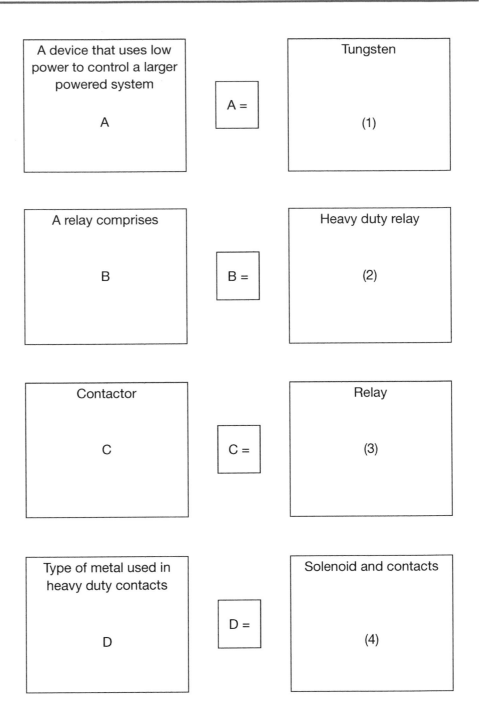

ELECT3/08 Electrical science and principles

Chapter checklist

Learning outcome	Assessment criteria	Page number
1. Understand the mathematical methods that are appropriate to electrical science and principles.	1.1 Apply mathematical methods relevant to electrical science and principles.	360
2. Understand the principles of electronic components and devices in electrotechnical systems.	2.1 Outline the operating principles of components and devices. 2.2 State the applications of components and devices in electrotechnical systems. 2.3 Describe the function of electronic components and devices in electrotechnical systems.	371 399 399
3. Understand the principles of alternating current circuits.	3.1 Determine electrical quantities in alternating current circuits. 3.2 Explain the relation between resistance, inductance, impedance and capacitance. 3.3 Determine power quantities in alternating current circuits. 3.4 Explain the relation between power quantities in alternating current circuits. 3.5 Calculate power factor. 3.6 Explain the term power factor correction. 3.7 Specify the methods of power factor correction. 3.8 Determine the neutral current in a three-phase four-wire system. 3.9 Calculate voltage and current in star and delta connected systems.	403 416 422 423 423 426 426 427 428
4. Understand the principles and applications of luminaires.	4.1 Understand the principles of the laws of illumination 4.2 Determine illumination quantities. 4.3 Explain the operating principles of luminaires. 4.4 Distinguish applications for luminaires.	431 433 438 438
5. Understand the principles and applications of direct current machines and alternating current motors.	5.1 State the basic types of direct current machines. 5.2 Describe the operating principles of direct current machines. 5.3 State the applications of direct current machines. 5.4 State the basic types of alternating current motors. 5.5 Describe the operating principles of single phase alternating current motors. 5.6 Describe the operating principles of three-phase alternating current motors. 5.7 State the applications of alternating current motors. 5.8 Describe the methods used for motor control.	453 455 455 462 462 459 459 464
6. Understand the operating principles of electrical components.	6.1 Specify the operating principles of electrical components 6.2 Distinguish the application of electrical components in electrical systems.	467 469
7. Understand the principles of electric heating.	7.1 Define the basic principles of electric space heating. 7.2 Define the basic principles of electric water heating. 7.3 Explain the operating principles of heating systems.	470 469 472

Answers to 'Test your knowledge' questions

Chapter 1

1. d	2. d	3. b	4.c	5. a, b, c, d
6. b, d	7. a	8. b	9. c	10. c
11. b	12. d	13. b, c	14. a, d	15. b
16. c	17. a	18. d	19. a, d	20. b, c
21. d	22. d	23. d	24. c	25. d
26. a, b	27. b	28. c	29. d	30. c
31. c	32. c	33. c	34. a	35. d
36. d				
37 to 51: Answers in text of Chapter 1.				

Chapter 2

1. d	2. c	3. b	4. c	5. a
6. a	7. b	8. a	9. a	10. c
11. b	12. a	13. d	14. d	15. b
16. c	17. d	18. c	19. c	20. c
21. b	22. d	23. c	24. c	25. c
26. a, b	27. a, b, c	28. b	29. c, d	30. d
31. c	32. a, b	33. a	34. b	35. a
36. a	37. c	38. a	39. b	40. a
41.b	42. a, b, d	43. c	44. a, d	45. a
46. c, d	47. b	48. a	49. c	50. a
51 to 59: Answers in text of Chapter 2.				

Chapter 3

1. c	2. a	3. c	4. d	5. a, c
6. d	7. d	8. a	9. c	10. d
11. a	12. b	13. a	14. d	15. a
16. c	17. b	18. a, b, d	19. d, e	20. d
21. a	22. b	23. b	24. c	25. c
26. c	27. a	28. d	29. d	30. a
31. b	32. a	33. a	34. d	35. a, b, d

36. c	37. c	38. b	39. c	40. a
41. c	42. b	43. d	44. c	45. a
46. d	47. b	48. c	49. b	50. c
51. c	52. a	53. c	54. a, d	55. c
56. b	57. c	58. a	59. a	60. d
61. b	62. b	63. b	64. d	65. b
66. a	67. d	68. c	69. b	70. c
71. a	72. c	73. b	74. d	75. c
76. c	77. a	78. c	79. a	80. a
81. d	82. a	83. c	84. c	85. b, c, d
86. b	87. c, d	88. b	89. a	90. c
91. b	92. b	93. a	94. b	95. b
96. b	97. a	98. c	99. d	100. a, d
101. b	102. c	103. d	104. c	105. b
106. a	107. a	108. c	109. b	110. d
111. b	112. a	113. a, b	114.c	115. d
116. b	117. d	118. d	119. a, b	120. c
121. d	122. b, c, d	123. a, b, d	124. b	
125 to 141: Answers in text of Chapter 3.				

Chapter 4

1. c	2. d	3. a	4. a	5. a
6. d	7. b	8. a	9. d	10. c
11. b	12. a	13. a, d	14. b, c, d	15. d
16. c	17. a, b, c	18. a, b	19. a	20. b, c
21. b	22. b, c, d	23. d	24. b	25. b
26. d	27. b	28. d	29. a	30. a, c
31. c	32. b	33. d	34. a	35. a
36. b, c	37. b, d	38. b	39. a, c	40. a, b, d
41. d	42. a, b	43. d	44. a	45. c
46. a, d				
47 to 53: Answers in text of Chapter 4.				

Chapter 5

1. d	2. c	3. c, d	4. c	5. d
6. b	7. a, b	8. a, c	9. a, c	10. a
11. d	12. a	13. d	14. a, d	15. b
16. b, c	17. b	18. a	19. a, c	20. a, b, c, d
21. c	22. a	23. b	24. b	25. a, b, c
26. a	27. a	28. a	29. b	30. b, d

31. a, c, d	32. b, d	33. b	34. c, d	35. a, b, c
36. b	37. c, d	38. b	39. a, c, d	40. b, c, d
41. c	42. d	43. c, d		

Chapter 6

1. b	2. a	3. a	4. b	5. d
6. d	7. b	8. c	9. a	10. c
11. d	12. c	13. a	14. d	15. d
16. d	17. c	18. c	19. d	20. a
21. b, c	22. b	23. b	24. c	25. b
26. d	27. a	28. d	29. a	30. b
31. c	32. d	33. b	34. c	35. a
36. c	37. d	38. a	39. b	40. c
41. c	42. c	43. b	44. a	45. a
46. c	47. c	48. b	49. c	50. b
51. b	52. a	53. a	54. c	55. b
56. a	57. b	58. b	59. a	60. a, c, d
61. c	62. a	63. a, d	64. a, b	65. c
66. a	67. b	68. c	69. a	70. c
71. d	72. b	73. c	74. a	75. c
76. b	77. c	78. a, d	79. b	80. b, c, d
81. c	82. b	83. d	84. d	85. c
86. a	87. b	88. b	89. a, c, d	90. a
91. b	92. c	93. d	94. c, d	95. c
96. c	97. c	98. b	99. c	100. a
101. d	102. c	103. d	104. c	105. c
106. a	107. b	108. a	109. a	110. a
111. c	112. c			
113 to 125: Answers in text of Chapter 6.				

Answers to classroom activities

Chapter 1

Environmental legislation

Clean Air Act 1993: Ensures that companies operating furnaces, boilers or incinerators use well-engineered combustion equipment A = 3

Pollution Prevention and Control Regulations 2000: Sets directives regarding controlling pollution from certain industrial activities B = 4

Waste Electrical and Electronic Equipment EU Directive 2007: Act that brings compliance regarding European obligation regarding recycling electrical waste C = 5

Environmental Protection Act 1990: Act that ensures that the best practicable environment option is employed to control pollution D = 1

Controlled Waste Regulations 1998: Act that ensures you transfer waste to an authorised company E = 2

Environmental terms and roles

Environmental Health Officer: Person who periodically tests the water quality of recreational water in swimming pools and private water supplies A = 2

Environment Agency: Organisation that has the power to seize vehicles involved in illegal waste disposal B = 1

Landfill: Waste that cannot be recycled C = 4

Built environment: Things we use to build with that can change the environment for everyone D = 5

Public nuisance: An act that interferes with the rights of the public by affecting life, health, property, morals or reasonable comfort or convenience E = 3

Chapter 2

Site paperwork

Time sheet: A record of the work activities of employees on a daily basis A = 3

Delivery note: A record of what material has been delivered on site B = 5

Day work sheet: A record that accounts for any materials and time spent on extra work C = 4

Job sheet: A record of the technical work required to be carried out D = 2

Variation order: A record of any extra work carried out E = 1

On-Site Guide symbols

2 way switch A = 2

Pull switch B = 1

Intermediate switch C = 3

Socket outlet D = 7

Fluorescent lamp: E = 6

Wall-mounted lamp F = 5

Earth G = 4

Chapter 3

Statutory and non-statutory regulations

BS7671 Wiring Regulations: Main non-statutory document that brings about an electrician's code of practice A = 5

On-Site Guide: Pocket version of BS 7671, although you can read earth loop impedance values directly without correction B = 2

Guidance Note 3: This non-statutory document gives specific *guidance* on inspection and testing of electrical installations C = 1

Electricity at Work Regulations 1989: A statutory document that details specific precautions when working with electricity at work D = 4

Health and Safety at Work Regulations 1974: A health and safety umbrella Act from which other statutory regulations were formed E = 3

Chapter 4

Testing Position

Measurement of Prospective Fault Current, values expected to be in KA: Measured at the supply origin to ensure its maximum value is documented to ensure that switch gear & protective devices are assessed for breaking capacity A = 1

Stated Loop impedance values have to be multiplied by 0.8: Reading Zs from BS7671 B = 4

Loop impedance values can be read directly as stated: Reading Zs from the On-Site-Guide or Guidance Note 3 C = 2

Measuring Earth Loop impedance: Measured at the furthest most point to ensure its maximum value is low enough to comply with shock protection requirements D = 3

Testing earth electrode resistance: Protective equipotential bonding is disconnected E = 6

Measuring Prospective Fault Current: Protective equipotential bonding is connected F = 5

Dead test values

0.05-0.08Ω - High fault current will flow and lead to a protective device operating quickly: Acceptable range of values for Continuity of Circuit Protective Conductor A = 2

1MΩ – Within limits but possible Latent fault: Acceptable value of Insulation Resistance but further investigation is recommended B = 3

2MΩ: Acceptable value of Insulation Resistance C = 4

500V: Insulation resistance meter setting for a domestic single phase supply D = 5

NULL: Built in test equipment facility that removes value of test leads from reading under test. Alternatively subtract meter leads from measured value E = 1

Chapter 5

Fault symptoms

Details and symptoms of the fault: Talk to the customer, job sheet, carry out functional test A = 2

Is the circuit deteriorating slowly? Compare previous testing data (schedule of test results), maintenance records regarding component replacement B = 1

Confirm correct size of circuit and protective conductors: Refer to BS7671, *On-Site Guide*, GN3 C= 5

Evaluate if equipment is operating operates correctly in different modes of operation: User guide D = 4

Technical advice on how to install equipment: Manufacturer's instructions E = 3

Symptoms and corrective action

No evidence of overcurrent but cable insulation is showing signs of deterioration (heat damaged): Possible overloading: increase size of conductor A = 2

Circuit cannot achieve disconnection times: install a supplementary bonding conductor B = 5

Assembly line motors are sluggish in operating after the circuit conductors were extended in length: Possible volt drop issue – increase size of assembly line motor conductors C = 1

Upstairs sockets circuit breaker trips when I stand near the bedroom door: Suspect loose floor board impacting on cable installed too close to the surface D = 3

Relay has burnt out: If a short circuit is not apparent, then check design current against relay rating E = 7

Incorrect rotation on three-phase motor: Two of the line conductors are cross connected F = 4

No indications of short circuit but circuit breaker operates as soon as power is applied: Check condition, type and rating of circuit breaker matches load requirements G = 6

Chapter 6

Motor control

Direct on line starter: Is used to start motors rated <5 kW A = 2

Star delta starter: Is used to start motors rated at between 5 kW to 20 kW but only when motor can be started against light or no load B = 3

Autotransformer starter: Starter that can be used to start heavy loads C = 4

Rotor resistance starter: Used to start wound rotor resistance D = 1

Electrical components

Relay: A device that uses low power to control a larger powered system A = 3

A relay comprises: Solenoid and contacts B = 4

Contactor: Heavy duty relay C = 2

Tungsten: Type of metal used in heavy duty contacts D = 1

Answers to design activities and design exercise

Extracting design data

1. State how we calculate design current: divide power rating by circuit voltage ($I = P/V$)

2. State how we determine In (size of protective device):
   Protective device needs to be equal to or greater than the design current ($In \geq Ib$).

3. Where can you find protective device ratings available:
   Appendix B, tables B1-B6, of the OSG, dependant on type and the rule mentioned above.

4. Write out the formula for tabulating the current-carrying capacity of a cable

   $$It \geq \frac{In}{Ca\ Cg\ Ci\ Cf} \quad \text{(Appendix F OSG)}$$

5. When using a BS 3036 fuse (Cf Correction) what correction factor should we use during the cable calculation stage = 0.725 (Appendix F OSG)

6. What correction factor should we use for 3 PVC/PVC twin core cables that are grouped together and installed using Reference Method C? (Look in Appendix F) = 0.79

7. What correction factor should we use for any type of cable that is surrounded by thermal insulation for 0.5m = 0.50 (Appendix F OSG)

8. What correction factor should we use for a PVC/PVC twin core cable that is exposed to an ambient temperature of 35° = 0.94 (Appendix F OSG)

9. $It \geq \dfrac{20}{0.8 \times 0.65}$ It = 38.46A

10. With reference to Appendix F of the OSG, what ambient temperature does not need correcting = 30° (correction factor = 1, dividing by 1 does not affect any number)

11. What is the correct description of Reference Method C = Clipped direct (Table 7.1 (ii) OSG)

12. Given the reference method above and using table F6 of the OSG, determine what size cable is required to carry 47A = 6mm²

13. State the tolerances for voltage drop regarding Lighting & Power? (Look in Appendix F of the OSG or volt drop in the index) = 3% for lighting (6.9V) & 5% for power (11.5V)

14. What is the formula used for calculating voltage drop? (Look in Appendix F of the OSG or volt drop from the index)

   $$\text{Voltage drop} = \frac{(mv/A/m) \times Ib \times L}{1000}$$

15. Using Table F6 of the OSG, the Reference Method at step 11, size of cable at step 12, a design current of 27A, a circuit that is 20m long, calculate if this Power circuit complies with volt drop requirements?

$$\text{Voltage drop} = \frac{(7.3) \times 27 \times 20}{1000}$$

VD = 3.942V well within 3%, 6.9V allowed for a lighting circuit.

16. What do we mean by shock protection:
Shock protection is ensuring that the earth loop impedance is low enough, to generate a large enough fault current to operate a protective device within a specified time.

17. Look in the OSG Table B6 and determine the maximum Zs value for a 40A Type B circuit Breaker = 0.88Ω

18. Do we have to correct the value by multiplying by 0.8: No values in the OSG can be read directly the same applies to GN3. BS7671 values however do have to be corrected.

19. Given the value above in step 17, if the measured loop impedance value Zs of a circuit is 0.91Ω, what does this mean regarding shock protection? It means shock protection has **NOT** been met since 0.91Ω > 0.88Ω in other words the measured value is greater than the tabulated value?

20. Given the Zs value above calculate the PFC?

$$I = \frac{V}{R} \qquad\qquad I = \frac{230}{0.91} \qquad\qquad I = 252.75A$$

21. What do we mean by Thermal constraint?
When an installation is installed with PVC flat profile cable which incorporates a cpc which is smaller than the live conductors. Thermal Constraint is a calculation to ensure that the size of the cpc is big enough to withstand a vey large fault current over a short period of time.

22. A 6mm^2 PVC flat profile cable has a 2.5mm^2 CPC and has a K constant of 115. If the PFC of the circuit is 1.7KA and the circuit needs to be disconnected in 0.4s, calculate if thermal constraint requirement is met.

$$S = \frac{\sqrt{I^2\, t}}{K}$$

$$S = \sqrt{\frac{1700^2 \times 0.4}{115}}$$

S= 9.34mm^2

A 6mm^2 PVC flat profile cable has a cpc of 2.5mm^2 (table 7.1 of the OSG). Therefore thermal constraint is NOT met because cpc needs to be 9mm^2 but circuit is fitted with 2.5mm^2 conductor.

Design exercise

A 10KW industrial oven is to be fed from a domestic property to a consumer unit which is located in a detached garage 15m away. The earthing system is TN-C-S and the wiring type to be used is PVC/PVC flat profile cable. This will be clipped direct (Installation method C) to the surface and will be directed away from any thermal insulation and routed through roof trusses. The circuit is to be protected by a Type B circuit breaker. The earth loop impedance Zs has been measured and recorded as 0.81Ω. Using Ohms law this means that the pfc is 283.9A. The cable is grouped with one other circuit in an area with an ambient temperature of 35 °C.

Work out the following:

1) Design Current Ib

$$I = \frac{10000}{230}$$

$$I = 43.47A$$

2) Size of protective device In

45A

3) Having established any correction factors, calculate the Tabulated current carrying capacity It

Correction factors Cg – 0.85, Ca – 0.94

Calculate the tabulated current carrying capacity (It)

$$It \geq \frac{45}{0.85 \times 0.94} \qquad It = 56.3A$$

4) Size of cable required

10mm²

5) Does volt drop comply?

$$\text{Voltage drop} = \frac{(4.4) \times 15 \times 43.47A}{1000}$$

VD = 2.87V complies with volt drop requirement (<11.5V)

6) Is shock protection requirements met?

OSG Table B6 maximum loop impedance value for Type B 45A breaker is 0.78. Shock protection requirements is **NOT** met given that 0.81 > 0.78 Ω (tabulated value).

7) Thermal Constraint ok?

Referring to table 7.1 (i) of the OSG (under the cable size column) a 10mm² cable has a 4mm² cpc. The circuit is a final circuit therefore will need to be disconnected in 0.4s.

$$S = \frac{\sqrt{283.95^2 \times 0.4}}{115}$$

S = 1.56 mm²

Thermal constraint requirement is met because 1.56 < 4mm²

APPENDIX
Abbreviations, symbols and codes

A

Abbreviations used in electronics for multiples and sub-multiples

T	Tera	10^{12}
G	Giga	10^{9}
M	Mega or Meg	10^{6}
k	Kilo	10^{3}
d	Deci	10^{-1}
c	Centi	10^{-2}
m	Milli	10^{-3}
μ	Micro	10^{-6}
n	Nano	10^{-9}
p	Pico	10^{-12}

Terms and symbols used in electronics

Term	Symbol
Approximately equal to	$\cong$
Proportional to	$\propto$
Infinity	∞
Sum of	Σ
Greater than	$>$
Less than	$<$
Much greater than	$\gg$
Much less than	$\ll$
Base of natural logarithms	e
Common logarithms of x	$\log x$
Temperature	θ
Time constant	T
Efficiency	η
Per unit	p.u.

APPENDIX
Formulas for electrical principles

Advanced Principles Formulas

1. **Trigonometry**

   $c^2 = a^2 + b^2$

   - Transpose for $c = \sqrt{a^2 + b^2}$
   - Transpose for $a = \sqrt{c^2 - b^2}$
   - Transpose for $b = \sqrt{c^2 - a^2}$

2. **Ohms Law**

   $V = I \times R,$

   - Transpose for $I = \dfrac{V}{R}$
   - Transpose for $R = \dfrac{V}{I}$

3. **Power** = Voltage (V) × I (current)

   - Transpose for $I = \dfrac{P}{V}$
   - Transpose for $V = \dfrac{P}{I}$

4. **Power** = $I^2 \times R$

   - Transpose for $I^2 = \dfrac{\text{Power}}{R}$
   - Transpose for $R = \dfrac{\text{Power}}{I^2}$

5. **Power** = $\dfrac{V^2}{R}$

   - Transpose for $V^2 = P \times R$
   - Transpose for $R = \dfrac{V^2}{R}$

6. **Resistors in series** Rt = R1 + R2 + R3

   - Transpose for R1 = Rt − R2 − R3

7. **Resistors in parallel** $\dfrac{1}{Rt} = \dfrac{1}{R1} + \dfrac{1}{R2} + \dfrac{1}{R3}$

   - Transpose for $\dfrac{I}{R3} = \dfrac{1}{Rt} - \dfrac{1}{R2} - \dfrac{1}{R3}$

8. **Left Hand Rule for motors**

   Force = β (Flux Density) x I (Current) x L (length of conductor)

   $F = \beta\, I\, L$

   - Transpose for $I = \dfrac{F}{\beta\, L}$

- Transpose for $L = \dfrac{F}{\beta\ I}$

- Transpose for $\beta = \dfrac{F}{I\ L}$

9. **Right Hand Rule for Generators**

e (electro motive force) = β (Flux Density) x L (length of conductor) x V (Velocity)

e = β L V

- Transpose for $L = \dfrac{e}{\beta\ V}$

- Transpose for $V = \dfrac{e}{\beta\ L}$

- Transpose for $\beta = \dfrac{e}{L\ V}$

10. **Phase Systems**

Star: VL = $\sqrt{3}$ VP

- Transpose for **VP** $= \dfrac{\text{VL}}{\sqrt{3}}$

Star IL = IP

Delta: IL = $\sqrt{3}$ IP

- Transpose for **IP** $= \dfrac{\text{IL}}{\sqrt{3}}$

11. **Single Phase Power Factor:**

Power (P) = V x I x Cos (PF)

- Transpose for $V = \dfrac{P}{I \times \text{Cos (pf)}}$

- Transpose for $I = \dfrac{P}{V \times \text{Cos (pf)}}$

12. **Three Phase Power Factor (V=400V):**

Power (P) = $\sqrt{3}$ x V x I x Cos (PF)

- Transpose for $I = \dfrac{P}{\sqrt{3} \times V \times \text{Cos (PF)}}$

- Transpose for $V = \dfrac{P}{\sqrt{3} \times I \times \text{Cos (PF)}}$

13. XL Inductive Reactance= 2π F (frequency) L (inductance in Henrys)

XL = 2π F L

- Transpose for F = $\dfrac{XL}{2\pi L}$

- Transpose for L = $\dfrac{XL}{2\pi F}$

14. XC Capacitive Reactance = $\dfrac{1}{2\pi F C}$

- Transpose for C = $\dfrac{1}{2\pi F\ XC}$

- Transpose for F = $\dfrac{1}{2\pi C\ XC}$

15. Impedance Z = $\sqrt{R^2 + XI^2}$ (Resistance & XL only)

16. Impedance Z = $\sqrt{R^2 + XC^2}$ (Resistance & XC only)

17. Impedance Z = $\sqrt{R^2 + (XC - XL)^2}$ (Resistance, XC & XL)

18. Extracting power factor (pf) from Impedance Triangle Pf = $\dfrac{R}{Z}$

- Transpose for R = pf X Z

- Transpose for Z = $\dfrac{R}{pf}$

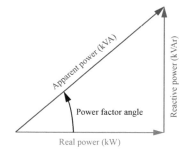

19. Extracting power factor (pf) from power Triangle: pf = $\dfrac{\text{Real Power}}{\text{Apparent power}}$

- Transpose for Real Power = pf x Apparent power

- Transpose for Apparent power = $\dfrac{\text{Real Power}}{Pf}$

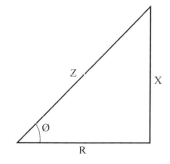

20. Inverse square law = $\dfrac{\text{Light Intensity}}{d^2}$

21. Cosine law = $\dfrac{\text{Light Intensity} \times \cos}{d^2}$

3 rules of Transposition

+ (positive)   – (negative)

X (multiply-Top) and (Divide-bottom) /

X^2 (Raise to a power) and $\sqrt{}$ (Square root)

Glossary of terms

Acceleration

Acceleration is the rate of change in velocity with time.

$$\text{Acceleration} = \frac{\text{Velocity}}{\text{Time}} = (\text{m/s}^2)$$

Accident

An accident may be defined as an uncontrolled event causing injury or damage to an individual or property.

Alarm call points

Manually operated alarm call points should be provided in all parts of a building where people may be present, and should be located so that no one need walk for more than 30 m from any position within the premises in order to give an alarm.

Alloy

An alloy is a mixture of two or more metals.

Appointed person

An appointed person is someone who is nominated to take charge when someone is injured or becomes ill, including calling an ambulance if required. The appointed person will also look after the first aid equipment, including re-stocking the first aid box.

Approved test instruments

The test instruments and test leads used by the electrician for testing an electrical installation must meet all the requirements of the relevant regulations. All testing must, therefore, be carried out using an 'approved' test instrument if the test results are to be valid. The test instrument must also carry a calibration certificate; otherwise the recorded results may be void.

Basic protection

Basic protection is provided by the insulation of live parts in accordance with Section 416 of the IET Regulations.

Bonding

The linking together of the exposed or extraneous metal parts of an electrical installation.

Bonding conductor

A conductor providing protective bonding.

Cable tray

Cable tray is a sheet-steel channel with multiple holes. The most common finish is hot-dipped galvanized but PVC-coated tray is also available. It is used extensively on large industrial and commercial installations for supporting MI and SWA cables which are laid on cable tray and secured with cable ties through the tray holes.

Capacitive reactance

Capacitive reactance (XC) is the opposition to an a.c. current in a capacitive circuit. It causes the current in the circuit to lead ahead of the voltage.

Centrifugal force

Centrifugal force is the force acting away from the centre, the opposite to centripetal force.

Centripetal force

Centripetal force is the force acting towards the centre when a mass attached to a string is rotated in a circular path.

Circuit protective conductor (CPC)

A protective conductor connecting exposed conductive parts of equipment to the main earthing terminal.

Cohesive or adhesive force

Cohesive or adhesive force is the force required to hold things together.

Compact fluorescent lamps (CFLs)

CFLs are miniature fluorescent lamps designed to replace ordinary GLS lamps.

Compressive force

Compressive force is the force pushing things together.

Conductor

A conductor is a material, usually a metal, in which the electrons are loosely bound to the central nucleus. These electrons can easily become 'free electrons' which allows heat and electricity to pass easily through the material.

Conduit	A conduit is a tube, channel or pipe in which insulated conductors are contained.
Corrosion	Corrosion is the destruction of a metal by chemical action.
Delivery notes	A delivery note is used to confirm that goods have been delivered by the supplier, who will then send out an invoice requesting payment.
Duty holder	This phrase recognizes the level of responsibility which electricians are expected to take on as a part of their job in order to control electrical safety in the work environment. Everyone has a duty of care, but not everyone is a duty holder. The person who exercises 'control over the whole systems, equipment and conductors' and is the electrical company's representative on-site is a duty holder.
Earth	The conductive mass of the earth, the electrical potential of which is taken as zero.
Earthing	The act of connecting the exposed conductive parts of an installation to the main protective earthing terminal of the installation.
Efficiency of any machine	The ratio of the output power to the input power is known as the efficiency of the machine. The symbol for efficiency is the Greek letter 'eta' (η). In general, $$\eta = \frac{\text{Power output}}{\text{Power input}}$$
Electric current	The drift of electrons within a conductor is known as an electric current, measured in amperes and given the symbol I.
Electric shock	Electric shock occurs when a person becomes part of the electrical circuit.
Electrical force	Electrical force is the force created by an electrical field.
Electrical industry	The electrical industry is made up of a variety of individual companies, all providing a service within their own specialism to a customer, client or user.
Emergency lighting	Emergency lighting is not required in private homes because the occupants are familiar with their surroundings, but in public buildings people are in unfamiliar surroundings. In an emergency people do not always act rationally, but well-illuminated and easily identified exit routes can help to reduce panic.
Emergency switching	Emergency switching involves the rapid disconnection of the electrical supply by a single action to remove or prevent danger.
Escape/standby lighting	Emergency lighting is provided for two reasons: to illuminate escape routes, called 'escape' lighting; and to enable a process or activity to continue after a normal lights failure, called 'standby' lighting.
Expansion bolts	The most well-known expansion bolt is made by Rawlbolt and consists of a split iron shell held together at one end by a steel ferrule and a spring wire clip at the other end. Tightening the bolt draws up an expanding bolt inside the split iron shell, forcing the iron to expand and grip the masonry. Rawlbolts are for heavy-duty masonry fixings.
Exposed conductive parts	The metalwork of an electrical appliance or the trunking and conduit of an electrical system which can be touched because they are not normally live, but which may become live under fault conditions.
Extraneous conductive parts	The structural steelwork of a building and other service pipes such as gas, water, radiators and sinks.
Faraday's law	Faraday's law states that when a conductor cuts or is cut by a magnetic field, an e.m.f. is induced in that conductor.
Fault protection	Fault protection is provided by protective bonding and automatic disconnection of the supply (by a fuse or circuit breaker, CB) in accordance with IET Regulations 411.3 to 6.
Ferrous	A word used to describe all metals in which the main constituent is iron.

Fire	Fire is a chemical reaction which will continue if fuel, oxygen and heat are present.
Fire alarm circuits	Fire alarm circuits are wired as either normally open or normally closed. In a *normally open circuit*, the alarm call points are connected in parallel with each other so that when any alarm point is initiated the circuit is completed and the sounder gives a warning of fire. In a *normally closed circuit*, the alarm call points are connected in series to normally closed contacts. When the alarm is initiated, or if a break occurs in the wiring, the alarm is activated.
First aid	First aid is the initial assistance or treatment given to a casualty for any injury or sudden illness before the arrival of an ambulance, doctor or other medically qualified person.
First aider	A first aider is someone who has undergone a training course to administer first aid at work and holds a current first aid certificate.
Flashpoint	The lowest temperature at which sufficient vapour is given off from a flammable substance to form an explosive gas–air mixture is called the flashpoint.
Flexible conduit	Flexible conduit manufactured to BS 731-1: 1993 is made of interlinked metal spirals often covered with a PVC sleeving.
Fluorescent lamp	A fluorescent lamp is a linear arc tube, internally coated with a fluorescent powder, containing a low-pressure mercury vapour discharge.
Force	The presence of a force can only be detected by its effect on a body. A force may cause a stationary object to move or bring a moving body to rest.
Friction force	Friction force is the force which resists or prevents the movement of two surfaces in contact.
Functional switching	Functional switching involves the switching on or off, or varying the supply, of electrically operated equipment in normal service.
Fuse	A fuse is the weakest link in the circuit. Under fault conditions it will melt when an overcurrent flows, protecting the circuit conductors from damage.
Gravitational force	Gravitational force is the force acting towards the centre of the earth due to the effect of gravity.
Hazard	A hazard is something with the 'potential' to cause harm, for example, chemicals, electricity or working above ground.
Hazard risk assessment	Employers of more than five people must document the risks at work and the process is known as hazard risk assessment.
Hazardous area	An area in which an explosive gas–air mixture is present is called a hazardous area, and any electrical apparatus or equipment within a hazardous area must be classified as flameproof to protect the safety of workers.
Heating, magnetic or chemical	The three effects of an electric current: when an electric current flows in a circuit it can have one or more of the following three effects: heating, magnetic or chemical.
Impedance	The total opposition to current flow in an a.c. circuit is called impedance and given the symbol Z.
Inductive reactance	Inductive reactance (X_L) is the opposition to an a.c. current in an inductive circuit. It causes the current in the circuit to lag behind the applied voltage.
Inertial force	Inertial force is the force required to get things moving, to change direction or stop.
Inspection and testing techniques	The testing of an installation implies the use of instruments to obtain readings. However, a test is unlikely to identify a cracked socket outlet, a chipped or loose switch plate, a missing conduit-box lid or saddle, so it is also necessary to make a visual inspection of the installation. All existing installations should be periodically

inspected and tested to ensure that they are safe and meet the IET Regulations (IET Regulations 610 to 634).

Instructed person (electrically)	An instructed person (electrically) is a person adequately advised or supervised by electrically skilled persons to be able to perceive risks and avoid the hazards which electricity can create.
Insulator	An insulator is a material, usually a non-metal, in which the electrons are very firmly bound to the nucleus and, therefore, will not allow heat or electricity to pass through it. Good insulating materials are PVC, rubber, glass and wood.
Intrinsically safe circuit	An intrinsically safe circuit is one in which no spark or thermal effect is capable of causing ignition of a given explosive atmosphere.
Intruder alarm systems	An intruder alarm system serves as a deterrent to a potential thief and often reduces home insurance premiums.
Isolation	Isolation is defined as cutting off the electrical supply to a circuit or item of equipment in order to ensure the safety of those working on the equipment by making dead those parts which are live in normal service.
Job sheets	A job sheet or job card carries information about a job which needs to be done, usually a small job.
Lamp	A lamp is a device for converting electrical energy into light energy.
Lever	A lever is any rigid body which pivots or rotates around a fixed axis or fulcrum. Load force × Distance from fulcrum = Effort force × Distance from fulcrum.
Levers and turning force	A lever allows a heavy load to be lifted or moved by a small effort.
Line conductor	A conductor in an a.c. system for the transmission of electrical energy other than a neutral or protective conductor. Previously called a phase conductor.
Luminaire	A luminaire is equipment which supports an electric lamp and distributes or filters the light created by the lamp.
Magnesium oxide	The conductors of mineral insulated metal sheathed (MICC) cables are insulated with compressed magnesium oxide.
Magnetic field	The region of space through which the influence of a magnet can be detected is called the magnetic field of that magnet.
Magnetic force	Magnetic force is the force created by a magnetic field.
Magnetic hysteresis	Magnetic hysteresis loops describe the way in which different materials respond to being magnetized.
Magnetic poles	The places on a magnetic material where the lines of flux are concentrated are called magnetic poles.
Maintained emergency lighting	In a maintained system the emergency lamps are continuously lit using the normal supply when this is available, and change over to an alternative supply when the mains supply fails.
Manual handling	Manual handling is lifting, transporting or supporting loads by hand or by bodily force.
Mass	Mass is a measure of the amount of material in a substance, such as metal, plastic, wood, brick or tissue, which is collectively known as a body. The mass of a body remains constant and can easily be found by comparing it on a set of balance scales with a set of standard masses. The SI unit of mass is the kilogram (kg).
Mechanics	Mechanics is the scientific study of 'machines', where a machine is defined as a device which transmits motion or force from one place to another.

Metallic trunking	Metallic trunking is formed from mild steel sheet, coated with grey or silver enamel paint for internal use or a hot-dipped galvanized coating where damp conditions might be encountered.
Mini-trunking	Mini-trunking is very small PVC trunking, ideal for surface wiring in domestic and commercial installations such as offices.
Mobile equipment	Electrical equipment which is moved while in operation or can be moved while connected to the supply. Previously called portable equipment.
Movement or heat detector	A movement or heat detector placed in a room will detect the presence of anyone entering or leaving that room.
Mutual inductance	A mutual inductance of 1 henry exists between two coils when a uniformly varying current of 1 ampere per second in one coil produces an e.m.f. of 1 volt in the other coil.
Non-ferrous	Metals which do not contain iron are called non-ferrous. They are non-magnetic and resist rusting. Copper, aluminium, tin, lead, zinc and brass are examples of non-ferrous metals.
Non-maintained emergency lighting	In a non-maintained system the emergency lamps are only illuminated if the normal mains supply fails.
Non-statutory regulations and codes of practice	Non-statutory regulations and codes of practice interpret the statutory regulations telling us how we can comply with the law.
Ohm's law	Ohm's law says that the current passing through a conductor under constant temperature conditions is proportional to the potential difference across the conductor.
Optical fibre cables	Optical fibre cables are communication cables made from optical-quality plastic, the same material from which spectacle lenses are manufactured. The energy is transferred down the cable as digital pulses of laser light, as against current flowing down a copper conductor in electrical installation terms.
Ordinary person	An ordinary person is a person who is neither a skilled person nor an instructed person.
Overload current	An overload current can be defined as a current which exceeds the rated value in an otherwise healthy circuit.
Passive infra-red (PIR) detectors	PIR detector units allow a householder to switch on lighting units automatically whenever the area covered is approached by a moving body whose thermal radiation differs from the background.
Perimeter protection system	A perimeter protection system places alarm sensors on all external doors and windows so that an intruder can be detected as he or she attempts to gain access to the protected property.
Person	A person can be described as ordinary, instructed or skilled depending upon that person's skill or ability.
Personal protective equipment (PPE)	PPE is defined as all equipment designed to be worn, or held, to protect against a risk to health and safety.
Phasor	A phasor is a straight line, having definite length and direction, which represents to scale the magnitude and direction of a quantity such as a current, voltage or impedance.
Plastic plugs	A plastic plug is made of a hollow plastic tube split up to half its length to allow for expansion. Each size of plastic plug is colour-coded to match a wood screw size.
Polyvinylchloride (PVC)	PVC used for cable insulation is a thermoplastic polymer.

Potential difference	The potential difference (p.d.) is the change in energy levels measured across the load terminals. This is also called the volt drop or terminal voltage, since e.m.f. and p.d. are both measured in volts.
Power	Power is the rate of doing work.

$$\text{Power} = \frac{\text{Work done}}{\text{Time taken}} \ \text{(W)}$$

Power factor	Power factor (p.f.) is defined as the cosine of the phase angle between the current and voltage.
Pressure or stress	Pressure or stress is a measure of the force per unit area.

$$\text{Pressure or stress} = \frac{\text{Force}}{\text{Area}} \ \text{(N/m}^2\text{)}$$

Primary cell	A primary cell cannot be recharged. Once the active chemicals are exhausted, the cell must be discarded.
Protective bonding	This is protective bonding for the purpose of safety.
PVC/SWA cable installations	Steel wire armoured PVC insulated cables are now extensively used on industrial installations and often laid on cable tray.
Reasonably practicable or absolute	If the requirement of the regulation is absolute, then that regulation must be met regardless of cost or any other consideration. If the regulation is to be met 'so far as is reasonably practicable', then risks, cost, time, trouble and difficulty can be considered.
Relay	A relay is an electromagnetic switch operated by a solenoid.
Resistance	In any circuit, resistance is defined as opposition to current flow.
Resistivity	The resistivity (symbol ρ – the Greek letter 'rho') of a material is defined as the resistance of a sample of unit length and unit cross-section.
Risk	A risk is the 'likelihood' of harm actually being done.
Risk assessments	Risk assessments need to be suitable and sufficient, not perfect.
Rubber	Rubber is a tough elastic substance made from the sap of tropical plants.
Safety first – isolation	We must ensure the disconnection and separation of electrical equipment from every source of supply and that this disconnection and separation is secure.
Secondary cells	A secondary cell has the advantage of being rechargeable. If the cell is connected to a suitable electrical supply, electrical energy is stored on the plates of the cell as chemical energy.
Secure supplies	A UPS (uninterruptible power supply) is essentially a battery supply electronically modified to provide a clean and secure a.c. supply. The UPS is plugged into the mains supply and the computer systems are plugged into the UPS.
Security lighting	Security lighting is the first line of defence in the fight against crime.
Shearing force	Shearing force is the force which moves one face of a material over another.
Shock protection	Protection from electric shock is provided by basic protection and fault protection.
Short-circuit	A short-circuit is an overcurrent resulting from a fault of negligible impedance connected between conductors.
SI units	SI units are based upon a small number of fundamental units from which all other units may be derived.
Silicon rubber	Introducing organic compounds into synthetic rubber produces a good insulating material such as FP200 cables.

Simple machines	A machine is an assembly of parts, some fixed, others movable, by which motion and force are transmitted. With the aid of a machine we are able to magnify the effort exerted at the input and lift or move large loads at the output.
Single PVC insulated conductors	Single PVC insulated conductors are usually drawn into the installed conduit to complete the installation.
Skilled person (electrically)	A skilled person (electrically) is a person with relevant education, training and practical skills, and sufficient experience to be able to perceive risks and to avoid the hazards which electricity can create.
Skirting trunking	Skirting trunking is a trunking manufactured from PVC or steel in the shape of a skirting board which is frequently used in commercial buildings such as hospitals, laboratories and offices.
Socket outlets	Socket outlets provide an easy and convenient method of connecting portable electrical appliances to a source of supply.
Sounders	The positions and numbers of sounders should be such that the alarm can be distinctly heard above the background noise in every part of the premises.
Space factor	The ratio of the space occupied by all the cables in a conduit or trunking to the whole space enclosed by the conduit or trunking is known as the space factor.
Speed	Speed is concerned with distance travelled and time taken.
Spring toggle bolts	A spring toggle bolt provides one method of fixing to hollow partition walls which are usually faced with plasterboard and a plaster skimming.
Static electricity	Static electricity is a voltage charge which builds up to many thousands of volts between two surfaces when they rub together.
Statutory regulations	Statutory regulations have been passed by Parliament and have, therefore, become laws.
Step-down transformers	Step-down transformers are used to reduce the output voltage, often for safety reasons.
Step-up transformers	Step-up transformers are used to increase the output voltage. The electricity generated in a power-station is stepped up for distribution on the National Grid network.
Switching for mechanical maintenance requirements	The switching for mechanical maintenance requirements are similar to those for isolation except that the control switch must be capable of switching the full load current of the circuit or piece of equipment.
Synthetic rubber	Synthetic rubber is manufactured, as opposed to being produced naturally.
Tensile force	Tensile force is the force pulling things apart.
Thermoplastic polymers	These may be repeatedly warmed and cooled without appreciable changes occurring in the properties of the material.
Thermosetting polymers	Once heated and formed, products made from thermosetting polymers are fixed rigidly. Plug tops, socket outlets and switch plates are made from this material.
Time sheets	A time sheet is a standard form completed by each employee to inform the employer of the actual time spent working on a particular contract or site.
Transformer	A transformer is an electrical machine which is used to change the value of an alternating voltage.
Trap protection	Trap protection places alarm sensors on internal doors and pressure pad switches under carpets on through routes between, for example, the main living area and the master bedroom.

Trunking

A trunking is an enclosure provided for the protection of cables which is normally square or rectangular in cross-section, having one removable side. Trunking may be thought of as a more accessible conduit system.

Velocity

In everyday conversation we often use the word velocity to mean the same as speed, and indeed the units are the same. However, for scientific purposes this is not acceptable since velocity is also concerned with direction.

Visual inspection

The installation must be visually inspected before testing begins. The aim of the visual inspection is to confirm that all equipment and accessories are undamaged and comply with the relevant British and European Standards, and also that the installation has been securely and correctly erected.

Weight

Weight is a measure of the force a body exerts on anything which supports it. Normally it exerts this force because it is being attracted towards the earth by the force of gravity.

Work done

Work done is dependent upon the force applied times the distance moved in the direction of the force. Work done = Force × Distance moved in the direction of the force (J). The SI unit of work done is the newton metre or joule (symbol J).

Index